Structural Health Monitoring of Large Structures Using Acoustic Emission–Case Histories

Structural Health Monitoring of Large Structures Using Acoustic Emission–Case Histories

Special Issue Editors

Kanji Ono
Tomoki Shiotani
Martine Wevers
Marvin A. Hamstad

MDPI • Basel • Beijing • Wuhan • Barcelona • Belgrade • Manchester • Tokyo • Cluj • Tianjin

Special Issue Editors

Kanji Ono
Department of Materials Science
and Engineering, HSSEAS
School of Engineering & Applied
Sciences, University of California
USA

Tomoki Shiotani
Laboratory, Dept. Civil & Earth
Resources Eng'g Grad. School
Eng'g, Kyoto University
Japan

Martine Wevers
Department of Materials
Engineering, University of
Leuven
Belgium

Marvin A. Hamstad
Department of Mechanical and
Materials Engineering,
University of Denver
USA

Editorial Office
MDPI
St. Alban-Anlage 66
4052 Basel, Switzerland

This is a reprint of articles from the Special Issue published online in the open access journal *Applied Sciences* (ISSN 2076-3417) (available at: https://www.mdpi.com/journal/applsci/special_issues/Structural_Health_Monitoring%E2%80%93Case_Histories).

For citation purposes, cite each article independently as indicated on the article page online and as indicated below:

LastName, A.A.; LastName, B.B.; LastName, C.C. Article Title. *Journal Name* **Year**, *Article Number*, Page Range.

ISBN 978-3-03928-474-0 (Hbk)
ISBN 978-3-03928-475-7 (PDF)

Contents

About the Special Issue Editors

Kanji Ono received a B.Eng. from the Tokyo Institute of Technology and a Ph.D. in materials science from Northwestern University. After a year of postdoctoral research at Northwestern, he joined the faculty of the University of California, Los Angeles, where he became professor of engineering in 1976 and professor emeritus in 2005. He was awarded the HM Howe Medal by ASM International, the Achievement Award by ASNT, the Gold Medal Award by AEWG, and the Kishinoue Award by JSNDI. He is the author of more than 250 publications in materials science, acoustic emission, and non-destructive testing, and was the founding editor of *Journal of Acoustic Emission*. His areas of research have included physical metallurgy, strength and fracture of metallic and composite materials, non-destructive testing, and physical acoustics.

Tomoki Shiotani has worked on AE and relevant NDT for the evaluation of civil engineering materials and structures. After graduate study at Tokushima University, he conducted research at a general contractor, then moved to Kyoto University in 2007. In 2014, he founded the Laboratory on Innovative Technology for Infrastructure at Kyoto University, and was promoted to Professor. He has served as chair of AEWG, USA, and of Research and Technical Committee on AE in JSNDI, Japan, and organized International Symposium of AE twice (2008 and 2016) in Kyoto. He was a co-founder of International Institute of Innovative AE. He received Gold Medal of AEWG in 2019, and Best Paper Award of Structural Faults and Repair (Scotland, U.K.) four times.

Martine Wevers, Engineer in Metallurgy and Materials Engineering KU Leuven 1981, PhD in Science and Engineering KU Leuven 1987, is a full Professor at the KU Leuven and coordinating the research group "Materials Performance and Non-Destructive Evaluation" at the Department of Metallurgy and Materials Engineering. She is Chair of the Department of Materials Engineering. She is responsible for the NDT research (quality control, defect detection, development of new NDT techniques and data analysis procedures) and the research on the mechanical behavior of materials (fracture mechanics and fatigue). She specializes in in-situ or on-line damage monitoring techniques (acoustic emission, optical fibers, and sensors for structural health monitoring) and the evaluation of microstructural damage in materials using X-ray computed tomography.

Marvin Hamstad (Ph.D. ME/Solid Mechanics, University of California, Berkeley) has been continuously involved in acoustic emission research and applications since 1971. This AE work has ranged from nondestructive testing applications to materials characterization. He has worked in the field of composite materials since the 1970s. His various research efforts have resulted in more than 140 papers. From 1971 to 1984, he was at Lawrence Livermore National Laboratory. From 1984 to 2010, Dr. Hamstad was Professor of Engineering at the University of Denver. He is currently Professor Emeritus. Since 1992, Dr. Hamstad has held positions working on AE at the National Institute of Standards and Technology, Boulder, CO. He was chairman of the Committee on Acoustic Emission from Reinforced Plastics for 16 years, where he initiated the AECM international symposiums (1983–1998). Dr. Hamstad was an associate editor of *J. Structural Health Monitoring*, and currently serves as a co-editor of *JAE*.

Editorial

Structural Health Monitoring of Large Structures Using Acoustic Emission–Case Histories

Kanji Ono

Department of Materials Science and Engineering, University of California, Los Angeles (UCLA),
Los Angeles, CA 90095, USA; ono@ucla.edu

Received: 13 October 2019; Accepted: 16 October 2019; Published: 29 October 2019

1. Introduction

Acoustic emission (AE) techniques have successfully been used for assuring the structural integrity of large rocket motorcases since 1963 [1], and their uses have been expanded to ever larger structures [2,3], especially since the structural health monitoring (SHM) of large structures has become an urgent task for engineering communities globally. The needs for advanced methods of AE monitoring are felt most keenly by those dealing with aging infrastructure. Many publications have appeared covering various aspects of AE techniques, but documentation of actual applications of AE techniques has been limited mostly to reports of successful results without technical details that allow objective evaluation of the results, except for some exceptions in the literature where special applications were detailed [4–6]. In this Special Issue of the Acoustics section of Applied Sciences, we sought contributions, like the exceptions cited here, which describe case histories of AE being applied to large structures. That is, papers that have achieved the goals of SHM and do so by giving adequate technical information supporting the success stories. Gathered here are 14 such articles that cover structures from aerospace and geological structures, bridges, buildings, factories, nuclear facilities, etc.

2. Acoustic Emission Applications

With the goal given above, this special issue was initiated to collect latest research on relevant subjects. There were 28 papers submitted to this special issue, and 14 papers were accepted (i.e., a 50% acceptance rate). A key review paper, authored by Manthei and Plenkars [7], covered AE applications for the SHM of mines and in various geological settings. Of these, underground repositories for nuclear waste are especially significant, and millimeter-size defects were located in a million cubic meter volume, demonstrating in situ AE monitoring is a useful tool to observe instabilities in rock long before any damage becomes visible. Moriya [8] presented the use of an AE method for determining the seismicity of a lake bottom following the massive 2011 Tohoku earthquake. Another review by Behnia et al. [9] discussed AE methods for evaluating the structural integrity of asphalt pavements located in cold regions. It amply shows that AE allows for relatively rapid and inexpensive characterization of pavement materials and can be used for enhancing pavement sustainability and resiliency to thermal loading.

Three papers dealt with the SHM of industrial structures, bridges, and masonry buildings. Elizarov et al. [10] presented a series of examples of AE monitoring of industrial facilities under load, both static and fatigue conditions. Some were one of a kind structures, forcing them to improvise various techniques. They successfully evaluated oil refinery tanks and towers, pipelines, rotary kilns, bridge structures, drag lines, etc. Swit [11] also covered AE detection of active destructive processes that are in progress on various structures, including steel bridges, steel columns, and a large suspension bridge. The use of pattern recognition analysis was effective in identifying active damage progression. Carpinteri et al. [12] detected AEs from seismically induced cracking in an ancient building.

Work by Esola et al. [13] was directed at the evaluation of aerospace composite parts via AE monitoring. They developed a multi-dimensional parts assessment method by taking advantage of AE data recorded during structural testing. For this goal, they tested 16 composite fixed-wing-aircraft spars using a structural loading sequence designed around a manufacturer-specified design limit load (DLL). While loading, the Felicity ratio was calculated. Along with specific AE data from post-processing, they deduced spar test classification in terms of apparent damage behavior.

In applications aimed toward the SHM of concrete structures, the Ziehl group presented two studies [14,15]. In the first one, Abdelrahman et al. [14] applied remote AE monitoring on an actively corroding concrete structure for one year and showed the feasibility of using AE for corrosion damage detection and classification. Here, studies of decommissioned components and control concrete parts were examined in parallel and comparative evaluation led to the conclusion. In the second study, Soltangharaei et al. [15] examined alkali-silicate reaction on large-scale concrete structures. They utilized an agglomerative hierarchical algorithm to classify the AE data based on energy-frequency based features. AE observations were correlated to confinement strains, allowing the assessment of structural degradation via AE signal features.

In all AE applications, source location is essential and Zhou et al. [16] proposed a new scheme for dealing with outliers, by introducing a preconditioned closed-form solution based on weight estimation. While this produced improvements, some real effects like damage-induced velocity variation still needs additional study, as noted by the authors. Zhong et al. [17] provided another approach for improving source location accuracy. This is called multiple signal classification (MUSIC), which adds directional scanning ability and easy arrangement of the sensor array. This approach takes advantage of Lamb wave propagation behavior, focusing on identified center frequency. This work further combined optimized ensemble empirical mode decomposition (EEMD) and a two-dimensional multiple signal classification algorithm for real-time impact localization on composite structures. This new method was validated on a cross-ply composite plate.

Another improvement in AE data analysis was given by Keshtgar et al. [18] through a statistical concept, Bayesian analysis. They considered AE data during fatigue monitoring of crack growth. Hong et al. [19] gave an analysis of transverse vibration on a beam that may be useful in preventing damages to high-rate systems.

Lastly, a review paper by the present author, 'Review on structural health evaluation with acoustic emission' [3] gave an overview of the current status of AE contribution to SHM. This article first examined signal attenuation, since any AE application seeks minimizing the number of sensors. It is shown that signal loss from geometrical spreading is a key issue in many shell-type structures. Another issue is general lack of attenuation data, found after an extensive survey of existing experimental reports. Since theory cannot provide such attenuation data, more effort is needed before quantitative SHM design procedures can be implemented. This is followed by discussion on source location, bridge monitoring, sensing and signal processing, pressure vessels and tanks, and special applications.

3. Future on AE-SHM

Although this special issue has been closed, more research and development work in AE technologies useful for SHM is in progress. It can be anticipated that future SHM will benefit from advanced AE applications that take advantage of various modeling tools and artificial intelligence technologies. Newer, smaller sensors, remote monitoring, feature and clustering analyses, and AE monitoring in extreme environments are some of anticipated topics to come. More effective AE methods are required more than ever for sustainable societies.

Acknowledgments: This issue was made successful by the contributions of all the talented authors. Also essential were gracious and professional reviewers, who gave the feedback, valuable comments, and suggestions. Thank you. I am particularly grateful to fellow Guest Editors, Marvin Hamstad, Tomoki Shiotani, and Martine Wevers for their contributions. Finally, thanks are due to the dedicated editorial team of Applied Sciences and special thanks to Daria Shi from MDPI Branch Office, Beijing, who helped guide this issue to its successful conclusion.

References

1. Green, A.T.; Lockman, C.S.; Steele, R.K. *Modern Plastics*; Breskin Publications Inc.: New York, NY, USA, 1964; Volume 41, pp. 137–139.
2. Ono, K. Application of Acoustic Emission for Structure Diagnosis. *Diagnostics* **2011**, *2*, 3–18.
3. Ono, K. Review on Structural Health Evaluation with Acoustic Emission. *Appl. Sci.* **2018**, *8*, 958. [CrossRef]
4. Anonymous. MONPAC Technology. *J. Acoust. Emiss.* **1986**, *8*, 1–34.
5. Hay, D.R.; Cavaco, J.A.; Mustafa, V. Monitoring the Civil Infrastructure with Acoustic Emission: Bridge Case Studies. *J. Acoust. Emiss.* **2009**, *27*, 1–9.
6. Gorman, M.R.; Modal, A.E. Analysis of Fracture and Failure in Composite Materials, and the Quality and Life of High Pressure Composite Pressure Vessels. *J. Acoust. Emiss.* **2011**, *29*, 1–28.
7. Manthei, G.; Plenkers, K. Review on In Situ Acoustic Emission Monitoring in the Context of Structural Health Monitoring in Mines. *Appl. Sci.* **2018**, *8*, 1595. [CrossRef]
8. Moriya, H. Acoustic Emission/Seismicity at Depth Beneath an Artificial Lake after the 2011 Tohoku Earthquake. *Appl. Sci.* **2018**, *8*, 1407. [CrossRef]
9. Behnia, B.; Buttlar, W.; Reis, H. Evaluation of Low-Temperature Cracking Performance of Asphalt Pavements Using Acoustic Emission: A Review. *Appl. Sci.* **2018**, *8*, 306. [CrossRef]
10. Elizarov, S.; Barat, V.; Terentyev, D.; Kostenko, P.; Bardakov, V.; Alyakritsky, A.; Koltsov, V.; Trofimov, P. Acoustic Emission Monitoring of Industrial Facilities under Static and Cyclic Loading. *Appl. Sci.* **2018**, *8*, 1228. [CrossRef]
11. Swit, G. Acoustic Emission Method for Locating and Identifying Active Destructive Processes in Operating Facilities. *Appl. Sci.* **2018**, *8*, 1295. [CrossRef]
12. Carpinteri, A.; Niccolini, G.; Lacidogna, G. Time Series Analysis of Acoustic Emissions in the Asinelli Tower during Local Seismic Activity. *Appl. Sci.* **2018**, *8*, 1012. [CrossRef]
13. Esola, S.; Wisner, B.; Vanniamparambil, P.; Geriguis, J.; Kontsos, A. Part Qualification Methodology for Composite Aircraft Components using Acoustic Emission Monitoring. *Appl. Sci.* **2018**, *8*, 1490. [CrossRef]
14. Abdelrahman, M.; ElBatanouny, M.; Dixon, K.; Serrato, M.; Ziehl, P. Remote Monitoring and Evaluation of Damage at a Decommissioned Nuclear Facility using Acoustic Emission. *Appl. Sci.* **2018**, *8*, 1663. [CrossRef]
15. Soltangharaei, V.; Anay, R.; Hayes, N.; Assi, L.; Le Pape, Y.; Ma, Z.; Ziehl, P. Damage Mechanism Evaluation of Large-Scale Concrete Structures Affected by Alkali-Silica Reaction Using Acoustic Emission. *Appl. Sci.* **2018**, *8*, 2148. [CrossRef]
16. Zhou, Z.; Rui, Y.; Zhou, J.; Dong, L.; Chen, L.; Cai, X.; Cheng, R. A New Closed-Form Solution for Acoustic Emission Source Location in the Presence of Outliers. *Appl. Sci.* **2018**, *8*, 949. [CrossRef]
17. Zhong, Y.; Xiang, J.; Chen, X.; Jiang, Y.; Pang, J. Multiple Signal Classification-Based Impact Localization in Composite Structures Using Optimized Ensemble Empirical Mode Decomposition. *Appl. Sci.* **2018**, *8*, 1447. [CrossRef]
18. Keshtgar, A.; Sauerbrunn, C.; Modarres, M. Structural Reliability Prediction Using Acoustic Emission-Based Modeling of Fatigue Crack Growth. *Appl. Sci.* **2018**, *8*, 1225. [CrossRef]
19. Hong, J.; Dodson, J.; Laflamme, S.; Downey, A. Transverse Vibration of Clamped-Pinned-Free Beam with Mass at Free End. *Appl. Sci.* **2019**, *9*, 2996. [CrossRef]

Review

Review on Structural Health Evaluation with Acoustic Emission

Kanji Ono

Department of Materials Science and Engineering, University of California, Los Angeles (UCLA), Los Angeles, CA 90095, USA; ono@ucla.edu

Received: 20 May 2018; Accepted: 8 June 2018; Published: 11 June 2018

Abstract: This review introduces several areas of importance in acoustic emission (AE) technology, starting from signal attenuation. Signal loss is a critical issue in any large-scale AE monitoring, but few systematic studies have appeared. Information on damping and attenuation has been gathered from metal, polymer, and composite fields to provide a useful method for AE monitoring. This is followed by discussion on source location, bridge monitoring, sensing and signal processing, and pressure vessels and tanks, then special applications are briefly covered. Here, useful information and valuable sources are identified with short comments indicating their significance. It is hoped that readers note developments in areas outside of their own specialty for possible cross-fertilization.

Keywords: acoustic emission; structural diagnosis; attenuation; source location; sensing; signal processing

---

1. Introduction

Nondestructive evaluation (NDE) of various structures, large and small, has been the primary target of acoustic emission (AE) technology along with its uses in materials research. The first success of AE technology was achieved at Aerojet for the inspection of Polaris missile chambers in the 1960s [1]. Further works continued for nuclear and chemical pressure vessels and tanks [2,3]. For industrial AE applications to fiber reinforced plastics (FRP) vessels, the Committee on Acoustic Emission for Reinforced Plastics (CARP) was instrumental in code development, culminating in ASME Boiler and Pressure Vessels Codes and ASTM standards. See Fowler et al. [4] and four following articles in the special issue of Journal of AE in 1989. A wide range of AE applications have been compiled in the AE volume of the *Nondestructive Testing Handbook* [5], while many articles appeared in conference proceedings and in Journal of AE [6]. Shiotani [7] and Bohse [8] reviewed various applications of AE to infrastructures and to structural diagnosis, respectively. Two recent review articles [9,10] on bridge and pressure-vessel inspection are noteworthy in connection to the topic of this introduction. Another review was published as a Sandia report, comparing AE with other methods for structural health monitoring (SHM) in evaluating damages to a full-scale wind turbine blade, and demonstrating the advantages of AE over others [11]. The present author also prepared survey papers on structural diagnosis [12] and on composites [13]. A comprehensive monograph by Giurgiutiu [14] on SHM of aerospace composites appeared recently and AE monitoring for SHM was covered in depth. AE uses in SHM have fully integrated acousto-ultrasonic methods, taking advantage of piezoelectric wafer active sensors (PWAS). A review paper by Mitra [15] covering the roles of guided waves in SHM is also useful. With the wealth of these available resources, this article will focus on those topics not addressed adequately elsewhere.

2. Signal Attenuation

An important issue during any structural inspection using AE methods is the decrease in signal intensity or attenuation. On the AE side, signal-to-noise ratio is important in signal detection. In SHM applications, quantitative system modeling is often desired and attenuation behavior needs to be a part of successful SHM system design [16]. The same phenomena at lower frequencies are referred to as structural damping and a recent review relative to composite materials [17] provides useful backgrounds. This topic is also indirectly related to the mechanical properties of polymers since most polymers exhibit viscoelastic behavior and strong acoustic damping [18,19].

2.1. Guided Wave Attenuation

In using AE to examine structures, we typically deal with two-dimensional wave propagation since thick-walled structures—like nuclear pressure vessels and concrete dams—are exceptional cases. For theoretical developments and early experimental results, see Viktorov's book from 1967 [20] or numerous other books that followed dealing with wave propagation, e.g., Rose [21]. In most instances, AE signals propagate along the surface as Rayleigh waves or through thin-walled structures as Lamb waves. These waves are collectively known as guided waves. When they spread on two-dimensional structures, signal loss occurs from geometrical spreading (with inverse square root dependence on travel distance, x, or $1/\sqrt{x}$ dependence). Additionally, attenuation is caused from material absorption and scattering with exponential decay, or $\exp(-ax)$ with attenuation coefficient a. Combined, signal amplitude decreases following $(1/\sqrt{x})\cdot\exp(-ax)$. For dispersive Lamb waves, signal loss also arises from dispersion or frequency-dependent wave speed, which spreads vibration energy over a longer period, reducing signal amplitude [13,20]. When the surfaces are covered with liquids or other damping matters, attenuation occurs from vibration energy leakage. These appear as additional a values. Mal et al. [22] gave an extension of wave propagation theory to anisotropic plates with dissipation given in terms of quality factors, Q. They presented several examples of attenuation in fiber reinforced composite plates.

Press and Healy [23] in 1957 gave theory and experimental confirmation for Rayleigh wave attenuation. Measured Rayleigh attenuation coefficients (a_R) for PMMA were nearly linear with frequency and was 55 dB/m at 100 kHz. Viktorov [20] provided a parametric equation for a_R and listed three more a_R values for aluminum (Dural or 2017 alloy), glass and polystyrene (15.4, 73.7, 422 dB/m at 1 MHz, respectively). In 1964, Zhukov et al. [24] derived theoretical expressions for Lamb wave attenuation coefficients, a_L. These guided wave attenuation coefficients, a_R and a_L, depend on the attenuation coefficients of longitudinal and transverse bulk waves, or a_p and a_t. Zhukov et al. [24] also measured a_L values for the 0th and 1st symmetric and asymmetric modes, or S_0, S_1, A_0, and A_1 modes on low carbon steel plates (C = 0.15%). The measured a_L values ranged from 3.8 to 5.7 dB/m at 1.3 MHz and were comparable to their longitudinal and shear wave attenuation coefficients, a_p = 3.7 and a_t = 3.9 dB/m. At zero thickness limit for S_0, $a_L = \sqrt{2}\,a_t$. Pressure-vessel and pipeline steels are known for their low bulk wave attenuation of a_p = 1 to 10 dB/m at 2 MHz as tabulated in Krautkramer's book [25], which are reduced further at lower frequencies. Such low attenuation values for Al and steels below 10 dB/m were reported by Mason and McSkimin [26], Roderick and Truell [27], Kamigaki [28], and Papadakis [29,30] among others at frequencies up to 20 MHz. These data were obtained using directly bonded, low-loss quartz transducers. Some of these tests also included a_t measurements. Thus, guided waves are expected to propagate with low loss when materials possess low a_p values.

Wave attenuation is characterized using several different parameters. In the ultrasonic and AE fields, attenuation coefficient a is commonly used to represent an exponential decay. We use two units for a; One is dB/m and the other Np/m with 8.686 dB = 1 Np. Np stands for Nepers, a non-dimensional unit, and is useful in numerical computation. This a is also related to damping (or loss) factor η by $\eta = a\lambda/\pi$, where λ is the wave length. Here, η is the ratio of energy dissipated per cycle to maximum energy stored per cycle, and is also equal to loss tangent, $\tan\delta$. This $\tan\delta$ is defined as the ratio of imaginary part (E″) to real part (E′) of a complex modulus, $E^* = E' - iE''$ with $i^2 = -1$. The damping factor (or loss tangent) is often used in dealing with vibration damping at lower frequencies. Another parameter is quality factor Q, defined as the inverse of η (or tan δ). See e.g., Kinsler et al. [31] and Cai et al. [32].

2.2. Attenuation Measurements on Large Metallic Structures

Graham and Alers [33] reported in 1975 the first quantitative study of AE signal attenuation on pressure vessels, showing that signal amplitude (expressed in dB scale) decreased linearly with distance of propagation, except near the signal source where the slope of the decrease was steeper. The frequency they examined was 100 to 850 kHz and the wall thickness of the pressure vessels was 100–130 mm. Thus, the signals examined were Rayleigh waves. Selected data was replotted against distance (in the logarithmic scale) as shown in Figure 1. Of the five plots, the data at 100 and 200 kHz followed the inverse square root distance dependence, indicating absorption loss was negligible. For the two higher frequency cases at 400 and 600 kHz, additional attenuation term of 1.7 or 2.1 dB/m accounted for the observed deviation from the $1/\sqrt{x}$ dependence. Data at 850 kHz was inadequate, but it seems to show even higher attenuation. For a given plate thickness, a limiting frequency exists, below which Rayleigh waves do not exist. For the 100-mm thick steel, it is 82 kHz, as shown in the Appendix A.

Pollock [34] reported AE signal attenuation for 30-kHz signals on a pipeline of nearly 300-m length. A replot of this data (with blue + symbols) in Figure 2a exhibits the same $1/\sqrt{x}$ behavior (indicated by a black line) with small additional attenuation of 0.17 dB/m. Data points for the $1/\sqrt{x}$ plus 0.17-dB/m attenuation are shown by red circles. Here, Lamb waves propagated on a relatively thin pipe wall (of a few cm thickness). Blackburn [35] reported attenuation data for large gas cylinders. These cylinders were 12-m length, 0.6-m diameter, and 14.3-mm wall thickness and made of heat-treated 4130 steel (quenched and tempered after fabrication). His data for a 3AAX tube at 150 kHz is plotted in Figure 2b. Most data points fit the $1/\sqrt{x}$ behavior except a few points deviated lower, suggesting possible effects of signal absorption at large distances. In these two studies, dispersion loss was minimal.

More attenuation measurements have been published recently. The data of Baran et al. [36] for 30-kHz signals on a pipeline up to 100-m length are shown in Figure 2c. Their results are similar to the Pollock case with a slightly higher attenuation of 0.34 dB/m. Sofer et al. [37] tested 150-kHz signals from pencil-lead breaks on steel sheet and plate, getting only the $1/\sqrt{x}$ behavior since the maximum distance was 2 m. Their cylindrical block did exhibit attenuation of 17 dB/m in addition to the $1/\sqrt{x}$ behavior. Thus, these observations fit with the theory and early experiments [20,24].

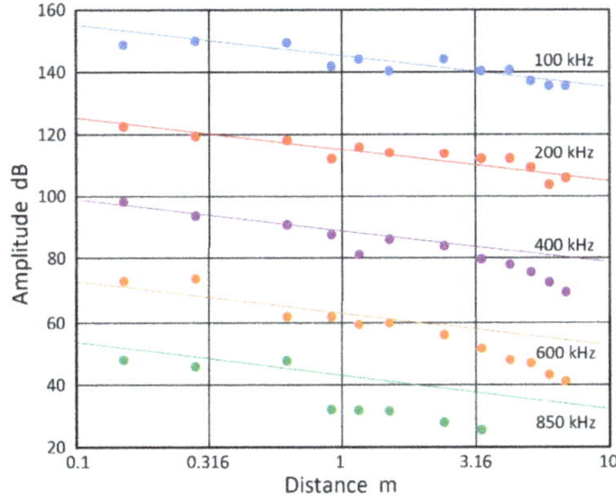

Figure 1. Attenuation data (amplitude in dB vs. distance in m) on steel pressure vessels at five frequencies from Graham-Alers [33]. From top to bottom: 100 kHz (blue), 200 kHz (red), 400 kHz (purple), 600 kHz (orange), and 850 kHz (green). Signal source was a white noise generator. Lines drawn represent $1/\sqrt{x}$ dependence.

Two other works were outside the above framework. CETIM group [38] examined large penstocks in a hydropower plant covering the length of 40 m. Their low and medium frequency attenuation studies resulted in the inverse-distance behavior. This may be due to many thick flanges for these particular penstocks that appear different from the common design of long pipe sections. El-Shaib [39] reported Lamb wave attenuation on a steel plate, but his signal energy values decreased with $1/\sqrt{x}$, implying negative amplitude attenuation. It is likely that his energy data was already converted to amplitude values.

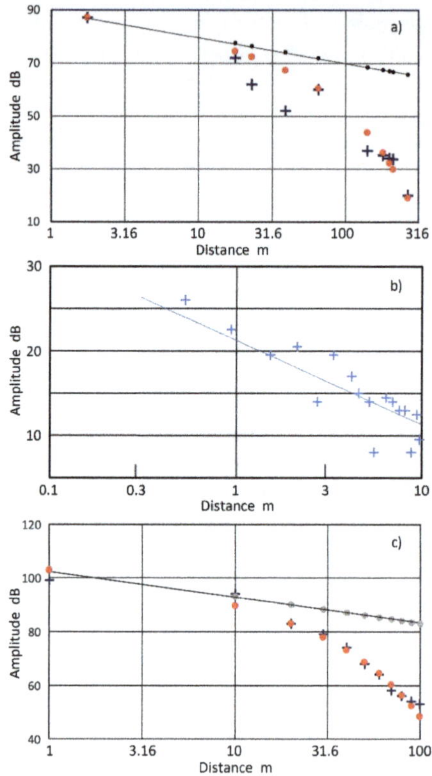

Figure 2. Attenuation data (amplitude in dB vs. distance in m) on steel structures. (**a**) Signal attenuation on a pipeline at 30 kHz by Pollock [34]; (**b**) Signal attenuation on a gas cylinder at 150 kHz by Blackburn [35]; (**c**) Signal attenuation on a pipeline at 30 kHz by Baran et al. [36]. Lines drawn represent $1/\sqrt{x}$ dependence. Measured data points are shown by + (blue) and modeled points are in filled circles (red) in (**a**,**c**).

2.3. Laboratory Attenuation Measurements

As a part of guided-wave sensor studies at UCLA [40,41], Lamb wave attenuation was measured for large aluminum (Al) plates (6.4-mm thick 1100 Al and 12.7-mm thick 6061 Al), steel (3.2-mm thick 410 stainless), soda-lime glass (4.7-mm thickness), PMMA (6.4-mm thickness), and polyvinyl chloride (PVC, 4.6-mm thickness). For the three metal and glass plates, the inverse-square root distance behavior prevailed with a few exceptions (when combined modes started to split at larger travel distances at some frequencies). The maximum travel distance was 500 mm and the frequency range was from 100 to 1500 kHz. The result of the $1/\sqrt{x}$ behavior for 410 stainless steel (SS) was surprising because this steel (along with pure Fe and pure Ni) was expected to show high bulk wave attenuation due to its magnetic properties [42]. However, Papadakis [43] reported only moderate attenuation for a comparable 416 stainless steel (a_p = 20 dB/m at 4 MHz). Papadakis [43] did find Ni to have a high a_p of 120 dB/m at 2 MHz. Drinkwater [44] calculated Lamb wave attenuation curves for glass, showing less than 0.1 dB/m attenuation below 0.5 MHz for a plate (3.9-mm thickness). Thus, the $1/\sqrt{x}$ behavior found here for glass confirms the calculation. In highly attenuating polymeric plates, signal levels were strongly reduced as shown in Figure 3. At 75 kHz on 6.4-mm thick PMMA (Figure 3a), S_0-mode waves propagated with symmetric excitation and showed the $1/\sqrt{x}$ behavior plus a_L of 25 dB/m. At 380 kHz on PMMA (Figure 3b), seven Lamb modes are expected with the group velocity of 0.8 to 1.1 mm/µs according to dispersion curve calculation. Since the duration of excited signals was about 50 µs, mode separation was not observed and the entire wave packets were analyzed. Attenuation a_L beyond the geometrical spreading was higher at 121 dB/m. A 4.6-mm thick PVC plate exhibited even higher attenuation, reflecting its higher bulk wave attenuation [45]. As shown in Figure 3c, 130-dB/m attenuation was observed beyond the geometrical spreading at 75 kHz. For PMMA, the values of a_p and a_t below 100 kHz were available from resonant ultrasonic spectroscopy [46,47]. Assuming that the damping factor (η) values of 0.035 and 0.025 reported for 50 and 60 kHz, respectively, hold at 75 kHz, a_p and a_t are found as 27.3 and 52.0 dB/m for PMMA at 75 kHz. Using the Zhukov theory for a_L and calculated coefficients given, the attenuation of S_0-mode waves was obtained as 33.7 dB/m for the frequency-thickness product of 0.48 MHz-mm for PMMA thickness of 6.4 mm. The observed a_L value for S_0-mode is 25 dB/m, so a_L matching is good between theory and experiment. Castaings and Hosten [48,49] obtained complex elastic moduli for PMMA and predicted Lamb wave attenuation for five lowest modes. At 75 kHz, a_L for S_0 was 21.5 dB/m, while a_L values ranged from 145 to 300 dB/m at 380 kHz. Thus, theory and experiment agree reasonably well also. When the observed a_p and a_t values are inserted to the Viktorov equation [20] for Rayleigh wave attenuation, a_R = A a_p + (1 − A) a_t = 51.0 dB/m as A = 0.05 with Poisson's ratio of 0.37 for PMMA. This is in good agreement with 47 dB/m at 75 kHz, obtained by Press and Healy [23].

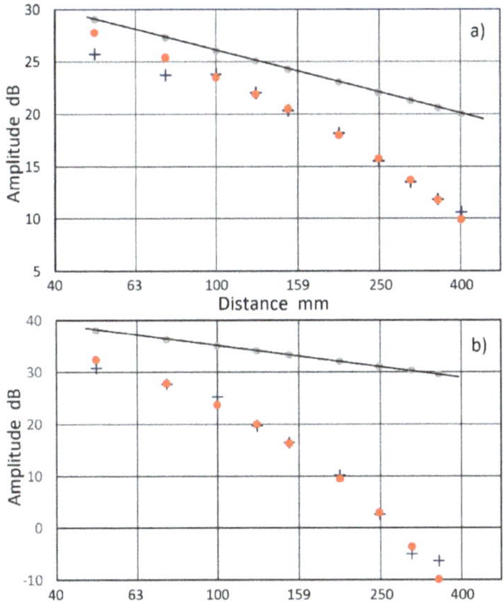

Figure 3. *Cont.*

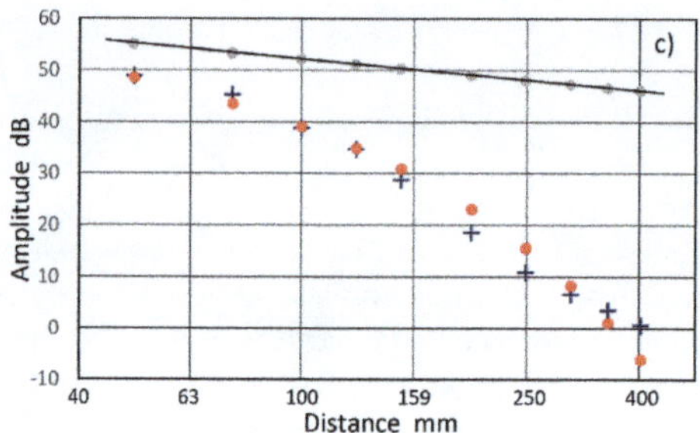

Figure 3. Lamb wave attenuation curves with amplitude in dB and distance in mm. Data symbols are identical to those in Figure 2. (**a**) S_0-mode propagation at 75 kHz on 6.4-mm thick PMMA. Red points are for the $1/\sqrt{x}$ behavior plus a_L of 25 dB/m; (**b**) mixed mode propagation at 380-kHz, fitting the $1/\sqrt{x}$ behavior plus a_L of 121 dB/m; (**c**) S_0-mode propagation at 75 kHz on 4.6-mm thick PVC matched the $1/\sqrt{x}$ behavior plus a_L of 130 dB/m.

2.4. Complex Elastic Moduli Measurements

Castaings et al. [50] developed an elegant ultrasonic method that is capable of determining the complex elastic moduli of PMMA discussed above. This method utilized iterative numerical inversion techniques and transmitted ultrasonic fields obtained for multiple incident angles on a plate sample. Either immersion or air-coupling technique can be used. This method yielded complex viscoelastic stiffness coefficients. The complex elastic moduli for PMMA were given in Castaings and Hosten [49], showing $\eta_1 = 0.023$, $\eta_2 = 0.056$, $\eta_{66} = 0.026$, and $\eta_{12} = 0.046$. These were measured at 0.3 MHz. The η data is tabulated in Table 1.

For PMMA, many tests have been reported for η and for a_p. Figure 4 shows a plot of damping factor vs. frequency data from [47,51–56]. The value of η starts to rise at 0.001 Hz and the peak η of 0.09 is reached at 3 Hz, then decreasing at higher frequencies. These low frequency tests were in torsional mode (corresponding to η_{66} or η_{12}) and yielded 40 to 60% higher values than Thakur's data [53], conducted in tension mode (equivalent to η_1), corresponding to a_t and a_p values, respectively, in terms of ultrasonic attenuation. Also plotted are η values converted from a_p measurements at higher ultrasonic frequencies between 0.5 to 6.4 MHz [54–56]. The η values from ultrasonic attenuation are relatively unchanged at approximately 0.01 and match Thakur's data near 1 MHz within 25%. In fact, Hartman's early η value of 0.089 is valid from 0.29 to 30 MHz [57]. If the trend at lower frequencies continues to the MHz range, it can be expected that a_t is 50% higher than a_p in the low MHz region. When all these η values are compared among them, it is evident that the transmission field inversion method [49] produced at least a factor of two larger results. Independent evaluation of this method seems advisable as their other complex elastic moduli data have been utilized by several other research groups as will be discussed below. Note also that η values for PMMA decrease by a factor of two from 10 kHz to 30 MHz. Given ultrasonic attenuation coefficient $a = \eta\pi/\lambda$, a values increase with frequency with their frequency dependence decreasing gradually. This behavior is expected in other engineering polymers.

From the above survey and results, we can estimate the attenuation of guided waves when we have bulk wave attenuation data. Unfortunately, values of a_p and a_t are unavailable at typical AE frequencies under 1 MHz for most engineering materials because large samples are needed for attenuation measurements. Even in usual ultrasonic frequencies above 1 MHz, it is difficult to find even a_p values for many materials. Still, high strength steels and Al alloys are qualitatively known to be good transmitters of ultrasounds. Thus, we can reasonably assume that AE signal attenuation follows the geometrical spreading and shows the $1/\sqrt{x}$ behavior. This assumption cannot be used when excess damping conditions exist from inside or outside contact with liquids or other lossy matters [58,59]. Then, we need to assess signal attenuation by traditional methods.

Table 1. Damping factors from complex elastic moduli measurements.

Ref. No. [1]	Material	Axis [2]	η_{11}	η_{22}	η_{33}	η_{44}	η_{55}	η_{66}	H_{12}	η_{13}	η_{23}	Freq (MHz)
[60]	Epoxy		0.01	0.01				0.011				
[49]	PMMA		0.023	0.056				0.026	0.046			0.3
[49]	GFRP-UD	3	0.047	0.043	0.033		0.050	0.038	0.033	0.036		0.3
[61]	GFRP-UD	3	0.051	0.038	0.056	0.100	0.070	0.091	0.014	0.106		0.05–0.2
[61]	CFRP-UD	3	0.027	0.05	0.075	0.06	0.063	0.11	0.037	0.1		0.05–0.2
[62]	CFRP-UD	1	0.061	0.024	0.046	0.084	0.062	0.055	0.114	0.105	0.029	
[63]	CFRP-UD	1	0.086	0.045				0.06	0.037	0.033		0.005–0.2
[64]	CFRP-UD	1	0.024	0.043	0.041	0.032	0.056	0.093	0.143	0.074		0.2–0.93
[64]	CFRP-QI	1	0.041	0.067	0.034	0.045	0.044	0.068	0.071	0.016		0.2–0.93
[65,66]	CFRP XP	1	0.020	0.018	0.001	0.010	0.017	0.026	0.018	0.016	0.001	0.01–1
[65,66]	CFRP-UD	1	0.015	0.099	0.099	0.010	0.045	0.045	0.015	0.015	0.261	0.01–1

[1]: Ref. No. refers to reference number. [2]: Axis indicates the coordinate direction along fiber direction (0° orientation in QI-quasi-isotropic or XP-cross-ply).

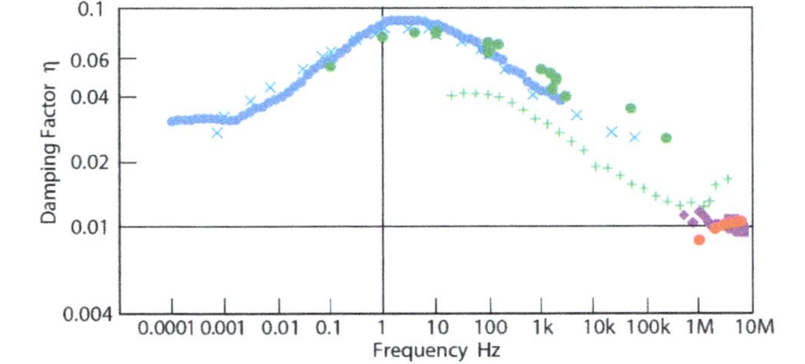

Figure 4. Damping factor of PMMA over 11 decades of frequency, peaking at 3 Hz, from the literature. Torsional damping: blue X [47], blue circle [51], green circle [52]; longitudinal damping: green + [53], red circle [54], purple square [55], purple diamond [56].

2.5. Survey of Ultrasonic Attenuation of Metallic Alloys

Table 2 lists ultrasonic attenuation of engineering alloys, noting material parameters given in the original articles. This list attempted to cover all the attenuation data available for engineering alloys, but some data were omitted for obvious errors in methods used. Large variations are sometimes seen for similar alloys, but detailed reevaluation of measurement methods is needed to explore their causes [67]. For example, Klinman et al. [68] obtained a_p of 60 dB/m at 5 MHz for an annealed 0.15%C steel with 20 μm grain size, but later Klinman et al. [69] reported for a similar sample a_p of 670 dB/m at 10 MHz. Papers from Harwell [70,71] also reported 500 to 1350 dB/m attenuation for annealed low C steels at 8–10 MHz. Such large differences in a_p at 5 and 10 MHz may arise from the frequency dependence of a_p. However, these appear to be strange, since Papadakis [29,72] obtained a_p = 12 dB/m for annealed 4150 steel and below 10 dB/m for a tempered martensitic bearing steel (52100 steel) over 1 to 10 MHz. Serious evaluation of these diverging results has not been conducted so far, but Papadakis' data are more reliable as he used directly bonded quartz disc transducers, while later studies used damped ultrasonic transducers and water immersion [68–71]. Magnetic effects could be a factor, but are less than 10 dB/m at 1 MHz and not large enough [73,74]. Most of the references dealt with the longitudinal wave attenuation. Recently, Hirao, Ogi, and Ohtani [75–79] have measured shear wave attenuation using non-contact electromagnetic sensors. For example, they found a_t = 116 dB/m at 5 MHz for 0.15%C steel with 49 μm grain size, almost doubling Klinman's a_p data at 5 MHz cited above [68]. Their a_t results in combination with longitudinal attenuation data allow one to estimate guided wave attenuation using the theories discussed in Viktorov [20]. The number of engineering alloys examined for a_t, however, is still limited.

In contrast to polymers, in which hysteretic effects due to molecular rearrangements are dominant [18], ultrasonic attenuation of crystalline metals and ceramics mainly comes from absorption, Rayleigh scattering, and stochastic scattering [26,80]. Absorption effect is similar to polymers with linear frequency dependence, though mechanisms are different. Both of the scattering contributions depend on frequency with a power law and the exponents are 4 and 2. Rayleigh scattering varies most with grain size. In Table 2, attenuation measurements that used low-loss quartz discs directly bonded to samples are marked with Q while non-contact electromagnetic measurements are marked with E. The rest utilized immersion, buffer rod, or direct contact methods. The unmarked group needs assumptions regarding signal loss at sample interfaces, where errors may be generated. Generazio [81] examined ultrasonic reflection coefficients and showed large variations and dependence on interface thickness and contact pressure. In view of low attenuation found in guided wave propagation dominated by the $1/\sqrt{x}$ dependence, it is likely that most structural steels used in pressure vessels, tanks, and pipelines have low values of a_p and a_t. Thus, high bulk wave attenuation of ultrasonic waves must be reexamined since high attenuation reports have mostly originated from immersion test procedures that included a plane-wave assumption in the analysis. This last point also needs further study because the sound fields ahead of a commonly used piston transducer suffer from divergence and diffraction [31,72].

Table 2. Ultrasonic attenuation of engineering alloys.

Material	Type	$\alpha\text{-p}$ [1]	$\alpha\text{-t}$ [1]	Method [2]	Condition [3]	Reference	Ref. No.
Al	1100 (2S)	0.16 (5), 1(10)		Q		Hikata (1957)	[82]
Al	Al (99.99%)		28 (6.3)	E	Anneal 200 °C 1 h	Hirao (2003)	[76]
	2017 (17ST)	5.6 (5)	13.1 (4)	Q	GS: 0.23 mm	Mason (1947)	[26]
	2017 (17ST)	2.6 (5)	7.9 (5)	Q	GS: 0.13 mm	Mason (1947)	[26]
Cu	Cu (99.99%)		88 (2.3)	E	GS: 35 μm	Hirao (2003)	[76]
	Brass (360)	45.7 (10)	286 (5)	Q	GS: 49 μm	Papadakis (1965)	[43]
Mg	–	32.3 (10)		Q	GS: 0.2 mm	Mason (1948)	[83]
	AZ31 (FS1)	6.9 (10)	4.1 (10)	Q	GS: 0.12 mm	Mason (1948)	[83]
Nb	Nb	280–510 (5)			GS: 43–124 μm	Zeng (2010)	[84]
Ni	Grade A	115 (3.5)	240 (2)	Q	GS: 55 μm	Papadakis (1965)	[43]
	Waspaloy		51 (4.5)	E	solution + aged	Ohtani (2004)	[75]
Fe	Fe-0.004%C	339 (8.2)			GS: 50 μm	Ahn (2000)	[85]
	Fe-0.02%C	300–1350 (10)			GS: 18–137 μm	Smith (1981)	[70]
	Fe-0.06%C	46–101 (5)			GS: 22–30 μm	Klinman (1980)	[68]
Steel	Fe-0.15%C	33–60 (5)			GS: 14–40 μm	Klinman (1980)	[68]
	Fe-0.15%C		116 (5)	E	GS: 49 μm	Hirao (2003)	[76]
	Fe-0.2%C	220–340 (5)			GS: 120–290 μm	Ahn (2000)	[85]
	Fe-0.25%C	45–91 (5)			GS: 12–29 μm	Klinman (1980)	[68]
	Fe-low C	83–237 (5)			GS: 50–120 μm	Klinman (1980)	[68]
	Fe-0.4%C	53 (5)			GS: 12–20 μm	Klinman (1980)	[68]
	Fe-0.4%C	86–127 (5)			GS: 50–65 μm	Klinman (1980)	[68]
	4150	78.7 (10)			RHC: 14	Roederick (1952)	[27]
	4150	19.7 (20)			RHC: 61	Roederick (1952)	[27]
	4145	14.5 (5)		Q	VHN: 746	Kamigaki (1957)	[28]
	4145	22.5 (5)		Q	VHN: 265	Kamigaki (1957)	[28]
	3140	21.2 (10)		Q	RHC: 45	Papadakis (1960)	[30]
	4150	21.2 (10)		Q	RHC: 54	Papadakis (1960)	[30]
	4150	12.2 (10)		Q	pearlie-bainite	Papadakis (1964)	[29]
	4150	6.4 (20)		Q	martensite (M)	Papadakis (1964)	[29]
	4150	11.0 (20)		Q	tempered M	Papadakis (1964)	[29]
	1Cr-Mo-V		33 (5)	E	tempered M	Ohtani (2002)	[77]
	2.25Cr-1Mo		25.6 (3.9)	E	annealed	Ohtani (2006)	[78]

Table 2. *Cont.*

Material	Type	α-p [1]	α-t [1]	Method [2]	Condition [3]	Reference	Ref. No.
	1075	10 (5)		Q	GS^c: 27 μm	Kamigaki (1957)	[28]
	1075	35 (5)		Q	GS^c: 38 μm	Kamigaki (1957)	[28]
	1075	68 (5)		Q	GS^c: 68 μm	Kamigaki (1957)	[28]
	1075	174 (5)		Q	GS^c: 174 μm	Kamigaki (1957)	[28]
	1075	73.5 (10)		Q	VHN: 696	Kamigaki (1957)	[28]
	1075	90 (10)		Q	VHN: 274	Kamigaki (1957)	[28]
	1075	174 (10)		Q	VHN: 270	Kamigaki (1957)	[28]
	1080	19.2–33.9 (10)		Q	spheroidized	Latiff (1974)	[83]
	1080	230–290 (5)			normalized	Ahn (2000)	[85]
	Rail steel	26–150 (10)			Quenched	Du (2014)	[86]
	52100	2.4–22 (10)	6.8–44 (10)	Q	RHC: 55–61	Papadakis (1970)	[72]
	Plain C	150–1000 (15)			GS: 15–33 μm	Smith (1983)	[71]
	Plain C	150–1690 (10)			GS: 10–31 μm	Klinman (1981)	[69]
	316L SS	230 (5), 360 (10)			GS: 37 μm	Wan (2017)	[87]
	316L SS		5.0 (5)	E	GS: 19.5 μm	Ohtani (2005)	[79]
	403 SS	26 (10)		Q	GS^c: 25 μm	Nadeau (1985)	[88]
	416 SS	27.8 (5)	159 (5)	Q	GS: 30 μm	Papadakis (1965)	[43]
	416 SS	348 (10)		Q	GS: 30 μm	Papadakis (1965)	[43]
	440C austenitic SS	313 (5)	814 (5)	Q	GS: 50 μm	Papadakis (1965)	[43]
	Cast austenitic SS	192 (0.5), 174 (1)			columnar 2.5 mm	Ramuhalli (2009)	[89]
	Cast austenitic SS	264 (0.5), 272 (1)			equiaxed 2.3 mm	Ramuhalli (2009)	[89]

[1] Attenuation coefficient is given in dB/m, followed by frequency in MHz in the parentheses. In many cases, results for additional frequencies are given as well. [2] Q refers to measurement method using quartz discs and E using electromagnetic transducer. [3] GS: grain size, GS^c: pearlite colony size or prior austenite grain size, VHN: Vickers hardness number, RHC: Rockwell hardness-C-scale, M: martensite.

2.6. Guided Wave Attenuation on Fiber-Reinforced Composites

Lamb wave experiments were also conducted at UCLA using fiber reinforced plastics (FRP). Figure 5a,b show two cases for an FRP with woven glass-fiber rovings and epoxy matrix (2.2 mm thickness, density 1.9 g/cm^3, 0.38 fiber fraction). Attenuation data for S_0 mode along the fiber direction at 60 and 100 kHz are plotted against (log) distance. The observed data can be attributed entirely to the $1/\sqrt{x}$ dependence at 60 kHz, while 100-kHz data matches the $1/\sqrt{x}$ dependence plus 10.8 dB/m attenuation. These attenuation coefficients are similar to a_p of 4.4 dB/m at 100 kHz for a unidirectional GFRP along the fiber (converted from η of 0.007) [90]. The above findings implies glass-fiber reinforced plastics (GFRP) behave similarly to common structural metallic alloys below 100 kHz. The same attenuation data can also be described conventionally with an attenuation coefficient of 21.2 dB/m (60 kHz) or 33.1 dB/m (100 kHz) including the geometrical spreading as shown in Figure 5c. These conventional attenuation coefficients are lower than a_p of 135 dB/m for another GFRP (with random mat) at 100 kHz, with its a_p increasing to 400 dB/m at 2 MHz [91]. However, this data was in the direction normal to fibers and not directly comparable. When attenuation due to absorption is high, it is convenient to include signal spreading in attenuation parameters in NDE applications.

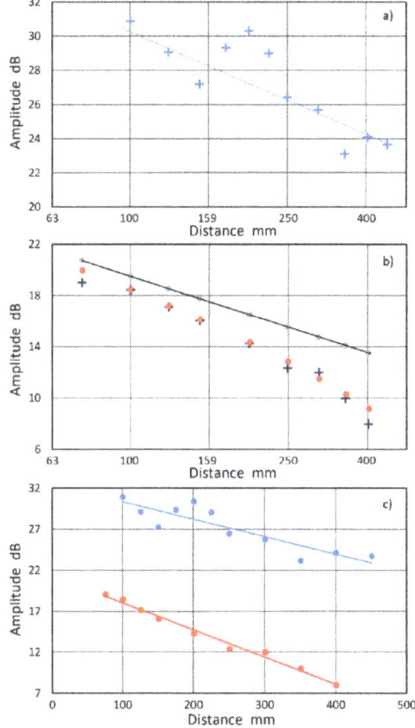

Figure 5. Lamb wave attenuation data for S_0 mode along the fiber direction of an FRP. (**a**) Data at 60 kHz match with $1/\sqrt{x}$ dependence; (**b**) data (blue +) at 100 kHz fit to the $1/\sqrt{x}$ dependence plus 10.8 dB/m attenuation, shown by red circles; (**c**) the same attenuation data (in dB) plotted against linear distance. Blue points: 60 kHz with the slope of 21.2 dB/m. Red points: 100 kHz with 33.1 dB/m. This FRP was used in an AE study [92].

Castaings and Hosten [48] used complex elastic moduli from [49] and predicted Lamb wave attenuation for unidirectional GFRP, giving S_0-mode a_L value at 100 kHz of 17 dB/m. For this case, Castaings and Hosten [49] reported $\eta_{11} = 0.047$, $\eta_{22} = 0.043$, $\eta_{33} = 0.033$, $\eta_{55} = 0.05$, $\eta_{66} = 0.038$, $\eta_{12} = 0.033$, and $\eta_{13} = 0.036$ (fiber direction is the 3-direction and fiber volume 0.6). Neau et al. [61]

reported another set of η_{ij} for a GFRP, measured by the Castaings method. This set showed the values of η_{33} and η_{66} to be 70 and 240% higher. These η values by the Castaings method (listed in Table 1) are higher than other published values of η, determined from conventional vibration damping methods at lower frequencies. Crane [93] thoroughly reviewed test methods and results. He also gave additional results on damping of composite materials. See also [60], which listed η values from the 1980s as in [93]. For unidirectional GFRP with 0.6-fiber volume fraction, longitudinal damping factor η in the fiber direction was 0.004–0.006, about six-times smaller than the ultrasonic data above [49,61]. Crane [93] also noted that epoxy matrix has η of 0.022. Vantomme [94] obtained even lower damping factor values, e.g., $\eta_1 = 0.002$ (or 0.0013 from [60]) and noted η for a single glass fiber to be 0.0015. Data in [60] gave η_2 of 0.008 and η_{12} of 0.011 for E-glass/DX210 epoxy. These GFRP data are mostly lower than the epoxy data in Table 1 [60]. These damping data were all taken under 1 kHz and increasing trends with frequency were reported. If the main contribution to damping comes from the matrix, as existing theories postulated [91], the increase with frequency is likely to be insignificant considering the decreasing trend of damping factor noted on PMMA [42]. In fact, the data of $a_p = 400$ dB/m at 2 MHz cited earlier [91] correspond to the damping factor of 0.02. Thus, the Castaings–Hosten determination of GFRP damping factors that averaged to 0.043 apparently suffers from overestimation of a factor of two or more.

Attenuation and damping studies for carbon-fiber reinforced plastics (CFRP) have been conducted since the 1970s. In his exhaustive review, Crane [93] showed that longitudinal damping factor in the fiber direction (η_1) was 0.001–0.005 for unidirectional CFRP with 0.6-fiber volume fraction. Transverse damping factor normal to the fiber direction (η_2) was approximately 0.01. CFRP data from [60] were similar to GFRP values given above and match with the Crane values. A newer study confirmed these results [95]. These damping studies were made at low frequencies below 20 kHz using flexural bending of long beam samples. A recent work also verified η_1 of ~0.001 using a longitudinal resonance technique with pultruded 0° samples [96]. At 2 MHz, η_1 or η_2 of 0.01 corresponds to ultrasonic attenuation coefficient (a_p) of 50 or 182 dB/m in the direction parallel or normal to fibers. Using quartz disc transducers, Kim [97] evaluated a_p and a_t of UD-CFRP (XA-S/1138) including those along the fiber direction over 1.8 to 9 MHz. At a fiber fraction of 0.6 and 2 MHz, α_p was 55 and 450 dB/m, parallel and normal to fibers, while α_t exceeded 1050 dB/m. Biwa et al. [98] made theoretical and experimental studies of attenuation for UD-CFRP (TR30/340) with epoxy matrix varying fiber volume fractions up to 0.6. At 2 MHz, a_p was 430 dB/m normal to fibers, about 10% higher than their theory. In both studies [97,98], the values of a_t were much higher than those of a_p. When the observed a_p of 450 dB/m is converted to η_2, we get η_2 of 0.025, about twice the low frequency data. An earlier work by Williams et al. [99] obtained η_1 and η_2 values of 0.014 and 0.049 along and normal to fibers at 2 MHz for a CFRP (AS/3501-6). They used multiple samples of square cross section (9.5 × 9.5 or 12.7 × 12.7 mm^2) and 3.8- to 127-mm length to get their η_1 and η_2 values. Even ignoring a diffraction correction [100,101], α_p along fibers at 2 MHz is 70 dB/m and is only 27% higher than Kim [97]. For a_p normal to fibers, it is 892 dB/m without diffraction correction and is about twice those of Kim [97] and Biwa et al. [98]. We also conducted an ultrasonic attenuation measurement at 2.25 MHz using an immersion method and obtained α_p normal to fibers of 704 dB/m for CFRP (G50/F584) in a cross-ply layup, corresponding to η_2 of 0.034. Along the fiber direction, the same CFRP yielded η_1 of 0.02 at 0.5 MHz. This CFRP sample was from our earlier AE study [102]. The differences in attenuation among CFRP are partly due to fibers and resins used, but may also be from methods used. Thus, we expect $\eta_1 = 0.01$~0.02 and $\eta_2 = 0.02$~0.05 for UD-CFRP at low-MHz ultrasonic frequencies.

Using a torsion pendulum method, Adams [103] evaluated the damping η for single carbon fibers. For polyacrylonitrile (PAN)-based fibers, η was 0.0013, while pitch-based fibers (after stretching) had η of 0.0028. Ishikawa et al. [104] examined η for single carbon fibers of 13 different combinations of tensile strength and elastic modulus. These were newer fibers with improved properties and both PAN- and pitch-based fibers were tested for torsional damping. For the low η group, the values of η were about 0.025 for low strength fibers (under 2 GPa tensile strength). The middle η group had

$\eta = 0.035 \pm 0.003$ and the tensile strength varied from 2.5 to 6 GPa, including both high strength PAN fibers and high modulus meso-phase pitch fibers. The last group, or the high η group, showed high damping with $\eta = 0.05$ to 0.08 and possessed high strength of 3.8–6.3 GPa. This high damping in the newer high-performance carbon fibers is surprising since these η values are three to four times larger than epoxy matrix and nearing those of some elastomers. Earlier damping results from the 1980s were on CFRP with fibers of low to medium strength in today's classification. Thus, these fibers had values of η under 0.04 and in combination with epoxy ($\eta = 0.022$), resultant η_2 is consistent with the observed η_2 of 0.02~0.05. Another check can be made by measuring ultrasonic attenuation. If the high damping values of 0.05 to 0.08 persist at low MHz frequencies, we should expect α_p of 900–1500 dB/m at 2 MHz (2300–3600 dB/m at 5 MHz) of ultrasonic attenuation. This level of α_p was reported on CFRP with T700 fiber. Olivier et al. [105] measured α_p of 3300 dB/m at 5 MHz for 8-ply unidirectional CFRP, but with 1.7% porosity. Our recent preliminary measurement on a CFRP with T700 fiber (2–10 mm thickness, made from Toray 2501 prepregs; fiber fraction of 0.4) showed $\alpha_p = 72$ or 600 dB/m at 0.45 MHz in the fiber direction or normal to fiber direction. These α_p values correspond to η_1 and η_2 of 0.05 and 0.13, respectively, and indeed imply that newer high-performance fibers have higher intrinsic damping in comparison to fibers made before the mid-1990s. The high damping behavior is beneficial for many engineering applications, but poses a challenge for SHM. Ishikawa et al. [104] related high torsional damping to amorphous carbon phase. Yet there seems to be no clear reason why newer high-performance carbon fibers possess high acoustic damping. Note that amorphous fused silica has very low damping. These ultrasonic studies should be repeated including the evaluation of neat resin and further research is needed for the origin of damping.

Complex elastic moduli of CFRP plates have been measured with the Castaings method. Four sets of η_{ij} values are tabulated in Table 1. Note that fibers are oriented along the 3-axis for Neau et al. [61], but along the 1-axis for Matt [62] and Calomfirescu [63,64]. In three cases for CFRP-UD samples, damping along the fiber axis averaged to 0.069 while shear damping normal to fibers 0.098. These damping values are again two or more times higher than η_2 values of comparable CFRP-UD samples. The complex elastic moduli were then used with higher order plate theory to calculate attenuation coefficients for S_0 and A_0 modes. Calomfirescu [63] obtained for a UD plate along the fibers ($0°$) α_L of 27 and 150 dB/m at 300 kHz (45 and 250 dB/m at 500 kHz) for S_0 and A_0 modes, respectively. Along the fiber normal ($90°$) direction, corresponding values were 75 and 108 dB/m at 300 kHz. These attenuation values are beyond geometrical spreading. A few other calculations resulted in much higher attenuation values [16,61,62] than Calomfirescu results.

The S_0 value along $0°$ of Calomfirescu [63] was much higher than our measurements on AS4/3506-1 CFRP plates [106]. In our CFRP study, three types of lay-ups (unidirectional, cross-ply, and quasi-isotropic) were used and complex attenuation behavior was found when the wave propagation direction shifted from the fiber orientations. For the UD plate, S_0-mode attenuation value along the fiber direction ($0°$) at 300 kHz was 4.8 dB/m, which was slightly less than that of Al plate (5.8 dB/m). Since geometrical spreading was not separated and receiver size was 8 mm, the same test was repeated using a KRN sensor of 1 mm size. Figure 6a gives attenuation data for S_0 and A_0 modes at 100 kHz, which are represented by attenuation coefficients of 20 dB/m for S_0 and 38 dB/m for A_0 mode. When the same data is plotted in a log-log graph, only geometrical spreading effect or $1/\sqrt{x}$ behavior is found without additional attenuation. No signal loss was observed for S_0-mode at 300 or 500 kHz, as shown in Figure 6b. This behavior appears to come from a sharp directivity for the UD plate observed previously [106]. In contrast, attenuation for A_0-mode at 300 or 500 kHz was substantial. Values of a_L were 78 and 178 dB/m beyond geometrical spreading for A_0-mode propagation along the fiber direction, as shown in Figure 7a,b. Thus, the a_L values for A_0 are within a factor of two with the Calomfirescu calculation [63]. However, a_L for S_0 at $0°$ vanished in our tests beyond geometrical spreading.

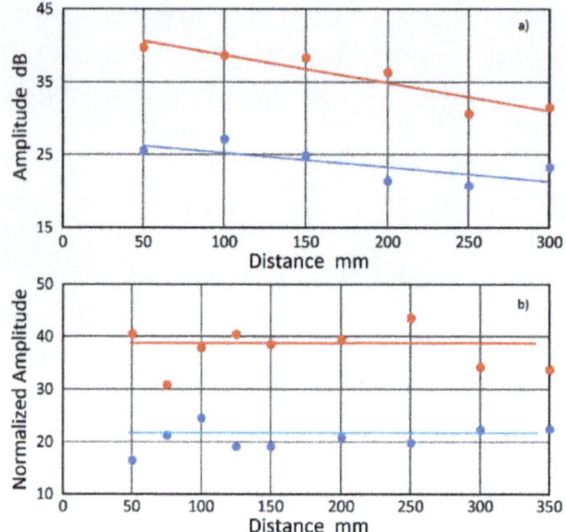

Figure 6. (**a**) UD-CFRP attenuation for S_0 (blue) and A_0 (red) modes at 100 kHz, amplitude in dB scale; (**b**) UD-CFRP attenuation for S_0 mode at 300 kHz (blue) and at 500 kHz (red), amplitude was calculated using the square root of the wave packet energy. Information on the UD-CFRP plate was given in [106].

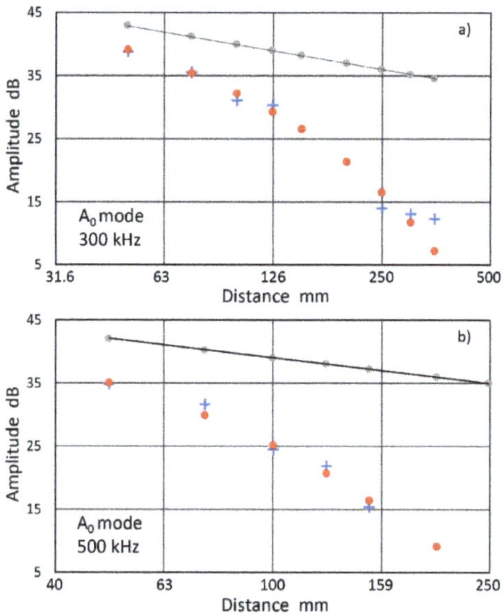

Figure 7. UD-CFRP attenuation for A_0 modes vs. (log) distance. Observed data (blue +), modeled attenuation of geometrical spread plus attenuation (red circle), geometrical spread (gray circle). Amplitude in dB scale. (**a**) Attenuation of 78 dB/m for 300 kHz; (**b**) attenuation of 178 dB/m for 500 kHz. Amplitude was calculated as in Figure 6.

Schmidt et al. [65] used another approach in characterizing CFRP properties. They measured Lamb wave propagation utilizing air-coupled ultrasonic techniques and deduced the dispersion curves and attenuation behavior. The data was then used to construct an analytic model. Their modeling relied on higher order plate theory and the attenuation utilized hysteretic model where the damping coefficients (η_{ij}) are independent of frequency. The values of η_{ij} were given in Schmidt et al. [66] and listed in Table 1. It appears that an UD plate is marked 45° considering C_{ij} values. Assuming this to be the case and noting that their η_{ij} stand for C''_{ij}, some of these are two-to-five times smaller than the corresponding values from the Castaings method [49]. However, off-fiber parameters of η_{22}, η_{33}, and η_{23} are grossly overestimated with $\eta_{23} = 0.26$ being five-times higher than η_{12} of PMMA. In dealing with different composite layups used, lamination theory was incorporated. An example given shows calculated attenuation coefficients for S_0 and A_0 modes on a quasi-isotropic CFRP plate. At 122 kHz, they obtained attenuation coefficients for S_0 and A_0 modes to be 4.3 and 50.2 dB/m beyond geometrical spreading. Another set of attenuation coefficients for S_0 and A_0 modes were reported graphically (CFRP layup was not given, but it seems to be UD). At 300 kHz, these were 31 and 140 dB/m, respectively. These α_L matched with their experimental values. These were higher than our data that included geometrical spreading effects [106], and our new test data discussed above showing no attenuation at 100 kHz. Only the A_0 data at 300 kHz matched within a factor of two. Schmidt et al. [65] included geometrical spreading in their modeling. This needs to be probed, however, since properly modeled calculations should predict $1/\sqrt{x}$ behavior from the viscoelastic parameters for quasi-isotropic CFRP, and a directional behavior for unidirectional CFRP. More comparative studies are obviously needed to obtain representative CFRP wave propagation characteristics.

2.7. Summaries

Section 2 considered signal attenuation, which often limits the use of AE inspection in many SHM applications. Since waves in most structures move as guided waves, Section 2.1 collected available theories for attenuation in isotropic media and described a general behavior, citing early experiments. Section 2.2 reviewed published wave propagation experiments, showing that available results can be rationalized well using the theories from Section 2.1. Section 2.3 reported Lamb wave propagation experiments conducted in laboratory scale for confirmation using elastic and viscoelastic plates. Results matched theoretical predictions. Section 2.4 discussed a new approach for attenuation studies using complex elastic moduli, which were obtained by the Castaings method. All the available results were collected in Table 1. Results for PMMA were then compared with damping factor determination, which had been accumulated over many years. It was found that the Castaings method prediction appears to overestimate the damping factor by a factor of two, suggesting the need for independent verification. Section 2.5 covered the ultrasonic attenuation of metals. For metallic alloys, structural attenuation behavior can be predicted when their attenuation coefficients, both longitudinal and transverse, are known. However, the attenuation data is limited and all the accessible values were tabulated in Table 2. More studies are needed, especially for commonly used structural alloys and for transverse attenuation coefficients that require special instrumentation. Section 2.6 dealt with the attenuation behavior of fiber composites. For these anisotropic media, only limited data sets are available and over short propagation distances. Calculations relied on higher order plate theory and complex elastic moduli from the Castaings method. This approach appears to be valid based on comparison with Lamb wave attenuation data and represents substantial advances in wave analysis. However, some of the damping factor data are apparently two or more times higher when compared to the corresponding values from ultrasonic attenuation measurements. Again, further studies are required to clarify the attenuation behavior of highly variable, anisotropic composite structures. Note that most available complex elastic moduli data sets lack manufacturing data on tested composite plates, making comparison difficult. Detailed material identification must accompany sophisticated mechanical characterization.

The above discussion demonstrated that modeling of guided waves for CFRP plates has advanced substantially. However, more refinements are needed in calculation procedures and validation of damping coefficients. While Rayleigh and hysteretic damping models produced realistic attenuation results that matched experiment [16,65], Kelvin–Voigt model as used in, e.g., [65], should be discarded for its physically unrealistic assumption. Additionally, the Kelvin–Voigt model introduces an arbitrary parameter (characteristic frequency) in damping calculation. Critical evaluation of the Castaings and Schmidt methods [49,66] for damping parameters is highly desirable to achieve a unified predictive method. As noted above, the Castaings method [49] gives damping factors twice higher than other methods and some η_{ij} were also excessive in Schmidt et al. [66].

These basic studies point to difficulties in practical testing of FRP structures except at lower frequencies. As in the Sandia report [11], Weihnacht et al. [107] discussed testing of large FRP components. In one of their tests, they needed to place 64 sensors over a 38-m long blade at 3 to 5 m sensor spacing. Such a sensor count demonstrates tough challenges facing real time composite inspection. Thus, the strategy introduced by CARP [4] of using low (60 kHz) and high (150 kHz) frequency sensors in tandem remains valid for global and local AE source location.

3. Source Location

Methods of AE source location came from seismology and have been refined to meet demands for accuracy, speed, and robustness against false location [5,108–112]. Actually, the source location or localization problem is a general topic of interest to broad segments of science and engineering. See a comprehensive review in connection to signal processing [113]. The AE location methods also evolved with available computational capabilities, starting with table look-up and first-hit or zonal location, as described by Hutton in 1972 [114]. In 1991, Crostak [115] developed a directional sensing method with four-sensor arrays to simplify source location, allowing a source location using two sets of sensors. More recently, three-sensor arrays worked just as well, despite reducing the sensor count [116,117]. New localization methods are adapted to AE field as reported in, e.g., [112,118].

For typical AE applications, currently available 2-D and 3-D location algorithms satisfy their basic needs. The algorithms utilize time differences of arrival (TDOA) and seek the intersection of a set of hyperboloids assuming constant wave propagation speed, usually relying on iterative processes. However, there are more requirements developing in the SHM field for faster calculations and less sensor placements [14,119]. In particular, smaller sized PWAS have added a new dimension to source location strategy as these allow mode discrimination and provide directional information [14] and phased array concept is also a useful addition [120]. Automation of AE detection will enable broader uses of AE technology and Holford et al. [112] discussed some examples developed, focusing on fatigue problems that still menace critical structural elements.

One of newer approaches relies on exact closed form solutions, which can reach the source position efficiently and accurately. This analytical method obtains the AE source as the intersection of spheres, the radii of which are related to TDOA [121–124]. Some are introducing artificial intelligence concepts for source location, including Gaussian process, support vector machine and deep learning [125–127]. Earlier, neural networks and genetic algorithm were also used [128]. For AE source location on thin plates, Gorman [129–131] introduced the Lamb wave velocity variation due to S_0 and A_0 modes, although the dispersion behavior was used in sensor frequency selection earlier [33,114]. Others followed his approach and showed that, for linear location, only a single sensor is needed [132,133]. For anisotropic plates, Kundu et al. [117] developed a method without solving a system of nonlinear equations. This method relies on three-sensor arrays, allowing simpler orientation detection. For 2-D location, this method can also find velocity values by itself with improvements [134]. Other methods can also be used for anisotropic plates [135]. In some composite layups, complex velocity patterns develop and a generic method was devised for such cases [118]. In this best-matched point search method, a structure is represented by points defined relative to sensor network. TDOA values of an AE event are matched with those of the previously defined points, quickly yielding the position of the AE event. Another method with TDOA, named delta-T, has also been advanced since its inception and now incorporate several signal-processing methods, becoming more reliable in dealing with complex geometries [136]. These methods have a root, tracing back to the table look-up methods [114], but now are highly refined for today's AE applications.

Another new approach utilizes time reversal methods. This concept originated from the need to focus on the origin through inhomogeneous media [137–139]. A time reversal method was applied to the analysis of a ribbed composite plate where Lamb wave mode conversion occurs adding an extra mode to the initial two-mode propagation [140]. Here, a matched filter is created that maximizes the ratio of output amplitude to the square root of the input energy. For source location, a number of waveforms recorded by a single sensor containing the impulse response of the medium are used, aided by scattering, mode conversions, and boundary reflections of the medium. As it uses a virtual focusing procedure, no iterative algorithms are used and no prior knowledge of the properties or anisotropic wave speed is needed. See also Douma et al. [141] and Robert et al. [142].

Other special topics are of interest. Liu et al. [143] examined source location errors when thick-walled vessels are tested. Traditionally, the accuracy limit of source location was empirically thought to be the thickness of a structure [5]. They defined errors in three different ways and conducted experiments using large concrete blocks with the thickness of 150 to 600 mm and sensor spacing of 4 to 5.5 m. Errors were found to be 2% to 10%, while the absolute errors were 5% to 30% of the thickness. Thus, results indicate that empirical estimates were realistic considering errors coming from the uncertainties in TDOA. A related report is on the inspection of heavy wall radioactive waste containers [144]. This does show we have to deal with thick-walled containers. Ozevin et al. [145] presented a source location method for open structures, like trussed towers or beams. Each element is modeled in 1-D and these elements are combined for complex spaced structures. A 2-D truss was modeled and compared to a laboratory scale bridge, confirming their approach to be valid. However, the practicality of such an approach depends on the attenuation of flaw AE signals and noise generated at joints inevitably present in large truss structures.

Section 3 covered various approaches to locate the origin of an AE signal, often referred to as an AE event. Basic location method developed from concepts used in seismology about 50 years ago, and was adapted to available instrumentation over the years. With advances in data processing capacity, newer approaches have sprung up in the past 20 years or so. Representative works were collected with brief description in this section.

4. Bridge Monitoring

AE monitoring on large bridges has a long history of successful outcome [5,9,12,146–148]. Initial emphasis was on steel truss structures as these contain many fatigue-prone joints in difficult to inspect locations. Fatigue damage of bridges has always presented technical challenge and remote monitoring capability with AE has become a viable solution [149]. Well-known fatigue damage in recent years was at the San Francisco–Oakland Bay Bridge, which led to a large-scale AE monitoring of the affected bridge segments following a successful demonstration of AE's capability [150,151]. For this project, 640 sensors of 60-kHz resonance frequency were used to monitor 384 eyebars covering over 6-km distance. The main aim was to detect cracks of 2.5-mm size. By the turn of the 21st century, bridge engineers' interests in acoustic (emission) monitoring of long-span suspension bridges finally reached the level for its practical implementation as shown by Hovhanessian [152]. Initial impetus came from the realization that a protection scheme of cable wire galvanizing (zinc coating) plus red lead (Pb_3O_4) paste was found inadequate on the Brooklyn Bridge and Williamsburg Bridge after nearly a century of continuous use. After 90–100 years, zinc coating was gone and many broken wires due to corrosion were found [153]. While these broken wires were repaired, more broken wires were also found in many newer bridges. On the Forth Road Bridge in the UK (a main span of 1006 m, opened in 1964), Colford [154] reported 8 out of 11,618 wires were broken at one of test locations in 2004. However, about 90% of the wires inspected exhibited Stage 3 (heavy) and 4 (severe) corrosion (Stage 5 being broken wire, according to NCHRP534 2004 [155]). After rehabilitation of the suspension cables along with the installation of a cable dehumidification system, fortified AE sensors were installed on the Forth Road Bridge in 2006 at 15 stations on each main cable (with about 140 m spacing over 2000 m span between anchorages). AE signals have been remotely monitored according to Hovhanessian [152], who earlier installed a similar AE monitoring system on the Bronx Whitestone Bridge (New York) in 1997 and Ancenis Bridge (France) in 2003. Additionally, M48 Severn Bridge (UK), Humber Bridge

(UK), Quincy Bayview Bridge (cable-stayed, US), Lane Memorial Bay Bridges (US), and a dozen others suspension bridges have AE monitoring systems installed [156].

The suspension-cable dehumidification system was invented for use on the Akashi Kaikyo Bridge (Japan) in 1998 and more than two dozen such systems have been installed globally for new bridges and for rehabilitated ones [156]. While it can keep wire corrosion to a minimum, perhaps eliminating the need for AE, suspension wires used with a conventional protection system will start to experience corrosion problems after 10–20 years. On such bridges, AE monitoring can detect serious deterioration on bridge elements. Hopefully, more AE installation can help reduce the number of structurally deficient bridges in the future.

Section 4 dealt with bridge monitoring using AE. This topic is usually discussed only among project participants or covered in civil engineering circles, but some of the reports provided technical details presented above.

5. Sensing and Signal Processing

Conventional piezoceramic sensors still dominate AE applications to structures, although PWAS has started to become smaller and more functional for potential aerospace uses [14]. Recently, Ono [41,157,158] reexamined sensor calibration methods for different types of wave propagation. These new approaches incorporated laser interferometry as the basis for calibration and Vallen [159] took an initial step for their standardization. It was also demonstrated that the so-called reciprocity calibration methods [160–162] have fundamental flaws. It was also found that most AE sensors fail to satisfy required reciprocity conditions [157,158].

Attempts to develop practical AE sensors based on optical fibers have continued [163,164], but cost and directionality problems still are obstacles. Yu et al. [165] presented a novel use of a fiber Bragg grating (FBG) sensor, a version that relies on phase shifts. This FBG sensor, connected by an optical fiber to a CFRP plate, has a broad bandwidth and allowed Lamb wave mode separation. Using FBG sensors for multiple channel operation needed in practical AE monitoring remains a challenge. Source location with arrays of FBG sensors are reported; e.g., Shrestha et al. [166] and Innes et al. [167].

On AE specific signal processing, Barat et al. [168] and Elizarov et al. [169] successfully developed new techniques for AE hit detection without relying on threshold crossing. They used an optimized combination of filters to detect the arrival of AE hit. This is possible due to the differences between noise processes and AE signals, which produce short-time perturbations. Their algorithms have been implemented into hardware. This approach traces back to high-order statistics and the use of short-time average and long-time average as used by Lokajícek [170]. Another method of hit time of arrival based on wavelet transform was given by Pomponi [171]. This method is conceptually easier to understand than those of high-order filtering and is based on a constraint imposed on wavelet decomposition. This comes from the rise time limit arising from the frequency characteristics of AE sensor used and an efficient signal denoising was achieved. Other methods were also reported [172,173]. Sagista et al. [174] applied a slightly different use of wavelet transform to amplitude distribution analysis defining b-values in terms of bandpass-filtered signal energy. Another innovative approach for pattern recognition analysis was explored by Godin and coworkers [175]. This method is based on genetic algorithm, which is a relatively new optimization method, pioneered by Goldberg [176]. See also [128] for its earlier application in source location. This approach can provide an alternate scheme is data clustering, which has relied primarily on the k-means method [177,178].

Another front of signal processing was AE tomography. Katsuyama et al. [179] presented the principle of this method in 1992, which was independently developed in 2004 by Schubert [180]. This utilizes AE hits (natural or artificial) received by multiple sensors to interrogate the conditions of wave paths. This imaging strategy can use attenuation, wave speed, or impedance mismatch. Shiotani and coworkers have applied it for concrete inspection and reported successful outcomes relying on wave speed variation [7,181,182]. Since concrete quality affects its sound velocity, special consideration was needed, including Kalman filtering [183]. Nishida et al. [184,185] applied this method to concrete bridge decks subjected to fatigue loading in laboratory and field use. A good correlation was demonstrated between fatigue cracking and wave speed loss. Takamine et al. [186] used a variation of this method in deck inspection.

Section 5 discussed AE signal detection or sensing and signal processing other than signal location, which was covered in Section 3. In sensing part, piezoelectric ceramics are still dominant, even for PWAS devices. Yet, their calibration methods lagged behind till recently. Despite high hopes held earlier, optical fiber-based detectors have not reached a breaking point, mainly from high component cost and from the lack of revolutionary concepts. Still, FBG sensors are advancing as discussed. Several new approaches of signal processing appeared in the last ten years as noted. AE tomography, developed originally in 1992, has made a significant progress of late.

6. Pressure Vessels and Tanks

Three AE applications have evolved in recent years at industrial scales. One was the inspection of small size consumer liquefied petroleum gas (LPG) tanks in Europe and was discussed by Tscheliesnig and coworkers [187]. The key to the success of this application was appropriate selection of inspection threshold parameters along with optimized sensor placement. Subsequently, AE testing was extended to larger than 13-m^3 tanks [188]. Next was the development of an examination method for high-pressure hydrogen tanks using AE, which led to the adaption of a special code case for ASME Boiler and Pressure Vessel Code, Section X [189,190]. Gorman [10] gave technical details of this approach, which was also discussed in [13]. The third AE application of significance is the life extension of self-contained breathing apparatus (SCBA), used by firemen and naval personnel. These are akin to SCUBA tanks, but have carbon-fiber wrapping over aluminum liners and are certified for the lifetime of 15 years. In April 2017, Gorman and coworkers at Digital Wave secured an exemption from the US Department of Transportation (DOT), allowing SCBA tanks to receive 15-year extension when these tanks pass AE inspection per DOT specifications. This permit was based on their studies conducted for US-DOT and for US Navy. DOT final report [191], which was completed in 2014, described technical details that led to evaluation criteria of SCBA tanks via AE testing. Anderson also provided the history of this inspection method and technical information that achieved the DOT approval [192]. The most critical aspect appears to be from fatigue and AE tests need to guarantee additional 15-year life. See [193,194] for a Navy report and recent specifications for SCBA testing from US DOT, issued 3 May 2018.

Stress rupture of structural members, also known as static fatigue, is an important topic and is central to predicting the remaining life of any structures for long term uses. Yet, it is extremely time consuming to conduct meaningful tests. For example, Digital Wave [191,192] relied on residual strength tests on damaged SCBA tanks, rather than long-term stress rupture testing, for assuring the remaining lifetime in SCBA testing. This topic was discussed in [13] in connection to pressure vessels made of fiber composites and the only known stress rupture life curves from Chang [195] were given. From this data, reproduced as Figure 8, one can see that Digital Wave's estimate was indeed within the life expectancy since they kept working pressure under 30% of rupture pressure for CFRP. The subject of lifetime prediction overlaps with the prediction of earthquakes and the underlying process is known as critical phenomenon. See [196] for discussion related to AE and earthquakes. For more than 10 years, Godin's group [197,198] has evaluated high temperature stress rupture response of SiC_f/Si-B-C composites. By conducting tests at 450–600 °C for up to 4000 h, they were able to identify a power-law behavior for the lifetime and to define AE-based parameters that relate to microscopic fracture processes; interfacial changes and fiber cracking. Figure 9 shows the stress rupture curves of SiC fiber bundles and SiC composites at 500 °C (see also related reports on ceramic matrix composites) [199–201].

Section 6 introduced three successful AE inspection technologies for LPG tanks, high pressure hydrogen tanks and SCBA tanks. Also discussed was stress rupture testing of a ceramic matrix composite at high temperatures. Life prediction with AE, if more studies become available in the future, might make a ground-breaking contribution to earthquake prediction.

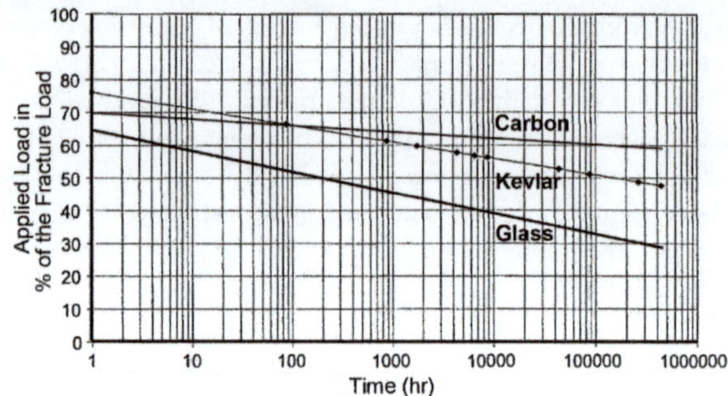

Figure 8. Applied stress vs. rupture time curves of composite-overwrapped pressure vessels with glass, carbon, and Kevlar fibers. Stress is expressed in percent of the short-time rupture pressure. Reproduced from [13], originally from Chang [195]. Reproduced with permission.

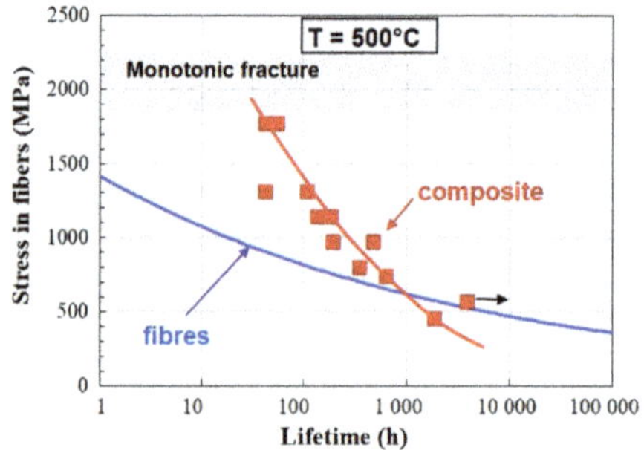

Figure 9. Applied stress vs. rupture time curves of SiC fiber bundles (blue) and SiC composites (red) at 500 °C. Shaded zone denoted monotonic fracture is for the short-time fracture stress of the SiC composite (1.85–2.25 GPa). From Godin et al. [198]. Reproduced with permission.

7. AE Inspection of Miscellaneous Processes

Some reports have appeared covering various processes. Since these are rare, yet useful, we list them below.

Tank-bottom inspection by AE has been widely practiced. For this testing, Papasalouros et al. [202] presented the statistical distribution data of their many inspection results. Naturally, such outcomes depend on specific sets of tested tanks, but this information should be of value to other inspectors and tank owners.

Kim et al. [203] monitored a blast furnace for steelmaking with AE for detecting cracks in its steel shell and leaks from hot air blower pipes. Burst and continuous signals were detected, respectively, providing a baseline data for devising an SHM system.

Serreti et al. [204] monitored the cold forming of aircraft wing panels using AE and seeking optimized forming parameters. The final aim is to automate the operation to get resultant panels equivalent to those made by qualified manual processes.

Zielke et al. [205] used AE to gain understanding of thermal spraying processes as AE signals are related to coating thickness achieved and to crack density of coatings. Different spray guns and their conditions must be selected properly and AE results provided guidelines to select the processes.

Manthei et al. [206] examined fracture of automobile windshields with AE and evaluated their fracture processes under static and dynamic loading together with finite element modeling. They identified the initiation point of fracture, which is crucial in proper modeling and support design.

Ravnik et al. [207] correlated low frequency AE during quenching of steel components to quenching cracks. Reliability of hydrophone detector was established and the system is ready for real component tests.

8. General Summary and Follow-Up Studies

Signal attenuation considered in Sections 2.1–2.3 led to the confirmation of the guided wave theories through reexamination of data from large structures as well as from laboratory-scale plate experiments. These theories for attenuation in isotropic media can be applied to predict signal loss in real structures of metallic alloys. For the distance of 10 m or so, the inverse square root distance (or $1/\sqrt{x}$) behavior is enough, while at longer propagation distances material absorption effects need to be added in the form of $(1/\sqrt{x}) \cdot \exp(-\alpha x)$. The guided wave attenuation coefficient can be calculated when reliable bulk wave attenuation coefficients are available. However, Section 2.5 revealed the lack of such data for most common alloys. For these cases, one needs to conduct attenuation tests as in [33–38]. It is hoped that more bulk wave attenuation data will be accumulated in the future, although shear wave attenuation tests are difficult.

A new approach for attenuation studies using complex elastic moduli was reviewed in Section 2.4. The damping factors for stiffness coefficients were obtained by the Castaings method. Since no direct verification is feasible, obtained damping factors for PMMA (in Section 2.4) and fiber-reinforced composites (in Section 2.6) were compared with damping factor determination from dynamic mechanical analysis and ultrasonic attenuation studies. The comparison showed Castaings' damping factors to typically be twice higher than those from other methods. Since the Castaings method was developed with sound theoretical foundation, some unaccounted factors appear to contribute to overestimation. A possible source is the common assumption made in most ultrasonic analysis, that is, the wave front to be planar. This was apparently a cause of higher bulk wave attenuation coefficients from immersion or buffer-rod methods as discussed in Section 2.5. In these ultrasonic methods, this point needs to be reevaluated. Clearly, the Castaings method is a valuable addition to viscoelastic analysis and making it verified will be beneficial to the field. Note that mechanical damping and ultrasonic attenuation studies are complementary, but these are usually separated. For example, Treviso's review [17] mentioned no complex elastic moduli works, and Castaings and Hosten [48,49] ignored low frequency damping studies. Still, an identical physical mechanism governs vibration energy loss in polymers like PMMA and epoxy [18,19].

Attenuation studies for fiber-reinforced composites considered in Section 2.6 show significant advances made in the past 15 years. Yet, the fact remains that attenuation in composites is higher than in homogeneous metallic alloys. This is especially true at higher frequencies above a few hundred kHz, where signal analysis must be conducted to characterize microscopic fracture mechanisms, as discussed in detail by Sause [208]. Even from practical engineering approach of using pattern recognition analysis, the availability of high frequency components in AE signals is an important element for success. In composite plate design, ply lay-up sequences are crucial parameters and these also affect attenuation characteristics. This part has not entered into AE or SHM consideration, but will eventually become necessary to incorporate it in attenuation analysis. Initial steps were made in examining the directivity of attenuation behavior [102,104,209], but more future works are needed, accompanied by modeling studies to avoid experimentation. Some of the past attenuation studies used a common AE or ultrasonic sensors [104,209] and this must be avoided in future works. As shown in Figure 6 above, a small aperture sensor of 1-mm diameter gives different attenuation behavior. While sensor aperture effects can be partly corrected by analysis [210], the sensor aperture must be smaller than the wavelength to minimize displacement cancellation effects. From the materials research side, high damping in newer high-strength carbon fibers and their composites is an interesting subject since there is no obvious mechanism for such a behavior. While not touched in this review, NDE of sandwiched structures is an important field by itself. Attenuation problems are far more serious issues and AE inspection has not progressed adequately. A general review [211] and a guided wave study [212] are cited as starters.

Section 3 surveyed algorithm developments for AE source location. Various approaches have been added since Ge's reviews [108,109] and each is expected to have strong and weak points. At this stage, it is hoped that some organizations or individuals formulate a signal generator software or a set of signal waveforms that can provide the basis for standard performance evaluation of an AE source location routine. It can start from a simple data set for testing location accuracy and speed. Eventually, an advanced data set should measure absolute/relative distance errors, robustness against noise and/or outliers, level of signal-to-noise ratio that can provide a certain level of location accuracy, as well as the minimum number of sensors needed for successful location. Without this type of standardized comparison, the objectivity of algorithm performance cannot be attained.

Section 4 covered AE applications to large bridges, which have been relatively unknown among AE engineers. It is hoped that more open reporting of success and failure promotes further technical developments for this crucial part of infrastructures.

Sensors and the rest of signal processing areas were discussed in Section 5. PWAS characterization is a logical next step in sensor calibration. Optical fiber sensor developments should be watched as a technical leap should come sooner or later. AE tomography has been applied primarily to concrete structures, but masonry can be another possible object for its application.

Section 6 highlighted four AE applications to structures, with three gaining commercial success. High temperature stress rupture monitoring is just as noteworthy technical success. With increasing uses of compressed gas in public and private transport—i.e., buses, trucks, and cars—the life extension of gas tanks can be a logical next step in AE applications.

Other AE applications worthy of mention were collected in Section 7. More applications of this type rarely appear in technical journals and one needs to peruse conference proceedings of AE and related fields.

9. Concluding Remarks

Limited aspects of AE uses have been reviewed, concentrating on the period of last several years. This review will hopefully provide an overview of various AE applications to large scale structures that will be discussed in this Special Issue on Structural Health Monitoring of Large Structures using Acoustic Emission Case Histories. However, many important issues were not covered here, especially on civil engineering side. On concrete issues, see a recent book edited by Ohtsu [213] and on SHM side, there are numerous books available. Another form of construction, masonry, has attracted increasing attention. Carpinteri et al. [214] examined historic masonry structures and identified internal crack distribution with AE source location methods. More recent reviews of AE studies of masonry structures were given by De Santis et al. [215] and Verstrynge et al. [216], again demonstrating the values of AE technology.

Author Contributions: The author has contributed from conception to writing in its entirety.

Funding: This research received no funding.

Acknowledgments: The author is grateful to N.G. for the use of Figure 9 and to UCLA Chapter of Materials Research Soc. SAMPE Competition team, N.S., team leader, for T700 CFRP plate samples. He also acknowledges numerous valuable discussions with AE colleagues in formulating opinions presented here, although any errors or omissions are his own.

Conflicts of Interest: The author declares no conflict of interest.

Appendix A   Limiting Frequency of Rayleigh Wave Propagation

Wave motion of Rayleigh waves decreases rapidly as the depth increases [20,21]. When the thickness of a plate, h, becomes smaller than three to four times the wavelength, wave propagation gradually shifts to Lamb wave modes. Hamstad [217] examined by experiment and by finite element analysis the limiting frequency of the Rayleigh wave propagation. Since this is of practical importance, some of his results are reproduced in Figure A1. He used a large steel plate (25.4-mm thick) in modeling Rayleigh wave propagation from a pencil-lead break. Wavelet transform spectrogram of calculated signal is shown in Figure A1a after 381-mm travel on the same side. Corresponding spectrogram for experimentally detected signal is given in Figure A1b. Both spectrograms show the strongest peak at the arrival time of Rayleigh waves (130 µs) and signal intensity diminishes below 250 kHz, shifting to Lamb wave S_0 and A_0 modes. His results of the low frequency limit of Rayleigh wave propagation, f_L, (in MHz) can be described by $f_L = 8.2/h$, as shown in Figure A1c. For a given thickness, h (in mm), Rayleigh waves exist only at frequencies above f_L. At h = 100 mm, this f_L is 82 kHz. This was the case for Graham–Alers attenuation study discussed in Section 2.2.

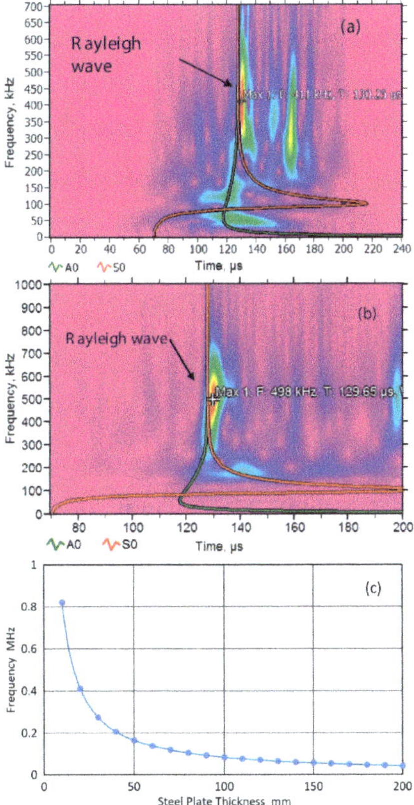

Figure A1. (**a**) Wavelet transform of calculated signal waveform, received at 381 mm from a pencil-lead break on the same surface of a steel plate of 25.4 mm thickness; (**b**) wavelet transform of experimentally received signal waveform, received at 381 mm from a pencil-lead break on the same surface of a steel plate of 24.4 mm thickness; (**c**) the low frequency limit of Rayleigh wave propagation (in MHz) vs. steel plate thickness (in mm). These figures were rearranged from Figures 13a,b and 19 by Hamstad [217]. Reproduced with permission.

References

1. Green, A.T.; Lockman, C.S.; Steele, R.K. Acoustic verification of structural integrity of Polaris chambers. *Mod. Plast.* **1964**, *41*, 137–139.
2. Liptai, R.G.; Harris, D.O.; Tatro, C.A. (Eds.) *Acoustic Emission*; STP-505; American Society for Testing and Materials: Philadelphia, PA, USA, 1972; 337p.
3. Spanner, J.C.; McElroy, J.W. *Monitoring Structural Integrity by Acoustic Emission*; STP-571; American Society for Testing and Materials: Philadelphia, PA, USA, 1975; 289p.
4. Fowler, T.J.; Blessing, J.A.; Conlisk, P.J.; Swanson, T.L. The MONPAC system. *J. Acoust. Emiss.* **1989**, *8*, 1–10.
5. Miller, R.K. (Ed.) Acoustic emission. In *Nondestructive Testing Handbook*, 3rd ed.; American Society for Nondestructive Testing: Columbus, OH, USA, 2005; Volume 6, 446p.
6. Ono, K. (Ed.) Journal Acoustic Emission. 1982–. Available online: www.aewg.org (accessed on 15 April 2017).
7. Shiotani, T. Recent advances of AE technology for damage assessment of infrastructures. *J. Acoust. Emiss.* **2012**, *30*, 76–99.
8. Bohse, J. Acoustic emission. In *Handbook of Technical Diagnostics: Fundamentals and Application to Structures and Systems*; Czichos, H., Ed.; Springer: Berlin, Germany, 2013; pp. 137–160.
9. Hay, D.R.; Cavaco, J.A.; Mustafa, V. Monitoring the civil infrastructure with acoustic emission: Bridge case studies. *J. Acoust. Emiss.* **2009**, *27*, 1–9.
10. Gorman, M.R. Modal AE analysis of fracture and failure in composite materials, and the quality and life of high pressure composite pressure vessels. *J. Acoust. Emiss.* **2011**, *29*, 1–28.
11. Rumsey, M.A.; Paquette, J.; White, J.R.; Werlink, R.J.; Beattie, A.G.; Pitchford, C.W.; van Dam, J. Experimental results of structural health monitoring of wind turbine blades, AIAA Paper AIAA-2008-1348. 2008. Available online: www.sandia.gov/wind/asme/AIAA-2008-1348.pdf (accessed on 25 April 2012).
12. Ono, K. Application of acoustic emission for structure diagnosis. *Diagnostyka Diagn. Struct. Health Monit.* **2011**, *2*, 3–18.
13. Ono, K.; Gallego, A. Research and applications of AE on advanced composites. *J. Acoust. Emiss.* **2012**, *30*, 180–229.
14. Giurgiutiu, V. *Structural Health Monitoring of Aerospace Composites*; Academic Press: New York, NY, USA, 2015; 470p.
15. Mitra, M.; Gopalakrishnan, S. Guided wave based structural health monitoring: A review. *Smart Mater. Struct.* **2016**, *25*, 053001. [CrossRef]
16. Gresil, M.; Giurgiutiu, V. Prediction of attenuated guided waves propagation in carbon fiber composites using Rayleigh damping model. *J. Intell. Mater. Syst. Struct.* **2015**, *26*, 2151–2169. [CrossRef]
17. Treviso, A.; Van Genechten, B.; Mundo, D.; Tournour, M. Damping in composite materials: Properties and models. *Compos. Part B Eng.* **2015**, *78*, 144–152. [CrossRef]
18. Jarzynski, J.; Balizer, E.; Fedderly, J.J.; Lee, G. *Acoustic Properties—Encyclopedia of Polymer Science and Technology*; Wiley: New York, NY, USA, 2003. [CrossRef]
19. Sinha, M.; Buckly, D.J. Acoustic properties of polymers. In *Physical Properties of Polymers Handbook*; Part X; Springer: Berlin, Germany, 2007; pp. 1021–1031.
20. Viktorov, I.A. *Rayleigh and Lamb Waves: Physical Theory and Applications*; Plenum: New York, NY, USA, 1967; p. 154.
21. Rose, J.L. *Ultrasonic Waves in Solid Media*; Cambridge University Press: Cambridge, UK, 1999; 476p.
22. Mal, A.K.; Bar-Cohen, Y.; Lih, S.-S. Wave attenuation in fiber reinforced composites. In *Mechanics and Mechanisms of Material Damping*; ASTM STP-1169; American Society for Testing and Materials: Philadelphia, PA, USA, 1992; pp. 245–261.
23. Press, F.; Healy, J. Absorption of Rayleigh waves in low-loss media. *J. Appl. Phys.* **1957**, *28*, 1323–1325. [CrossRef]
24. Zhukov, K.V.; Merkulov, L.G.; Pigulevskii, E.D. Normal mode damping in a plate with free edges. *Sov. Phys. Acoust.* **1964**, *10*, 133–136.
25. Krautkramer, J.; Krautkramer, H. *Ultrasonic Testing of Materials*; Springer: Berlin, Germany, 1969; p. 521.
26. Mason, W.J.; McSkimin, H.J. Attenuation and scattering of high frequency sound waves in metals and glasses. *J. Acoust. Soc. Am.* **1947**, *19*, 464–473. [CrossRef]

27. Roderick, R.L.; Truell, R. The measurement of ultrasonic attenuation in solids by the pulse technique and some results in steel. *J. Appl. Phys.* **1952**, *23*, 267–279. [CrossRef]

28. Kamigaki, K. Ultrasonic attenuation in steel and cast iron. *Sci. Rep. Res. Inst. Tohoku Univ. Ser. A* **1957**, *9*, 48–77.

29. Papadakis, E.P. Ultrasonic attenuation and velocity in three transformation products in steel. *J. Acoust. Soc. Am.* **1963**, *35*, 1884. [CrossRef]

30. Papadakis, E.P. Ultrasonic attenuation in SAE 3140 and 4150 steel. *J. Acoust. Soc. Am.* **1960**, *32*, 1628–1639. [CrossRef]

31. Kinsler, L.E.; Frey, A.R.; Coppens, A.B.; Sanders, J.V. *Fundamentals of Acoustics*, 3rd ed.; Wiley: New York, NY, USA, 1982.

32. Cai, C.; Zheng, H.; Khan, M.S.; Hung, K.C. Modeling of Material Damping Properties in ANSYS. 2002 International ANSYS Conference Proceedings. Available online: https://support.ansys.com/staticassets/ANSYS/staticassets/resourcelibrary/confpaper/2002-Int-ANSYS-Conf-197.PDF (accessed on 19 May 2018).

33. Graham, L.J.; Alers, G.A. Acoustic emission in the frequency domain. In *Monitoring Structural Integrity by Acoustic Emission*; STP-571; American Society for Testing and Materials: Philadelphia, PA, USA, 1975; pp. 11–39.

34. Pollock, A.A. Classical wave theory in practical AE testing. In *Progress in AE III*; The Japanese Society for Non-Destructive Inspection: Tokyo, Japan, 1986; pp. 708–721.

35. Blackburn, P.R. Acoustic emission from fatigue cracks in Cr-Mo steel cylinders. *J. Acoust. Emiss.* **1988**, *7*, 49–56.

36. Baran, I.; Lyasota, I.; Skrok, K. Acoustic emission testing of underground pipelines of crude oil of fuel storage depots. In Proceedings of the 32nd European Conference on Acoustic Emission Testing, Prague, Czech Republic, 7–9 September 2016; pp. 15–26.

37. Šofer, M.; Crha, J.; Zengerle, H. From near field to far field and beyond. In Proceedings of the 32nd European Conference on Acoustic Emission Testing, Prague, Czech Republic, 7–9 September 2016; pp. 475–483.

38. Bryla, P.; Walaszek, H.; Herve, C.; Catty, J. Real time & long term acoustic emission monitoring: A new way to use acoustic emission—Application to hydroelectric penstocks and paper machine. In Proceedings of the 31st European Conference on Acoustic Emission Testing, Dresden, Germany, 3–5 September 2014.

39. El-Shaib, M.N. Predicting Acoustic Emission Attenuation in Solids Using Ray-Tracing within a 3D Solid Model. Ph.D. Thesis, Heriot-Watt University, Edinburgh, UK, 2012; 230p.

40. Ono, K.; Hayashi, T.; Cho, H. Bar-wave calibration of acoustic emission sensors. *Appl. Sci.* **2017**, *7*, 964. [CrossRef]

41. Ono, K. On the piezoelectric detection of guided ultrasonic waves. *Materials* **2017**, *10*, 1325. [CrossRef] [PubMed]

42. Zhang, J.; Perez, R.J.; Lavernia, E.J. Documentation of damping capacity of metallic, ceramic and metal-matrix composite materials. *J. Mater. Sci.* **1993**, *28*, 2395–2404. [CrossRef]

43. Papadakis, E.P. Ultrasonic attenuation caused by scattering in polycrystalline metals. *J. Acoust. Soc. Am.* **1965**, *37*, 711–717. [CrossRef]

44. Drinkwater, B.W.; Castaings, M.; Hosten, B. The interaction of Lamb waves with solid-solid interfaces. *AIP Conf. Proc.* **2003**, *657*, 1064–1071. [CrossRef]

45. Zhang, F.; He, G.; Xu, K.; Wu, H.; Guo, S. The damping and flame-retardant properties of poly(vinyl chloride)/chlorinated butyl rubber multilayered composites. *J. Appl. Polym. Sci.* **2015**, *132*. [CrossRef]

46. Povolo, F.; Goyanes, S.N. Amplitude-dependent dynamical behavior of poly(methyl methacrylate). *Polym. J.* **1994**, *26*, 1054–1062. [CrossRef]

47. Lee, T.; Lakes, R.S.; Lal, A. Resonant ultrasound spectroscopy for measurement of mechanical damping: Comparison with broadband viscoelastic spectroscopy. *Rev. Sci. Instrum.* **2000**, *71*, 2855–2861. [CrossRef]

48. Castaings, M.; Hosten, B. The use of electrostatic, ultrasonic, air-coupled transducers to generate and receive Lamb waves in anisotropic, viscoelastic plates. *Ultrasonics* **1998**, *36*, 361–365. [CrossRef]

49. Castaings, M.; Hosten, B. Air-coupled measurement of plane wave, ultrasonic plate transmission for characterising anisotropic, viscoelastic materials. *Ultrasonics* **2000**, *38*, 781–786. [CrossRef]

50. Castaings, M.; Hosten, B.; Kundu, T. Inversion of ultrasonic, plane-wave transmission data in composite plates to infer viscoelastic material properties. *NDT E Int.* **2000**, *33*, 377–392. [CrossRef]

51. Capodagli, J.; Lakes, R. Isothermal viscoelastic properties of PMMA and LDPE over 11 decades of frequency and time: A test of time–temperature superposition. *Rheol. Acta* **2008**, *47*, 777–786. [CrossRef]

52. Wang, Y.C.; Ko, C.C.; Wu, H.K.; Wu, Y.T. Pendulum-type viscoelastic spectroscopy for damping measurement of solids. *J. Jpn. Soc. Exp. Mech.* **2013**, *13*, s137–s142.

53. Thakur, V.K.; Vennerberg, D.; Madbouly, S.A.; Kessler, M.R. Bio-inspired green surface functionalization of PMMA for multifunctional capacitors. *RSC Adv.* **2014**, *4*, 6677–6684. [CrossRef]

54. Treiber, M.; Kim, J.Y.; Jacobs, L.J. Correction for partial reflection in ultrasonic attenuation measurements using contact transducers. *J. Acoust. Soc. Am.* **2009**, *125*, 2946–2953. [CrossRef] [PubMed]

55. Carlson, J.E.; van Deventer, J.; Scolan, A.; Carlander, C. Frequency and temperature dependence of acoustic properties of polymers used in pulse-echo systems. In Proceedings of the 2003 IEEE Symposium on Ultrasonics, Honolulu, HI, USA, 5–8 October 2003; pp. 885–888.

56. Pouet, B.F.; Rasolofosaon, N.J.P. Measurement of broadband intrinsic ultrasonic attenuation and dispersion in solids with laser techniques. *J. Acoust. Soc. Am.* **1993**, *93*, 1286–1292. [CrossRef]

57. Hartmann, B.; Jarzynski, J. Ultrasonic hysteresis absorption in polymers. *J. Appl. Phys.* **1972**, *43*, 4304–4312. [CrossRef]

58. Merkulov, L.G. Damping of normal mode in a plate immersed in a liquid. *Sov. Phys. Acoust.* **1964**, *10*, 169–173.

59. Shehadeh, M.F. Monitoring of Long Steel Pipes Using Acoustic Emission. Ph.D. Thesis, Heriot-Watt University, Edinburgh, UK, 2006; 204p.

60. European Coop. Space Standardization (ECSS). In *Space Engineering, Structural Materials Handbook, Part I, Overview and Materials Properties and Applications*; EDSS-E-HB-32-20 Part 1A; Sec. 5.4; ECSS: Noordwijk, The Netherlands, 2011; 535p.

61. Neau, G.; Lowe, M.J.S.; Deschamps, M. Propagation of Lamb waves in anisotropic and absorbing plates: Theoretical derivation and experiments. *AIP Conf. Proc.* **2002**, *615*, 1062–1069. [CrossRef]

62. Matt, H.M. Structural Diagnostics of CFRP Composite Aircraft Components by Ultrasonic Guided Waves and Built-in Piezoelectric Transducers. Ph.D. Thesis, University of California, San Diego, CA, USA, 2006; 242p.

63. Calomfirescu, M.; Herrmann, A.S. On the propagation of Lamb waves in viscoelastic composites for SHM applications. *Key Eng. Mater.* **2007**, *347*, 543–548. [CrossRef]

64. Calomfirescu, M. Lamb Waves for Structural Health Monitoring in Viscoelastic Composite Materials. Ph.D. Thesis, Universitat Bremen, Bremen, Germany, 2008; 128p.

65. Schmidt, D.; Sadri, H.; Szewieczek, A.; Sinapius, M.; Wierach, P.; Siegert, I.; Wendemuth, A. Characterization of Lamb wave attenuation mechanisms. In *Health Monitoring of Structural and Biological Systems*; SPIE: San Diego, CA, USA, 2013; Volume 8695, p. 869503. [CrossRef]

66. Schmidt, D.; Wierach, P.; Sinapius, M. Mode selective actuator-sensor system for Lamb wave-based structural health monitoring. In Proceedings of the 7th European Workshop on SHM, Nantes, France, 8–11 July 2014.

67. Papadakis, E.P. The measurement of ultrasonic attenuation. In *Methods of Experimental Physics*; Edmonds, P.D., Ed.; Academic Press: New York, NY, USA, 1981; Chapter 3; pp. 108–156.

68. Klinman, R.; Webster, G.R.; Marsh, F.J.; Stephenson, E.T. Ultrasonic prediction of grain size, strength, and toughness in plain carbon steel. *Mater. Eval.* **1980**, *38*, 26–32.

69. Klinman, R.; Stephenson, E.T. Ultrasonic prediction of grain size and mechanical properties in plain carbon steel. *Mater. Eval.* **1981**, *39*, 1116–1120.

70. Smith, R.L.; Reynolds, W.N.; Wadley, H.N.G. Ultrasonic attenuation and microstructure in low-carbon steels. *Met. Sci.* **1981**, *15*, 554–558. [CrossRef]

71. Smith, R.L.; Rusbridge, K.L.; Reynolds, W.N.; Hudson, B. Ultrasonic attenuation, microstructure, ductile to brittle transition temperature in Fe-C alloys. *Mater. Eval.* **1983**, *41*, 219–222.

72. Papadakis, E.P. Ultrasonic Attenuation and velocity in SAE 52100 steel quenched from various temperatures. *Met. Trans.* **1970**, *1*, 1053–1057.

73. Latiff, R.H.; Fiore, N.F. Ultrasonic attenuation in spheroidized steel. *J. Appl. Phys.* **1974**, *45*, 5182–5186. [CrossRef]

74. Coronel, V.F.; Beshers, D.N. Magnetomechanical damping in iron. *J. Appl. Phys.* **1988**, *64*, 2006–2015. [CrossRef]

75. Ohtani, T.; Ogi, T.; Hirao, M. Ultrasonic attenuation peak during creep of a nickel-base superalloy with electromagnetic acoustic resonance. *J. Soc. Mater. Sci. Jpn.* **2004**, *53*, 692–698. [CrossRef]

76. Hirao, M.; Ogi, H. *EMATS for Science and Industry, Noncontacting Ultrasonic Measurements*; Kluwer: Boston, MA, USA, 2003; p. 369.

77. Ohtani, T.; Ogi, H.; Hirao, M. Change of ultrasonic attenuation and microstructure evolution in crept 2.25%Cr-1%Mo steels. *J. Soc. Mater. Sci. Jpn.* **2002**, *51*, 195–201. (In Japanese) [CrossRef]

78. Ohtani, T.; Ogi, H.; Hirao, M. Electromagnetic acoustic resonance to assess creep damage in Cr–Mo–V steel, Japan. *J. Appl. Phys.* **2006**, *45*, 4526–4533. [CrossRef]

79. Ohtani, T.; Takei, K. Change of ultrasonic attenuation and microstructure evolution during creep of SUS316L austenite stainless steels. *J. Soc. Mater. Sci. Jpn.* **2005**, *54*, 607–614. (In Japanese) [CrossRef]

80. Reynolds, W.N.; Smith, R.N. Ultrasonic wave attenuation spectra in steels. *J. Phys. D Appl. Phys.* **1984**, *17*, 109–116. [CrossRef]

81. Generazio, E.R. *The Role of the Reflection Coefficient in Precision Measurement of Ultrasonic Attenuation*; NASA Technical Memorandum 83788; NASA Lewis Research Center: Cleveland, OH, USA, 1984; 32p.

82. Hikata, A.; Truell, R. Frequency dependence of ultrasonic attenuation and velocity on plastic deformation. *J. Appl. Phys.* **1957**, *28*, 522–523. [CrossRef]

83. Mason, W.P.; McSkimin, H.J. Energy losses of sound waves in metals due to scattering and diffusion. *J. Appl. Phys.* **1948**, *19*, 940–946. [CrossRef]

84. Zeng, F.; Agnew, S.R.; Raeisinia, B.; Myneni, G.R. Ultrasonic attenuation due to grain boundary scattering in pure niobium. *J. Nondestruct. Eval.* **2010**, *29*, 93–103. [CrossRef]

85. Ahn, B.; Lee, S.S. Effect of microstructure of low carbon steels on ultrasonic attenuation. *IEEE Trans. Ultrason. Ferroelectr. Freq. Control* **2000**, *47*, 620–629. [PubMed]

86. Du, H.; Turner, J.A. Ultrasonic attenuation in pearlitic steel. *Ultrasonics* **2014**, *54*, 882–887. [CrossRef] [PubMed]

87. Wan, T.; Naoe, T.; Wakui, T.; Futakawa, M.; Obayashi, H.; Sasa, T. Effects of grain size on ultrasonic attenuation in type 316L stainless steel. *Materials* **2017**, *10*, 753. [CrossRef] [PubMed]

88. Nadeau, F.; Bussiere, J.F.; Van Drunen, G. On the relation between ultrasonic attenuation and fracture toughness in type 403 stainless steel. *Mater. Eval.* **1985**, *43*, 101–107.

89. Ramuhalli, P.; Good, M.S.; Diaz, A.A.; Anderson, M.T.; Watson, B.E.; Peters, T.J.; Dixit, M.; Bond, L.J. *Ultrasonic Characterization of Cast Austenitic Stainless Steel Microstructure: Discrimination between Equiaxed- and Columnar-Grain Material—An Interim Study*; PNNL-18912; Pacific Northwest National Laboratory: Richland, WA, USA, 2009; 84p.

90. Jennes, J.R., Jr.; Kline, D.E. The dynamic mechanical properties of some *epoxy* matrix composites. *J. Appl. Polym. Sci.* **1973**, *17*, 3391–3422. [CrossRef]

91. Dokun, O.D.; Jacobs, L.J.; Haj-Ali, R.M. Ultrasonic techniques to quantify material degradation in FRP composites. In *Review of Progress in Quantitative Nondestructive Evaluation*; Springer: Boston, MA, USA, 1999; pp. 1365–1371.

92. Roman, I.; Ono, K. Acoustic emission characterization of failure mechanisms in Woven roving glass-epoxy composites. In *Progress in Acoustic Emission II*; Japanese Society for Non-Destructive Inspection: Tokyo, Japan, 1984; pp. 496–503.

93. Crane, R.M. *Vibration Damping Response of Composite Materials*; US Navy Report; DTRC-SME-91/12; April 1991 AD A235 614; David Taylor Research Center: Bethesda, MD, USA, 1991; 303p.

94. Vantomme, J. A parametric study of material damping in fibre-reinforced plastics. *Composites* **1995**, *26*, 147–153. [CrossRef]

95. Yim, J.H.; Gillespie, J.W., Jr. Damping characteristics of 0° and 90° AS4/3501-6 unidirectional laminates including the transverse shear effect. *Compos. Struct.* **2000**, *50*, 217–225. [CrossRef]

96. Jalili, M.M.; Mousavi, S.Y.; Pirayeshfa, A.S. Investigating the acoustical properties of carbon fiber-, glass fiber-, and hemp fiber-reinforced polyester composites. *Polym. Compos.* **2014**, *35*, 2103–2111. [CrossRef]

97. Kim, H.C.; Park, J.M. Ultrasonic wave propagation in carbon fibre-reinforced plastics. *J. Mater. Sci.* **1987**, *22*, 4536–4540. [CrossRef]

98. Watanabe, Y.; Biwa, S.; Ohno, N. Experimental investigation of ultrasonic attenuation behavior in carbon fiber reinforced epoxy composites. *J. Soc. Mater. Sci. Jpn.* **2002**, *51*, 451–457. [CrossRef]

99. Williams, J.H., Jr.; Lee, S.S.; Nayeb-Hashemi, H. Ultrasonic wave propagation loss factor in composite in terms of constituent properties. *J. Nondestruct. Eval.* **1980**, *1*, 191–199. [CrossRef]

100. Seki, H.; Granato, A.; Truell, R. Diffraction effects in the ultrasonic field of a piston source and their importance in the accurate measurement of attenuation. *J. Acoust. Soc. Am.* **1956**, *28*, 230–238. [CrossRef]

101. Papadakis, E.P. Ultrasonic velocity and attenuation: Measurement methods with scientific and industrial applications. In *Physical Acoustics*; Academic Press: New York, NY, USA, 1976; Volume XII, pp. 277–375.

102. Ono, K. Acoustic emission behavior of flawed unidirectional carbon fiber-epoxy composites. *J. Reinf. Plast. Compos.* **1988**, *7*, 90–105. [CrossRef]

103. Adams, R.D. The dynamic longitudinal shear modulus and damping of carbon fibres. *J. Phys. D Appl. Phys.* **1975**, *8*, 738. [CrossRef]

104. Ishikawa, M.; Kogo, Y.; Koyanagi, J.; Tanaka, F.; Okabe, T. Torsional modulus and internal friction of polyacrylonitrile and pitch-based carbon fibers. *J. Mater. Sci.* **2015**, *50*, 7018–7025. [CrossRef]

105. Olivier, P.A.; Marguerès, P.; Mascaro, B.; Collombet, F. CFRP with voids: Ultrasonic characterization of localized porosity, acceptance criteria and mechanical characteristics. In Proceedings of the 16th International Conference on Composite Materials, Kyoto, Japan, 3–8 July 2007.

106. Ono, K.; Gallego, A. Attenuation of Lamb waves in CFRP plates. *J. Acoust. Emiss.* **2012**, *30*, 109–123.

107. Weihnacht, B.; Schulze, E.; Frankenstein, B. Acoustic emission analysis in the dynamic fatigue testing of fiber composite components. In Proceedings of the 31st European Working Group on Acoustic Emission, Dresden, Germany, 3–5 September 2014.

108. Ge, M. Analysis of source location algorithms, Part I: Overview and non-iterative methods. *J. Acoust. Emiss.* **2003**, *21*, 14–28.

109. Ge, M. Analysis of source location algorithms, Part II: Iterative methods. *J. Acoust. Emiss.* **2003**, *21*, 29–51.

110. Summerscales, J. *Acoustic Emission Source Location in Fibre-Reinforced Composite Materials*; Advanced Composites Manufacturing Centre at Plymouth University: Plymouth, UK, 2013; 36p, ISBN 978-1-870918-04-6.

111. Kundu, T. Acoustic source localization. *Ultrasonics* **2014**, *54*, 25–38. [CrossRef] [PubMed]

112. Holford, K.M.; Eaton, M.J.; Hensman, J.J.; Pullin, R.; Evans, S.L.; Dervilis, N.; Worden, K. A new methodology for automating acoustic emission detection of metallic fatigue fractures in highly demanding aerospace environments: An overview. *Prog. Aerosp. Sci.* **2017**, *90*, 1–11. [CrossRef]

113. Li, X.; Deng, Z.D.; Rauchenstein, L.T.; Carlson, T.J. Contributed Review: Source-localization algorithms and applications using time of arrival and time difference of arrival measurements. *Rev. Sci. Instrum.* **2016**, *87*, 041502. [CrossRef] [PubMed]

114. Hutton, P.H.; Jolly, W.D.; Vetrano, J.B. Acoustic emission for periodic and continuous flaw detection in pressure vessels. In Proceedings of the U.S.-Japan Joint Symposium on Acoustic Emission, Tokyo, Japan, 4–6 July 1972.

115. Crostack, H.A.; Böhm, P. Monitoring of a pressure vessel (ZB2) by means of acoustic emission. *J. Acoust. Emiss.* **1990**, *9*, 29–36.

116. Aljets, D.; Chong, A.; Wilcox, S.; Holford, K. Acoustic emission source location in plate-like structures using a closely arranged triangular sensor array. *J. Acoust. Emiss.* **2010**, *28*, 85–98.

117. Kundu, T.; Nakatani, H.; Takeda, N. Acoustic source localization in anisotropic plates. *Ultrasonics* **2012**, *52*, 740–746. [CrossRef] [PubMed]

118. Scholey, J.J.; Wilcox, P.D.; Wisnom, M.R.; Friswell, M.; Pavier, M.; Aliha, M.R. A generic technique for acoustic emission source location. *J. Acoust. Emiss.* **2009**, *27*, 291–298.

119. Boller, C.; Chang, F.K.; Fujino, Y. (Eds.) *Encyclopedia of Structural Health Monitoring*; Wiley: Hoboken, NJ, USA, 2009.

120. He, T.; Pan, Q.; Liu, Y.; Liu, X.; Hu, D. Near-field beamforming analysis for acoustic emission source localization. *Ultrasonics* **2012**, *52*, 587–592. [CrossRef] [PubMed]

121. Li, X.; Dong, L. An efficient closed-form solution for acoustic emission source location in three-dimensional structures. *AIP Adv.* **2014**, *4*, 027110. [CrossRef]

122. Spencer, S.J. Closed-form analytical solutions of the time difference of arrival source location problem for minimal element monitoring arrays. *J. Acoust. Soc. Am.* **2010**, *127*, 2943–2954. [CrossRef] [PubMed]

123. Liu, H.; Milios, E. Acoustic positioning using multiple microphone arrays. *J. Acoust. Soc. Am.* **2005**, *117*, 2772–2782. [CrossRef] [PubMed]

124. Mellen, G.; Pachter, M.; Raquet, J. Closed-form solution for determining emitter location using time difference of arrival measurements. *IEEE Trans. Aerosp. Electron. Syst.* **2003**, *39*, 1056–1058. [CrossRef]

125. Hensman, J.J.; Mills, R.; Pierce, S.G.; Worden, K.; Eaton, M.J. Locating acoustic emission sources in complex structures using Gaussian processes. *Mech. Syst. Signal Process.* **2010**, *24*, 211–223. [CrossRef]

126. Ebrahimkhanlou, A.; Salamone, S. A probabilistic framework for single-sensor acoustic emission source localization in thin metallic plates. *Smart Mater. Struct.* **2017**, *26*, 095026. [CrossRef]
127. Jiang, M.; Lu, S.; Sai, Y.; Sui, Q.; Jia, L. Acoustic emission source localization technique based on least squares support vector machine by using FBG sensors. *J. Mod. Opt.* **2014**, *61*, 1634–1640. [CrossRef]
128. Worden, K.; Staszewski, W.J. Impact location and quantification on a composite panel using neural networks and a genetic algorithm. *Strain* **2000**, *36*, 61–70. [CrossRef]
129. Gorman, M.R. Plate wave acoustic emission. *J. Acoust. Soc. Am.* **1991**, *90*, 358–364. [CrossRef]
130. Ziola, S.M.; Gorman, M.R. Source location in thin plates using cross-correlation. *J. Acoust. Soc. Am.* **1991**, *90*, 2551–2556. [CrossRef]
131. Gorman, M.R.; Prosser, W.H. AE source location by plate wave analysis. *J. Acoust. Emiss.* **1990**, *9*, 283–288.
132. Yamada, H.; Mizutani, Y.; Nishino, H.; Takemoto, M.; Ono, K. Lamb wave source location of impact on anisotropic plates. *J. Acoust. Emiss.* **2000**, *18*, 51–60.
133. Maji, A.K.; Satpathi, D.; Kratochvil, T. Acoustic emission source location using Lamb wave modes. *J. Eng. Mech.* **1997**, *123*, 154–161. [CrossRef]
134. Park, W.H.; Packo, P.; Kundu, T. Acoustic source localization in an anisotropic plate without knowing its material properties—A new approach. *Ultrasonics* **2017**, *79*, 9–17. [CrossRef] [PubMed]
135. Ebrahimkhanlou, A.; Salamone, S. AE source localization in thin metallic plates: A single sensor approach based on multimodal edge reflections. *Ultrasonics* **2017**, *78*, 134–145. [CrossRef] [PubMed]
136. Al-Jumaili, S.K.; Pearson, M.R.; Holford, K.M.; Eaton, M.J.; Pullin, R. Acoustic emission source location in complex structures using full automatic delta T mapping technique. *Mech. Syst. Signal Process.* **2016**, *72–73*, 513–524. [CrossRef]
137. Dorme, C.; Fink, M. Focusing in transmit-receive mode through inhomogeneous media: The TR matched filter approach. *J. Acoust. Soc. Am.* **1995**, *98*, 1155–1162. [CrossRef]
138. Tanter, M.; Thomas, J.L.; Fink, M. Time reversal and the inverse filter. *J. Acoust. Soc. Am.* **2000**, *108*, 223–234. [CrossRef] [PubMed]
139. Ing, R.-K.; Quieffin, N.; Catheline, S.; Fink, M. In solid localization of finger impacts using acoustic time-reversal process. *Appl. Phys. Lett.* **2005**, *87*, 204104. [CrossRef]
140. Ciampa, F.; Meo, M. Impact detection in anisotropic materials using a time reversal approach. *Struct. Health Monit.* **2011**, *11*, 43–49. [CrossRef]
141. Douma, J.; Niederleithinger, E.; Snieder, R. Locating events using time reversal and deconvolution: Experimental application and analysis. *J. Nondestruct. Eval.* **2015**, *34*, 2. [CrossRef]
142. Robert, E.; Jurg, D. Acoustic emission source detection using the time reversal principle on dispersive waves in beams. In Proceedings of the 2013 International Congress on Ultrasonics (ICU 2013), Singapore, 2–5 May 2013. [CrossRef]
143. Liu, F.; Ding, S.; Hu, D.; Wang, Q.; Xu, Y.; Kong, S.; Zheng, M. Sound source location error analysis of acoustic emission technique for thick-wall pressure vessel inspection. In Proceedings of the ASME 2009 Pressure Vessels and Piping Conference, Prague, Czech Republic, 26–30 July 2009; pp. 485–491. [CrossRef]
144. Iliopoulos, S.; Aggelis, D.G.; Pyl, L.; Vantomme, J.; Van Marcke, P.; Coppens, E.; Areias, E. Detection and evaluation of cracks in the concrete buffer of the Belgian nuclear waste container using combined NDT techniques. *Constr. Build. Mater.* **2015**, *78*, 369–378. [CrossRef]
145. Ozevin, D. Geometry-based spatial acoustic source location for spaced structures. *Struct. Health Monit.* **2010**, *10*, 503–510. [CrossRef]
146. Nair, A.; Cai, C.S. Acoustic emission monitoring of bridges, review and case studies. *Eng. Struct.* **2010**, *32*, 1704–1714. [CrossRef]
147. Washer, G. Nondestructive evaluation methods for bridge elements. In *Bridge Engineering Handbook*, 2nd ed.; Construction and Maintenance; CRC Press: Boca Raton, FL, USA, 2014; pp. 301–335.
148. Kosnik, D.E.; Hopwood, T.; Corr, D.J. Acoustic emission monitoring for assessment of steel bridge details. *AIP Conf. Proc.* **2011**, *1335*, 1410–1417.
149. Attanayake, U.; Aktan, H.; Hay, R.; Catbas, N. *Remote Monitoring of Fatigue-Sensitive Details on Bridges*; Report MDOT RC-1629; Michigan Department of Transport: Kalamazoo, MI, USA, 2015; 49p.
150. Johnson, M.B.; Ozevin, D.; Washer, G.; Ono, K.; Gostautas, R.; Tamutus, T. *Acoustic Emission Method for Real-Time Detection of Steel Fatigue Crack in Eyebar*; National Academy of Sciences: Washington, DC, USA, 2012; pp. 72–79.

151. Ley, O.; Gostautas, R.; Godinez-Azcuaga, V.F. Recent advances in structural health monitoring using acoustic emission. In *Progress in Acoustic Emission XVIII*; Japanese Society for Non-Destructive Inspection: Tokyo, Japan, 2016; pp. 153–156.

152. Hovhanessian, G.; Laurent, E. Instrumentation and monitoring of critical structural elements unique to suspension bridges. In *Advances in Cable-Supported Bridges*; Taylor & Frances: London, UK, 2006; Chapter 8; pp. 111–119.

153. Sluszka, P. *Studies on the Longevity of Suspension Bridge Cables*; Transportation Research Record 1290; National Academy of Sciences: Washington, DC, USA, 1990; pp. 272–278.

154. Colford, B. The maintenance of long span bridges. In Proceedings of the 8th International Cable Supported Bridge Operators Conference, Edinburgh, UK, 3–5 June 2013.

155. Mayrbaurl, R.M.; Camo, S. *Guidelines for Inspection and Strength Evaluation of Suspension Bridge Parallel Wire Cables*; National Cooperative Highway Research Program Report; NCHRP 534; National Academy of Sciences: Washington, DC, USA, 2004.

156. Beabes, S.; Faust, D.; Cocksedge, C. Suspension bridge main cable dehumidification—An active system for cable preservation, Sustainable bridge structures. In Proceedings of the 8th New York City Bridge Conference, New York, NY, USA, 24–25 August 2015; pp. 3–18.

157. Ono, K. Calibration methods of acoustic emission sensors. *Materials* **2016**, *9*, 508. [CrossRef] [PubMed]

158. Ono, K. Critical examination of ultrasonic transducer characteristics and calibration methods. *Res. Nondestruct. Eval.* **2017**, 1–46. [CrossRef]

159. Vallen, H. Proposal for an absolute AE sensor calibration setup. In Proceedings of the World Conference on Acoustic Emission, Xi'an, China, 11–13 October 2017.

160. NDIS 2109-91. *Method for Absolute Calibration of Acoustic Emission Transducers by Reciprocity Technique*; The Japanese Society for Non-Destructive Inspection: Tokyo, Japan, 1991; 13p.

161. Hatano, H.; Watanabe, T. Reciprocity calibration of acoustic emission transducers in Rayleigh-wave and longitudinal-wave sound fields. *J. Acoust. Soc. Am.* **1997**, *101*, 1450–1455. [CrossRef]

162. Monnier, T.; Friedrich, P.; Zhang, F. Primary calibration of acoustic emission sensors by the method of reciprocity-industrial exploitation of the calibration bench. In Proceedings of the 31st European Working Group on Acoustic Emission, Dresden, Germany, 3–5 September 2014.

163. Wild, G.; Hinckley, S. Acousto-ultrasonic optical fiber sensors: Overview and state-of-the-art. *IEEE Sens. J.* **2008**, *8*, 1184–1193. [CrossRef]

164. Teixeira, J.G.V.; Leite, I.T.; Silva, S.; Frazão, O. Advanced fiber-optic acoustic sensors. *Photonic Sens.* **2014**, *4*, 198–208. [CrossRef]

165. Yu, F.; Okabe, Y.; Wu, Q.; Shigeta, N. Damage type identification based on acoustic emission detection using a fiber-optic sensor in carbon fiber reinforced plastic laminates. In Proceedings of the 32nd European Conference on Acoustic Emission, Prague, Czech Republic, 7–9 September 2016; pp. 543–550.

166. Innes, M.; Davis, C.; Rosalie, C.; Norman, P.; Rajic, N. Acoustic emission detection and characterisation using networked FBG sensors. *Procedia Eng.* **2017**, *188*, 440–447. [CrossRef]

167. Shrestha, P.; Kim, J.H.; Park, Y.; Kim, C.G. Impact localization on composite wing using 1D array FBG sensor and RMS/correlation based reference database algorithm. *Compos. Struct.* **2015**, *125*, 159–169. [CrossRef]

168. Barat, V.; Borodin, Y.; Kuzmin, A. Intelligent AE signal filtering methods. *J. Acoust. Emiss.* **2010**, *28*, 109–119.

169. Elizarov, S.; Barat, V.; Shimansky, A. Nonthreshold acoustic emission data registration principles. In Proceedings of the 31st European Working Group on Acoustic Emission, Dresden, Germany, 3–5 September 2014.

170. Lokajícek, T.; Klíma, K. A first arrival identification system of acoustic emission (AE) signals by means of a high-order statistics approach. *Meas. Sci. Technol.* **2006**, *17*, 2461. [CrossRef]

171. Pomponi, E.; Vinogradov, A. Wavelet based approach to acoustic emission phase picking. In Proceedings of the 31st European Working Group on Acoustic Emission, Dresden, Germany, 3–5 September 2014.

172. Manhertz, G.; Csicso, G.; Gardonyi, G.; Por, G. Real-time acoustic emission event detection with data evaluation for supporting material research. In Proceedings of the 31st European Working Group on Acoustic Emission, Dresden, Germany, 3–5 September 2014.

173. Kharrat, M.; Ramasso, E.; Placet, V.; Boubakar, L. Acoustic emission in composite materials under fatigue tests: Effect of signal-denoising input parameters on the hits detection and data clustering. In Proceedings of the 31st European Working Group on Acoustic Emission, Dresden, Germany, 3–5 September 2014.

174. Sagasta, F.A.; Zitto, M.; Piotrokowski, R.; Gallego, A.; Benavent-Climent, A. A novel wavelet b-value of acoustic emissions to evaluate local damage in RC frames subjected to earthquakes. In Proceedings of the 31st European Working Group on Acoustic Emission, Dresden, Germany, 3–5 September 2014.

175. Sibil, A.; Godin, N.; R'Mili, M.; Maillet, M.; Fantozzi, G. Optimization of acoustic emission data clustering by a genetic algorithm method. *J. Nondestruct. Eval.* **2012**, *31*, 169–180. [CrossRef]

176. Goldberg, D.E. *Genetic Algorithms in Search, Optimization, and Machine Learning*; Addison Wesley: Reading, MA, USA, 1989; 412p.

177. Anastassopoulos, A.A.; Philippidis, T.P. Clustering methodology for the evaluation of acoustic emission from composites. *J. Acoust. Emiss.* **1995**, *13*, 11–22.

178. Godin, N.; Huguet, S.; Gaertner, R.; Salmon, L. Clustering of acoustic emission signals collected during tensile tests on unidirectional glass/polyester composite using supervised and unsupervised classifiers. *NDT E Int.* **2004**, *37*, 253–264. [CrossRef]

179. Katsuyama, K.; Seto, M.; Kiyama, T.; Utagawa, M. Three-dimensional AE tomography for image processing of the deteriorated material. *Saf. Eng. (Anzen Kogaku)* **1992**, *31*, 321–326. (In Japanese)

180. Schubert, F. Principle of AE tomography. *J Acoust. Emiss.* **2004**, *22*, 147–158.

181. Kobayashi, Y.; Shiotani, T. Computerized AE tomography. In *Innovative AE and NDT Techniques for On-Site Measurement of Concrete and Masonry Structures*; Ohtsu, M., Ed.; Springer: Dordrecht, Germany, 2016; pp. 47–68.

182. Kobayashi, Y.; Shiotani, T.; Oda, K. Three-dimensional AE-tomography with accurate source location technique. In Proceedings of the Structural Faults and Repair Conference, London, UK, 8–10 July 2014.

183. Kobayashi, Y.; Shiotani, T.; Oda, K. System identification for three-dimensional AE-tomography with Kalman filter. In Proceedings of the 31st European Working Group on Acoustic Emission, Dresden, Germany, 3–5 September 2014.

184. Nishida, T.; Shiotani, T.; Asaue, H.; Maejima, T.; Kobayashi, Y. Damage evaluation for CR bridge deck under wheel loading test by means of AE tomography. In *Progress in Acoustic Emission XVIII*; The Japanese Society for Non-Destructive Inspection: Tokyo, Japan, 2016; pp. 111–116.

185. Fukuda, M.; Shiotani, T.; Nishida, T.; Asaue, H.; Watabe, K.; Kobayashi, Y. Damage evaluation for in-field bridge deck by AE tomography. In *Progress in Acoustic Emission XVIII*; The Japanese Society for Non-Destructive Inspection: Tokyo, Japan, 2016; pp. 481–486.

186. Takamine, H.; Watabe, K.; Miyata, H.; Asaue, H.; Nishida, T.; Shiotani, T. Efficient damage inspection of deteriorated RC bridge deck with rain-induced elastic wave. *Constr. Build. Mater.* **2018**, *162*, 908–913. [CrossRef]

187. Lackner, G.; Tscheliesnig, P. Requalification of LPG tanks in Europe: Verifying the structural integrity by monitoring the pressure test with acoustic emission. In Proceedings of the 19th World Conference on Non-Destructive Testing, Munich, Germany, 13–17 June 2016.

188. Di Fratta, C.; Ferraro, A.; Tscheliesnig, P.; Lackner, G.; Correggia, V.; Altamura, N. AT on buried LPG tanks over 13 m^3: An innovative and practical solution. In Proceedings of the 31st European Working Group on Acoustic Emission, Dresden, Germany, 3–5 September 2014.

189. Newhouse, N.L.; Rawls, G.B.; Rana, M.D.; Shelley, B.F.; Gorman, M.R. Development of ASME section X code rules for high pressure composite hydrogen pressure vessels with non-load sharing liners. In Proceedings of the 2010 ASME Pressure Vessels and Piping Conference, PVP 2010, Bellevue, WA, USA, 18–22 July 2010.

190. *ASME Boiler and Pressure Vessel Code, Sec. X, Fiber-Reinforced Plastic Pressure Vessels*; Mandatory Appendix 8-620 and NB10-0601, Supplement 9; The American Society of Mechanical Engineers: New York, NY, USA, 2017; 323p.

191. Digital Wave Corp. *Use of Modal Acoustic Emission (MAE) for Life Extension of Civilian Self-Contained Breathing Apparatus (SCBA) DOT-CFFC Cylinders*; Final Report; US DOT Contract DTPH56-13-P-000029; Digital Wave Corp: Centennial, CO, USA, 2014; 204p. Available online: www.phmsa.dot.gov/sites/phmsa.dot.gov/files/docs/technical-resources/56376/finalreport-june20-2014.pdf (accessed on 20 May 2018).

192. Anderson, M. *A Review of Composite SCBA Cylinders and DOT Life Extension*; Digital Wave Corp: Centennial, CO, USA, 2016; 7p, Available online: www.digitalwavecorp/scba-life-extension-1 (accessed on 1 May 2018).

193. Flage, D.; Hunter, D.; Robinson, S.; Ceres, R.; Gorman, M.; Ziola, S. *Self-Contained Breathing Apparatus (SCBA) Cylinder Life-Extension Study*; Final Report; NAVSEA Contract N00024-11-C-4314; CACI International: Fairfax, VA, USA, 2012; 23p, Available online: navy-self-contained-breathing-apparatus-scba-composite-cylinder-life-extension-research-project.pdf (accessed on 20 May 2018).

194. US DOT Pipeline and Hazardous Materials Safety Administration. *Modal Acoustic Emission (MAE) Examination Specification for Requalification of Composite Overwrapped Pressure Vessels (Cylinders and Tubes)*; Tech. Report; US Dept. Transportation: Washington, DC, USA, 3 May 2018; 23p. Available online: www.phmsa.dot.gov/technical-resources/hazmat-technical-resources/technical-reports (accessed on 20 May 2018).

195. Chang, J.B. *Implementation Guidelines for ANSI/AIAA S-081: Space Systems Composite Overwrapped Pressure Vessels*; Aerospace Report No. TR-2003(8504)-1; AD A413531; Aerospace Corp: El Segundo, CA, USA, 2003; 83p.

196. Carpinteri, A.; Lacidogna, G. (Eds.) *Acoustic Emission and Critical Phenomena: From Structural Mechanics to Geophysics*; CRC Press: Boca Raton, FL, USA, 2008; 282p.

197. Maillet, E.; Godin, N.; R'Mili, M.; Reynaud, P.; Lamon, J.; Fantozzi, G. Analysis of acoustic emission release during static fatigue tests at intermediate temperatures on ceramic matrix composites: Towards rupture time prediction. *Compos. Sci. Technol.* **2012**, *72*, 1001–1007. [CrossRef]

198. Godin, N.; Reynaud, P.; R'Mili, M.; Fantozzi, G. Identification of a critical time with acoustic emission monitoring during static fatigue tests on ceramic matrix composites: Towards lifetime prediction. *Appl. Sci.* **2016**, *6*, 43. [CrossRef]

199. Dassios, K.G.; Aggelis, D.G.; Kordatos, E.Z.; Matikas, T.E. Cyclic loading of a SiC-fiber reinforced ceramic matrix composite reveals damage mechanisms and thermal residual stress state. *Compos. Part A* **2013**, *44*, 105–113. [CrossRef]

200. Tracey, J.; Waes, A.; Daly, S. A new experimental approach for in situ damage assessment in fibrous ceramic matrix composites at high temperature. *J. Am. Ceram. Soc.* **2015**, *98*, 1896–1906. [CrossRef]

201. Li, L.B. Damage development in fiber-reinforced ceramic-matrix composites under cyclic fatigue loading using hysteresis loops at room and elevated temperatures. *Int. J. Fract.* **2016**, *199*, 39–58. [CrossRef]

202. Papasalouros, D.; Bollas, K.; Kourousis, D.; Anastasopoulos, A. Acoustic emission tank floor testing: A study on the data-base of tests and follow-up inspections. In Proceedings of the 31st European Working Group on Acoustic Emission, Dresden, Germany, 3–5 September 2014.

203. Kim, D.H.; Lee, S.B.; Yang, B.S.; Bae, D.M. Structural health monitoring of blast furnace in steel mill using acoustic emission technique. In *Progress in Acoustic Emission XVIII*; The Japanese Society for Non-Destructive Inspection: Tokyo, Japan, 2016; pp. 37–42.

204. Seretti, A.; Boyer, L.; Adam, L.; Proust, A. Acoustic emission monitoring of cold forming automated operation on airplane wing panels. In Proceedings of the 31st European Working Group on Acoustic Emission, Dresden, Germany, 3–5 September 2014.

205. Zielke, R.; Tillmann, W.; Abdulgader, M.; Sievers, N.; Wang, G. Process control of thermal spraying. In Proceedings of the 31st European Working Group on Acoustic Emission, Dresden, Germany, 3–5 September 2014.

206. Manthei, G.; Alter, C.; Kolling, S. Localization of initial cracks in laminated glass using acoustic emission analysis—Part I. In Proceedings of the 31st European Working Group on Acoustic Emission, Dresden, Germany, 3–5 September 2014.

207. Ravnik, F.; Grum, J. Identification of machine components cracking with sound emission during steel quenching. In Proceedings of the 32nd European Working Group on Acoustic Emission, Prague, Czech Republic, 7–9 September 2016; pp. 419–431.

208. Sause, M.G.R. *In Situ Monitoring of Fiber-Reinforced Composites*; Springer: Berlin, Germany, 2016; 633p.

209. Asamene, K.; Hudson, L.; Sundaresan, M. Influence on acoustic emission signals in carbon fiber reinforced composites. *Ultrasonics* **2015**, *59*, 86–93. [CrossRef] [PubMed]

210. Schubert, K.; Herrmann, A.S. On attenuation and measurement of Lamb waves in viscoelastic composites. *Compos. Struct.* **2011**, *94*, 177–185. [CrossRef]

211. Ibrahim, M.E. Nondestructive evaluation of thick-section composites and sandwich structures: A review. *Compos. Part A* **2014**, *64*, 36–48. [CrossRef]

212. Schaal, C.; Mal, A. Core-skin disbond detection in a composite sandwich panel using guided ultrasonic waves. *ASME J. Nondest. Eval.* **2017**, *1*, 011006. [CrossRef]

213. Ohtsu, M. (Ed.) *Acoustic Emission and Related Non-Destructive Evaluation Techniques in the Fracture Mechanics of Concrete*; Woodhead: Cambridge, UK, 2015; 318p.

214. Carpinteri, A.; Lacidogna, G.; Niccolini, G. Multidimensional approaches to study of Italian seismicity. In *Acoustic Emission and Critical Phenomena: From Structural Mechanics to Geophysics*; CRC Press: Boca Raton, FL, USA, 2008; pp. 245–270.

215. De Santis, S.; Tomor, A.K. Laboratory and field studies on the use of acoustic emission for masonry bridges. *NDT E Int.* **2013**, *55*, 64–74. [CrossRef]
216. Verstrynge, E.; De Wilder, K.; Drougkas, A.; Voet, E.; Van Balen, K.; Wevers, M. Crack monitoring in historical masonry with distributed strain and acoustic emission sensing techniques. *Constr. Build. Mater.* **2018**, *162*, 898–907. [CrossRef]
217. Hamstad, M.A. Some observations on Rayleigh waves and acoustic emission in thick steel plates. *J. Acoust. Emiss.* **2009**, *27*, 114–136.

Review

Review on In Situ Acoustic Emission Monitoring in the Context of Structural Health Monitoring in Mines

Gerd Manthei [1,*] **and Katrin Plenkers** [2]

[1] Department of Mechanical and Energy Engineering, THM University of Applied Sciences, D-35390 Gießen, Germany
[2] GMuG mbH (Gesellschaft für Materialprüfung und Geophysik mbH), D-61231 Bad Nauheim, Germany; k.plenkers@gmug.eu
* Correspondence: gerd.manthei@me.thm.de; Tel.: +49-152-0154-2896

Received: 31 July 2018; Accepted: 24 August 2018; Published: 9 September 2018

Featured Application: **This article describes various applications of in situ acoustic emission (AE) monitoring in the context of structural health monitoring (SHM) in mines and extensively presents the state of the art in this field.**

Abstract: A major task in mines and even more in underground repositories for nuclear waste is to investigate crack formation for evaluation of rock mass integrity of the host rock. Therefore, in situ acoustic emission (AE) monitoring are carried out in mines as part of geomechanical investigations regarding the stability of underground cavities and the integrity of the rock mass. In this work, the capability of in situ AE monitoring in the context of structural health monitoring (SHM) in mines and in various geological settings will be reported. SHM pointed out, that the AE network is able to monitoring AE activity in rock with a volume up to 10^6 cubicmeter and distances up to 200 m (e.g., 100 m $\times$ 100 m $\times$ 100 m) in the frequency range of 1 kHz to 150 kHz. Very small AE events with source size in approximately centimeter to millimeter scale are detected. The results show that AE activity monitors rock deformation in geological boundaries due to convergence of the rock. In addition, high AE activity occurs in zones of dilatancy stress in homogenous rock. In conclusion in situ AE monitoring is a useful tool to observe instabilities in rock long before any damage becomes visible.

Keywords: in situ acoustic emission (AE) monitoring; structural health monitoring; mines; host rock

---

1. Introduction

For the safe use of various structures, a regular inspection is required. For critical infrastructures in particular, like aircraft, pressure vessels, and bridges as well as underground structures like mines, monitoring of their conditions is necessary. Therefore, methods for a reliable and automated monitoring are of great interest. Under these circumstances, structural health monitoring (SHM) has been an intriguing research topic in recent decades.

SHM is known as the continuous or periodical and automated method for monitoring and evaluating the condition of a monitoring subject. It is part of condition monitoring according to the International Organization for Standardization (ISO) 17359 [1]. This standard gives an overview of the basic procedures of condition monitoring and diagnostics programs for all types of machinery. The standard considers parameters such as vibration, temperature, flow rates, contamination, performance, and rotation speed, which are typically related to operation, state, and quality criteria. A concept is introduced of condition monitoring with the so-called root cause failure modes, which shows basic guidance on setting warning and alarm criteria, making diagnoses and predictions, and increasing their reliability.

A traditional means of monitoring big structures is visual inspection by trained personnel. Although simple, visual inspection is not successful for realizing all sources of damage, so a need exists for more reliable methods. A wide range of methods is now available for SHM of large structures such as bridges. These methods can be broadly classified as local methods for machine condition monitoring and global methods for monitoring of the whole structure (see Table 1) [2]. Machine condition monitoring is not strictly concerned with structural health (SH), but it requires information about the internal condition of the machine to be obtained externally. Vibration analysis is the most commonly applied method, and the analysis techniques have advanced greatly over the historic observation that if a machine is vibrating more than normal then it is likely to be faulty. These kinds of measurements are also used as predictive maintenance (PdM), which are performed to determine the condition of equipment in order to predict when maintenance should be done. One expects to save costs over routine or time-interval maintenance, because maintenance is carried out only when necessary or warranted. In order to identify wear particles or chemical contaminants, it is common to apply lubricant measurements, while the use of thermography to identify temperature anomalies is increasing.

Table 1. Type, aim, and characteristics of structural health monitoring (SHM) in machine condition, monitoring, and in monitoring of whole structures.

Type of Measurements	Aim of Measurements	Characteristics
SHM in machine condition monitoring		
Vibration measurements Lubricant measurements Thermography Acoustic emission (AE) measurements	Observation of machine vibration Identification wear particles and chemical contaminants Identification of temperature anomalies Identification of cracks	• infer the health of a whole machine • condition of rotating machines • mainly passive measurements • preventive and predictive maintenance • continuous and intermittent monitoring • small number of sensors are required • applied to e.g., large turbo machines, bearing housing
SHM in monitoring of entire structures		
Vibration measurements	Observation of vibration of a whole structure	• use of resonance frequencies for damage prediction • detection of local damage with full structure coverage • large number of sensors are required • applied to large structures, e.g., bridges, off-shore oil platforms
Fiber optics measurement	Sensing mechanisms of optical fibers are based on intensity, wavelength, and interference of the light waves	• applied to a wide range of civil structures such as building, bridges, pipelines, tunnels and dams
Load monitoring	Observation of strain, traffic or wind	• applied to large structures such as bridges
Ultrasonic measurements	Observation of inhomogeneities in the material induce changes to the propagating waves	• transducers are used to introduce high-frequency waves into a specimen and receive the pulses • determination of position and size of flaw • very time consuming and expensive • active measurement • applied to large structures of airplanes,
In situ AE monitoring	Observation of elastic waves arising from the rapid release of energy inside material, e.g., from crack initiation and crack propagation	• very high sensitivity • In situ location of damage • passive measurement • small number of sensors are required for monitoring the whole structure • applied to pressure vessels, tank, building, bridges, pipelines, and to underground structures like mines

For SHM of entire structures, several non-destructive evaluation/testing (NDE/NDT) techniques are available. These techniques do not involve the destruction of the structure during testing, as the name implies. Most commonly used non-destructive techniques are based on the use of elastic waves e.g., ultrasonic and acoustic emission (AE) measurements, and fiber optics measurements. Further details on these methods can be found in [3–5].

Vibration measurements usually give the global figure, indicating damage in the entire structure, and can also locate and assess the damage. The basis of the principle is that the changes in the global properties e.g., mass, stiffness, and damping of a structure cause a change in its modal properties such as natural frequencies and mode shapes. The modal properties such as modal flexibility and strain energy are used for the identification of damage [6–8]. These global methods usually use accelerometers to measure the vibration of the structure for calculating the modal properties. But the use of a vibration-based method in large structures such as bridges can be uncertain, just then, if damage may only cause negligible change in dynamic properties and thus may go unnoticed. Moreover, in order to find the exact location of damage, local methods are often better alternatives.

AE measurements are applied for machine condition monitoring as well as SHM of large structures. In engineering materials, some common sources of AE are initiation and growth of cracks, yielding, failure of bonds, fiber failure, and pullout in composites. It should be noted that only active or growing cracks emitted elastic waves. If cracks are present but do not grow, no AE is emitted. Ono [9,10] gives in his papers a review on structural integrity evaluation and SH evaluation using AE.

Another application of SHM in the broadest sense is in situ AE monitoring in mines. This monitoring involves permanent AE measurements of microcracking in parts of the mine or the entire mine. A major task in underground storage like repositories for nuclear waste and deep mines is the safety assessment. Therefore, several geotechnical monitoring methods, for instance, micro-seismic measurements are used in many mines. This study demonstrates the capability of In situ AE monitoring for SHM in mines. Section 2 deals with fundamentals about in situ AE monitoring. After some comprehensive consideration of the scale between earthquakes and AE events, a list of various applications of in situ AE monitoring found in the literature (Section 2.2) is given. Sections 3–5 show examples on in situ AE monitoring in salt mines, gold mines, and underground research laboratories.

2. In Situ Acoustic Emission (AE) Monitoring—Fundamentals

2.1. Comprehensive Consideration of the Scale

An interesting analogy can be made to AE events by comparing it to earthquakes, which can be regarded as largest natural occurring emission sources. The principal mechanism does not depend on the scale: The rapid release of elastic energy by processes of crack growth or deformation within the rock generates transient pulses of elastic wave energy as AE events. In an even larger scale, the same is true for earthquakes. Similar to AE sources, seismic waves propagate through the earth, and are detected with a global network of seismometers located around the world. Therefore, most of the theory of earthquakes can be transferred to AE sources [11,12]. Figure 1 shows the dependences between measured corner frequency f_C, the moment magnitude M_W, the seismic moment M_0, and the source radius r_0 for studies in earthquake seismology, microseismicity, in situ AE monitoring, and AE in laboratory (see more details in Box 1) [13]. In addition, in situ AE monitoring is reported from rock face monitoring e.g., on construction sites [14,15].

The largest and thereby the longest events, namely earthquakes, are found at the lowest end of the frequency scale (Figure 1). The focal length and displacement of an earthquake can amount to more than several hundred kilometers and up to several meters, respectively. Whereas global seismology exploits frequencies less than 1 Hz, local seismology focuses on the analysis of frequencies of hundreds of Hertz or even 1000 Hz. Recordings of microseismicity in dense local (e.g., borehole) networks or underground are limited to frequencies up to a few kilohertz. In these cases pendulum-based geophones or seismometers are utilized. The corner frequency in the field of microseismic measurements ranges between 5 Hz and some hundred Hertz with magnitudes from approximately 4.0 down to −2.0. Above these frequencies, the range of in situ AE monitoring in rock begins with applications in mines. In the frequency rang of about 1 kHz up to 200 kHz accelerometers or special piezoelectric AE sensors were developed for in situ application in mines. Source-receiver distances realized depend strongly on rock type, but can reach up to 150 m in rocks with very low wave attenuation. The covered areas may have linear dimensions of up to 200 m. The expected magnitudes are in the range between −2.0 and −5.0.

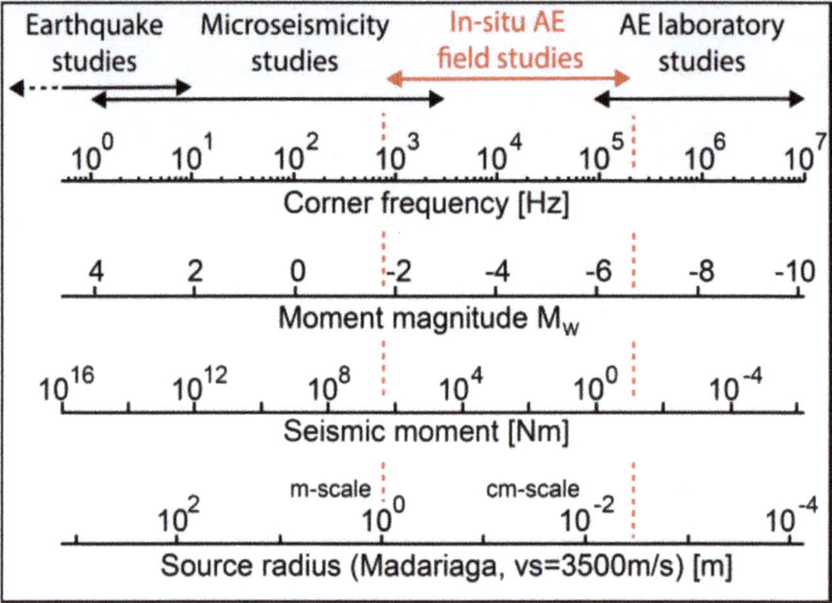

Figure 1. Relations between measured corner frequency f_C, moment magnitude M_W, seismic moment M_0, and the source radius r_0 for studies in earthquake seismology, microseismicity, In situ AE monitoring and AE in laboratory (see more details in Box 1) (modified from [13]).

On the other hand, the highest frequencies up to several MHz are utilized in laboratory studies. These frequencies are generated by AE events in the microscopic region, for instance by microcracking of rock or dislocation movement. Only sensitive piezoelectric sensors for laboratory applications are able to measure such high frequencies. In the laboratory, the source radius may extend to some micrometers only and the displacement (Burgers vector) is to be measured in nanometers. The expected magnitude of such small AE events is below −6.0 [11]. A detailed study on the theoretical limits on detection and analysis of small earthquakes in dependency on the sensitivity and frequency coverage of the monitoring network was published by Kwiatek and Ben Zion [16].

Figure 2 taken from Bohnhoff et al. [17] shows the co-seismic stress drop plotted with key earthquake source parameters over the entire bandwidth of observed rupture processes, extending from large natural earthquakes to AEs in the laboratory. The source parameters were calculated from individual data sets of natural earthquakes, induced seismicity in mines and reservoirs, and volcano seismicity [18–22]. AE data are unpublished results from the rock-deformation laboratory in the German research Centre for Geosciences, Potsdam, (GFZ)-Section 4.2. The Madariaga [23] circular source model is assumed to calculate source radii and the lines of constant static stress drop from 0.01 MPa up to 100 MPa and $v_s = 3500$ m/s (see Box 1).

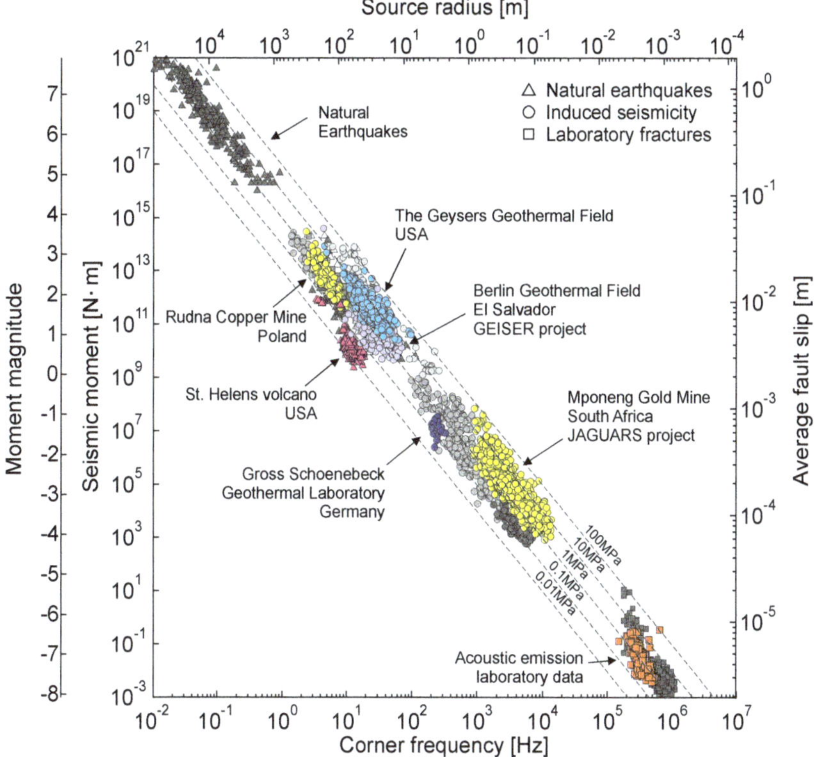

Figure 2. Dependence between seismic moment, moment magnitude, source radius, average fault slip, and corner frequency for natural earthquakes (gray rectangles) in comparison with data from other studies in the low-magnitude range. The Madariaga [18] circular source model is assumed to calculate source radii and the lines of constant static stress drop from 0.01 MPa up to 100 MPa and $v_s = 3500$ m/s (see Box 1). This figure is an updated figure from Kwiatek et al. [16] and was first published in [17]. Reproduced with permission.

Box 1. Estimation of source radius, static stress drop, and moment magnitude.

Seismic moments M_0 and corner frequencies f_c in Figure 2 were estimated from the spectral level of ground velocity or displacement spectra corrected for instrument response and wave propagation effects, as described in detail in the references. It follows that the dependency in Figure 2 between seismic moment and corner frequency are based on data-driven observables. The seismic moment is a measure of how much "work" an earthquake does in sliding when rock slips off other rock. It is necessary to take into account that the physical unit of M_0 is given in N·m (corresponds to 10^7 dyne·cm).

The estimation of source radii, the lines of constant static stress drop, and the average fault slip are based on the model of Madariaga [23] and Brune [24]. Both modeled a circular fault and use the corner frequency f_c to calculate the source radius r_0 and the static stress drop $\Delta\sigma$. The stress drop and source radius is defined as $\Delta\sigma = 7/16 \cdot M_0/r_0^3$ and $r_0 = c \cdot v_s/2\pi f_c$, respectively ($c = 2.34$ for Brune's model, $c = 1.32$ for Madariaga's modell). The moment magnitude M_W is calculated using the standard relation for tectonic events developed by Hanks et al. [25] $M_W = (log_{10}M_0 - 9.1)/1.5$. The formula is empirically established for tectonic earthquakes only. Extrapolation over so many orders of magnitude seems questionable, but is useful for visualization.

The question to which extent AE events represent double-couple shear events or events with a dominant isotropic component remains open. Owing to the AE sensor calibration problem (see Box 2), few studies exist that investigate reliably the source mode of AE events e.g., calculating the full moment tensor. Results from hydraulic fracturing studies suggest that AE events may represent sources with a major double-couple component [26,27], whereas studies related to stress-induced AE events find indications for isotropic components or mixed mode events [28–31].

2.2. Applications and Characteristics of In Situ AE Monitoring

Many examples of application of in situ AE monitoring in mines as found in the literature are given in Table 2. This table shows the test site of the project, the resulting publications, the type of AE networks with source–receiver distance and mode of recording, and the rock type with number of recorded events. Above the frequencies of microseismic measurements, the lower frequency of the in situ AE method begins at 1 kHz and ranges up to approximately 200 kHz. Due to the high sensitivity of the AE sensors, the in situ AE method allows to monitor fractures in the millimeter scale to decimeter scale (see Figure 2). This is important information because such small events indicate weakening in the rock long before macroscopic fractures occur [32–34]. Due to the high frequencies, the signals of AE events attenuate more than signals of microseismic events [35,36]. Thus, in situ AE monitoring is limited to a few tens of meters in hard rock or a few meters only in soft rock like clay rock or sedimentary rock. However, if the rock is very homogeneous and the attenuation of the seismic waves is low, larger rock volumes over 150 m in length can be monitored [12,33–38]. The applications in Table 2 can be roughly divided into three groups. The first group shows applications from underground laboratories in Switzerland and Sweden. The second group deals with the application of in situ AE monitoring in gold mines in South Africa. The third and biggest group shows applications in salt mines in Germany. These in situ measurements in rock salt often focused on monitoring AE activity caused by work like backfilling [39,40], excavation [41] or gas and fluid injection [42]. Early projects in in situ AE monitoring were limited to small rock volumes.

2.3. Method of In Situ AE Monitoring

With the in situ AE method, a network of AE sensors records very small events with low seismic energy in the kilohertz range. The frequency range of in situ AE monitoring is 1 to approximately 200 kHz. In contrast to seismic and micro-seismic sensors, AE sensors do not measure ground movement based on the principle of spring-mass or a pendulum, but detect stress changes purely based on the piezoelectric effect. AE sensors are accordingly piezoelectric-based sensors, which are much more sensitive in the kHz frequency range than spring-mass based accelerometers or pendulum-based geophones/seismometers as shown in Plenkers et al. [36] and Zang et al. [43]. But, in situ AE measurements are often missing absolute calibration of the AE sensors (see Box 2). The differences between those sensor types are discussed in more detail in Box 3.

Box 2. Annotations about characterization of AE sensors.

An important issue in today's works on in situ AE measurements is the often missing absolute calibration of the AE sensors. These piezoelectric sensors (not to be confused with piezoelectric accelerometers, see Box 3) work mostly in resonant mode and do not have a flat sensor response. For this reason, the AE sensors are very sensitive at the resonant frequencies, but the exact sensor response necessary for magnitude estimations (especially based on waveform amplitudes) or source type analysis is normally unknown. By contrast, AE sensors used at frequencies f much higher than 100 kHz and used in the laboratory are successfully calibrated [44–51]. The calibration of AE sensors used in situ remains difficult due to the longer wavelength [45]. In addition, in situ coupling of AE sensors can have a severe influence on the sensor recording and need to be taken into account. Today, mostly two characterization methods are used in situ. sensor characterization by signal deconvolution [16,52,53] or by regression analysis [27,54,55]. An absolute calibration technique for reliable in situ calibration is still missing.

For this reason, in the case of in situ AE measurements in rock, magnitudes are often listed nominal magnitudes [56] or relative magnitudes [54]. These magnitudes are useful to gain insights into the relative event size, but may not be compared *directly* to other seismic magnitudes or in between events from different regions, different source types or different source–receiver distances as the effect of the sensors resonant response is not corrected for. Few studies exist of in situ AE events, where reliable source parameters (seismic moments) were estimated after careful sensor characterization and correction for the sensor response in situ [16,27,53,55].

Many authors applied the in situ AE method in order to detect AE events in mines. This method was applied to most applications as shown in Table 2. Due to the high measurement frequencies and the very high number of AE events, fast data acquisition systems are required. A multi-channel transient recorder (often 16 or 32 channels) running in trigger mode does the digitization. This means, a limited time window (e.g., 32 ms) is stored, when the threshold was passed. In this time window, all signals modes like P wave and S wave should be included. In order to detect the complete waveforms during the whole measurements, some measuring system also allows continuous recording of the measuring signals, where the waveform is recorded without interruption.

Box 3. Comparison of accelerometer and piezoelectric AE sensor.

A combination of AE sensors and triaxial accelerometers were used in the gold mines in South Africa, salt mines in Germany, and underground hard rock laboratories in Switzerland and Sweden. These AE sensors manufactured by Gesellschaft für Materialprüfung und Geophysik (GMuG) uses a piezoelectric disk of PZT ceramic, which is sensitive in the frequency range from approximately 1 kHz to approximately 200 kHz. On the other hand, the commercial triaxial piezoelectric accelerometer composed of Wilcoxon 736T has a flat frequency response between 100 Hz and 25 kHz with a sensitivity of 100 mV/g and a resonance frequency at 60 kHz. Zang et al. [43] concluded, "that the Wilcoxon accelerometers were not able to record AE events detected with the in situ AE sensors, despite the fact that AE events were present in the frequency range of the accelerometer". Note that the in piezoelectric accelerometers the piezoelectric disk is solely used to measure the movement of the seismic mass, whereas for AE sensors the piezoelectric effect is exploited to measure stress changes.

AE sensors are piezoelectric sensors that are much more sensitive in the kilohertz frequency range than a spring-mass based accelerometer. In the case of a constant acceleration, the acceleration force is in equilibrium with the restoring force of the spring deflected by x with the spring constant c. $F = m \cdot a = c \cdot x$. The measuring sensitivity S of the system results in $S = x/a = m/c$, if x is proportional to the output voltage. Accordingly, a large mass and a low spring stiffness lead to a high measuring sensitivity. In the dynamic case, damping forces and inertial forces have to be taken into account in addition to the spring force. The essential damping force is proportional to the velocity $\dot{x}$ and is described with the attenuation coefficient p. The inertial force is proportional to the acceleration. The resulting equation describes a resonant system: $F = m \cdot a = c \cdot x + p \cdot \dot{x} + c \cdot \ddot{x}$. Starting from a negligible damping ($p \approx 0$), it has a resonance frequency $\omega_0 = \sqrt{c/m}$. Thus, according to the above equation, the measurement sensitivity S is firmly linked to the resonance frequency ω_0 in the following way: $\omega_0^2 \cdot S = 1$. This means that a twice as high resonance frequency must be paid for with sensitivity reduced to the factor 1/4. Such spring-mass systems show only sufficiently below their resonance frequency a sufficiently constant proportionality between the measured variable and the deflection.

Table 2. Characteristics and applications of in situ AE monitoring in mines.

Test Site/Project	Keyword	Publications, Year	Network/Source-Receiver Distance R/Mode of Recording	Rock Type/No. of AE Events
Underground tunnel, Japan	Hydraulic fracturing	Sasaki et al. [57], 1987 Ohtsu [58], 1991	17 AE sensors (up to 100 kHz), 17 accelerometers, R ≈ 1 m, trigger mode with waveforms	Siliceous sandstone 200 AEs during four hydraulic fracturing tests (including microseismic events)
Underground research laboratory (URL), Canada, TSX Project	Excavation/tunnel sealing	Young & Collins [59], 1999 Young et al. [60], 2000 Young & Collins [61], 2001 Collins & Pettitt [62], 2002 Young & Collins [63], 2004	16 AE sensors (40 to 400 kHz), 16 accelerometers, R ≈ 10 m trigger mode with waveforms	Granite 15,350 AEs in 5 months approximately 400 m depth
Salt mine Asse, Germany	Cavity stability/heating	Eisenblätter et al. [41], 1998 Dahm & Manthei, [64], 1998	29 AE sensors (100 kHz), R ≈ 100 m, trigger mode with waveforms	Salt rock, 250,000 AEs in 11 months
Salt mine Bernburg, Germany	Hydraulic fracturing	Manthei et al. [65], 1998 Dahm et al. [66], 1999	8 AE sensors (up to 250 kHz) R ≈ 10 m	Salt rock 1500 AEs during eleven hydraulic fracturing tests
		Manthei et al. [26], 2001 Manthei et al. [67], 2003	trigger mode with waveforms 8 AE sensors (up to 250 kHz) and hydraulic fracturing tool, R ≈ 5 m trigger mode with waveforms	Salt rock 15,000 AEs during four hydraulic fracturing tests
Äspö Hard Rock Laboratory (HRL), Sweden	Excavation	Pettitt et al. [68], 2002	24 AE sensors (35 to 350 kHz) R ≈ 10 m, trigger mode with waveforms	Dioritic granite 884 AEs in 24 hours
Salt mine Morsleben, Germany (southern part)	-	Spies et al. [69], 2004 Manthei et al. [28], 2007	24 AE sensors (up to 100 kHz) R ≈ 100 m, trigger mode	Salt rock 50,000 AEs in one month approximately 400 m depth
Salt mine Morsleben, Germany (central part)	Backfilling Cavity stability	Spies & Eisenblätter [37], 2001 Manthei et al. [70], 2001 Spies et al. [39], 2005 Manthei et al. [35], 2006 Köhler et al. [71], 2009 Becker et al. [40], 2010 Becker et al. [72], 2014	48 AE sensors (up to 100 kHz) R ≈ 200 m, trigger mode	Salt rock 100,000 AEs in one month approximately 400 m depth
Mponeng gold mine, Carletonville, South Africa JAGUARS project	Pillar stress loading	Nakatani et al. [73], 2008 Yabe et al. [33], 2009 Plenkers et al. [36], 2010 Kwiatek et al. [55], 2010 Plenkers et al. [52], 2011 Kwiatek et al. [16], 2011 Naoi et al. [74], 2011 Davidsen et al. [75], 2013 Kwiatek & BenZion [30], 2013 Davidsen et al. [76], 2012 Ziegler et al. [77], 2015 Yabe et al. [78], 2015 Kozlowska et al. [79], 2015	8 AE sensors (1 to 200 kHz) and 1 triaxial accelerometer, R ≈ 10 m to 200 m, trigger mode with waveforms	Quarzite/Gabbro more than 500,000 AEs in 2 years approximately 3200 m depth

Table 2. *Cont.*

Test Site/Project	Keyword	Publications, Year	Network/Source-Receiver Distance R/Mode of Recording	Rock Type/No. of AE Events
Salt mine Merkers, Germany	Gas loading	Doerner et al. [80], 2012 Manthei et al. [42], 2012 Popp et al. [81], 2015 Plenkers et al. [82], 2018	8 AE sensors (1 to 150 kHz) and 4 AE sensors (1 to 80 kHz), $R \approx 10$ m to 60 m, trigger mode with waveforms	Salt rock more than 5,000,000 AEs in 2 years approximately 300 m depth
Mont Terri URL, St Ursanne, Switzerland	Excavation	Le Gonidec et al. [83], 2012	16 AE sensors (unknown) and 4 AE sensors (2 kHz to 60 kHz) $R \approx 0.3$ m to 6.5 m, trigger mode with waveforms	Opalinus clay more than 20,000 AEs in 2 weeks (2127 located), 300 m depth
Cooke 4 gold mine, South Africa, SATREPS project	Mining stress	Naoi et al. [53], 2013 Naoi et al. [84], 2015 Naoi et al. [85], 2015 Naoi et al. [86], 2015 Moriya et al. [34], 2015	24 AE sensors (1 to 50 kHz) and 6 triaxial accelerometers (50 Hz to 10 or 25 kHz), $R \approx 0$ m to 180 m, trigger mode with waveforms	Quartzite 365,237 AEs in approximately 3 months 1000 m depth
Salt mine Asse, Germany	Cavity stability	Philipp et al. [38], 2015	16 AE sensors (1 to 100 kHz) $R \approx 0$ m to 180 m, trigger mode with waveforms	Salt rock more than 100,000 AEs in 10 month, 300 m depth
Äspö Hard Rock Laboratory (HRL), Sweden	Hydraulic fracturing	Zang et al. [43], 2017 López et al. [87], 2017 Kwiatek et al. [27], 2018	11 AE sensors (1 to 100 kHz) and 4 accelerometers (50 Hz to 25 kHz), $R \approx 10$ m to 30 m, trigger mode with waveforms and continuous recording	Granodiorite/Diorite-gabbro/Granite 196 located AE events during six hydraulic fracturing tests, (more than 4000 AEs during one hydraulic fracturing fracturing test in continuous data) 400 m depth
Grimsel Test Site (GTS), Switzerland	Hydraulic fracturing	Gischig et al. [54], 2018 Jalali et al. [88], 2018	28 AE sensors (1 to 100 kHz) and 4 accelerometers (50 Hz to 25 kHz), $R \approx 9$ m to 30 m, trigger mode with waveforms and continuous recording	Granodiorite, 2,000 AEs during three hydraulic fracturing tests 400 m to 500 m depth
Salt mine Merkers, Germany	Brine loading	Plenkers et. al. [82], 2018	8 AE sensors (1 to 100 kHz) and 4 AE sensors (1 to 150 kHz) and 4 AE sensors (1 to 80 kHz), $R \approx 5$ m to 30 m, trigger mode with waveforms and continuous recording	Salt rock approximately 300 m depth

Often real-time processing (P- and S-wave onset picking and localization) of events is implemented in trigger mode recording. As data needs to be processed and stored on computer hard drives some dead time in recording occurs in between different windows. In the trigger mode approximately 10 AE events per second can be recorded. The signals of AE events are recorded when a specified threshold is crossed on one or more channels. As mentioned above, continuous recording of the waveform is possible in the last few years, as large computer storage is now available. Typical sampling rates are 500 kHz to 1 MHz. The P-wave and S-wave onsets are automatically picked after band-pass filtering of the traces by applying an adapted, speed-optimized short-term-average to long-term-average (STA-to-LTA) trigger algorithm [89]. After picking of the onsets, a least-square algorithm based on a gradient method is used to determine the location of the AE events. A location is valid if a sufficient number of P-wave and S-wave onsets are used for source location and the time residuals are small enough. With this location procedure working, noise without discernible onsets can be eliminated. For post-analysis, the digitized waveforms and location results are stored on hard disk. With fast Internet access, the AE measuring system can be remotely controlled. Transient noise of anthropogenic or electronic origin on the other hand is often localized and can pollute the seismic catalog, especially in active mines [36,38,87].

3. In Situ AE Monitoring in Salt Mines

3.1. Salt Mine Asse in Germany

The first example we present in this review is the study of Philipp et al. [38], which demonstrates structural monitoring in cavity roofs of the salt mine Asse II. In the Asse II mine in Lower Saxony there is the risk of permanent brine inflows due to failure processes in pillars and tunnels in the southern flank and in the adjacent overburden [90,91]. In order to monitor these failure processes, in situ AE measurements are carried out to detect microcracking in the roof of two cavities (that are subject to work operations and noise). It is expected, that AE events outline weakening in rock, structural damage due to dilatation and other dynamic processes long before significant damage is visible and in areas that are not accessible.

Figure 3a displays a top view onto the two chambers, which are monitored. The contours of the chambers are indicated by rectangles. North is on top. The monitoring system consists of two networks with 16 AE sensors, which are installed in short borehole of 1 m to 3 m length in the chamber's roof. The positions of the AE sensors of the western and eastern galleries are marked by grey and black dots, respectively. The dashed lines in Figure 3a indicate the positions of Crosscut 1 to 4 with a width of 12 m. Figure 3b shows a perspective view of the two test sites. The vertical short lines mark the boreholes for the bottom-view AE sensors, which are especially sensitive in the frequency range 1 kHz to 150 kHz.

Data is recorded in trigger-mode (1 MHz sampling rate) and automatically processed i.e., that events recorded are localized in near-real time. For this P- and S-wave onset picking is performed using a picking algorithm based on the Hilbert transform. The network geometry of both networks (network dimension 37 m $\times$ 31 m $\times$ 5 m and 46 m $\times$ 39 m $\times$ 4.5 m) differs owing to the actual usage of the two chambers that define the accessibility of the roof for sensor installation e.g., the roof could not be accessed above three large brine ponds. The monitoring project has been ongoing for several years, but the study considers data of a 10-month period, namely the time period 4 February to 31 November 2013. In this time period more than 100,000 AE events were recorded that populate a rock volume of approximately 250 m $\times$ 250 m $\times$ 160 m outlining dynamic processes not only in the chambers roof, but also in the salt dome flank and in the upper salt dome.

Figure 4 displays a geological cross section (sketch) through the Asse salt dome. The sketch shows that the southern flank (left-hand side) has a steep slip.

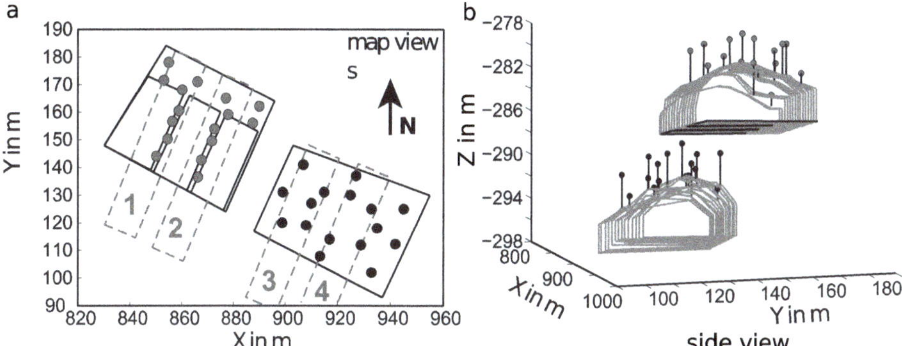

Figure 3. (**a**). Top view onto the monitored chambers. The contours of the chambers are indicated by squares. North is on top. The positions of the AE sensors are marked by grey and black dots. (**b**) Perspective view of the two test sites. The vertical short lines mark the boreholes for the bottom-view AE sensors (modified from [38]).

In this cross section, the chambers in Figure 3 are marked by a rectangle at the left-hand side in Figure 4. This figure gives a good overview of the approximate location of the four main clusters C1, C2, C3a, and C3b of AE events, which are marked by framed zones. More than 70% of the located AE events occur in the Cluster C1 along the so-called southern flank of the salt mine Asse.

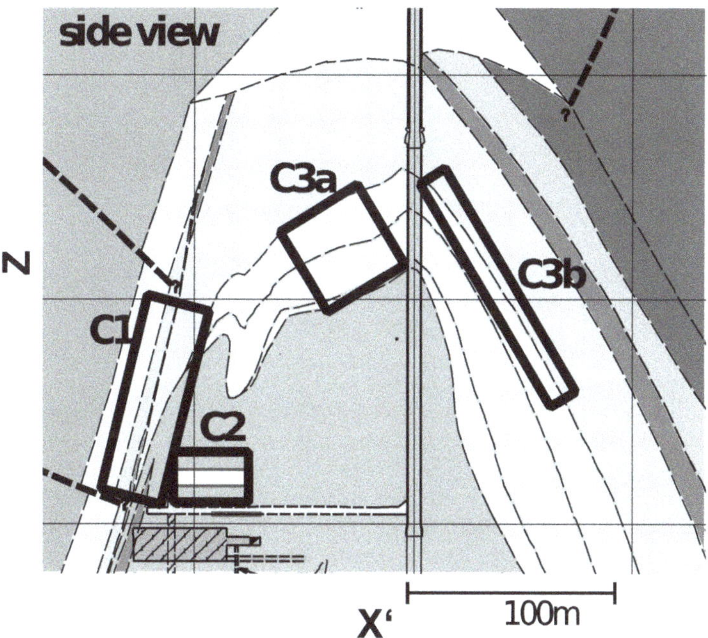

Figure 4. Geologic cross section (sketch) of the Asse salt mine with the four AE clusters C1, C2, C3a, and C3b (framed zones) [38].

In addition to these events, high AE activity occurs in zones above the roof of the chambers. At the roof of the western and eastern chamber (Cluster C2) more than 5000 and 10,000 AE events are located, respectively. In addition, in the upper part of the salt dome and the northern flank (Clusters 3a and 3b) two clusters of AE events are located. Because of the low wave attenuation of salt rock, the AE networks are very sensitive, so that source–receiver distances greater than 150 m are possible. Figure 5 displays AE activity above the roof of the chambers and in the southern flank above the two chambers in a rotated coordinate system. The location of the AE events are marked as black dots in the four crosscuts as shown in Figure 3. In Crosscut 1 and 2 (left-hand side) and Crosscut 3 and 4 (right-hand side), the AE events were recorded by the AE system in the western and eastern chamber, respectively. The width of the crosscut is about 12 m.

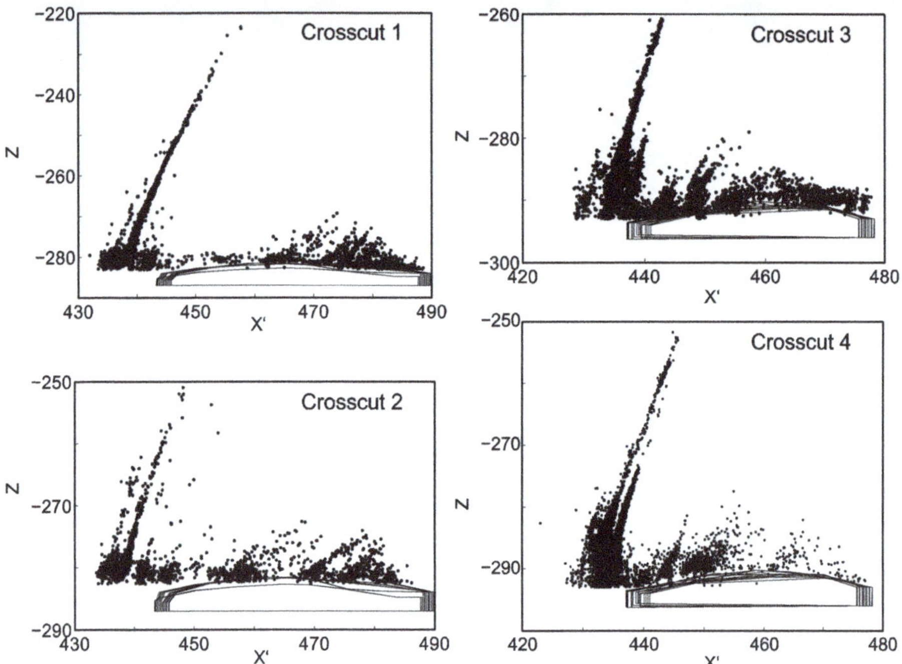

Figure 5. Location of AE events (black dots) in projection onto four vertical crosscuts in a rotated coordinate system above the chamber and along the anhydrite-host rock boundary at the southern flank [38]. All dimensions are given in meter.

The observed AE activity is not homogenously distributed but clusters clearly, in this way outlining planes of activity. In the chamber's roof most events outline planes oriented east–west and dipping to the south according to the salt's layering. The events extend up to 15 m from the roof into the salt rock, but at greater depth no activity is recorded, which demonstrates that currently no active damage process is occurring. In the roof of the chambers, all events occur in a homogenous part of the younger salt rock (Leine formation) and most likely correspond to damage processes owing to stress re-distribution. A geomechanical survey confirmed an increased permeability in the area of AE clusters, but did not show macroscopic damage.

The strongest AE activity is observed on the southern flank of the salt dome. The events outline a plane of activity that is oriented roughly east–west and dipping to the south. The AE events seem to follow the rock salt–anhydrite-sandstone rock boundary, which is subject to significant geomechanical processes including the loss of integrity owing to a barrier thickness of only 15 m in the upper mine. Events are observed as far as 70 m above the network. Within this plane, AE events cluster on vertical structures.

Overall, the authors conclude that although a singular AE event is too small to have a damage potential, the AE events are clearly able to outline in high-resolution areas and, even more precisely, the exact position, extension, and orientation of potential damage zones.

3.2. Salt Mine Merkers in Germany

From another salt mine in Germany, SHM during a borehole loading experiment is reported by Manthei et al. [42], Popp et al. [81], and Plenkers et al. [82]. In the first experiment that took place from January 2010 to January 2012, the rock response of four different stages were monitored using in situ AE monitoring: (1) pre-excavation; (2) drilling of wide-diameter (1.3 m diameter, 60 m extension) borehole; (3) partial backfill with MgO concrete to create gas tight seal; and (4) during borehole loading with compressed air. In the second experiment, that took place October 2017 to Summer 2018, in the same wide-diameter borehole, the rock response to brine loading was monitored [82].

The monitoring system of the first experiment consisted of 12 AE sensors installed in four monitoring boreholes in 12 m to 15 m distance to the large injection borehole. The sensors were equally spaced along the borehole and therefore, monitor the whole borehole with a similar recording sensitivity. The AE events are recorded in trigger mode. The dimension of the network is 28 m × 25 m × 27 m monitoring a rock volume of approximately 40 m × 40 m × 90 m.

Figure 6 shows at the left-hand side and in the middle results of the located AE events during excavation and cementation in side view (projection onto the x-z plane). The right-hand side of Figure 6 shows the number of located events per day during start of excavation (blue vertical arrow) and during stepwise cementation (red vertical arrows). In total, more than six million AE events were recorded and localized. Highest event rates are found during excavation and cementation, when more than 170,000 events are recorded and located per day (blue vertical arrow). It is shown that the activity starts in formerly inactive homogenous salt rock as soon as excavation is starting (left-hand side of Figure 6). After excavation and also after cementation is finished, the activity rate dropped nearly to the level of background seismicity rates (approximately 100 to 300 events/day). During cementation, the AE activity was limited to a zone approximately 0.5 m from the newly created borehole outlining the excavation damage zone [82]. The AE activity was extending outwards with time, but was also limited within a zone approximately 2 m from the borehole boundary. During excavation, AE events outlined the migration of humidity and temperature from the cementation into the rock.

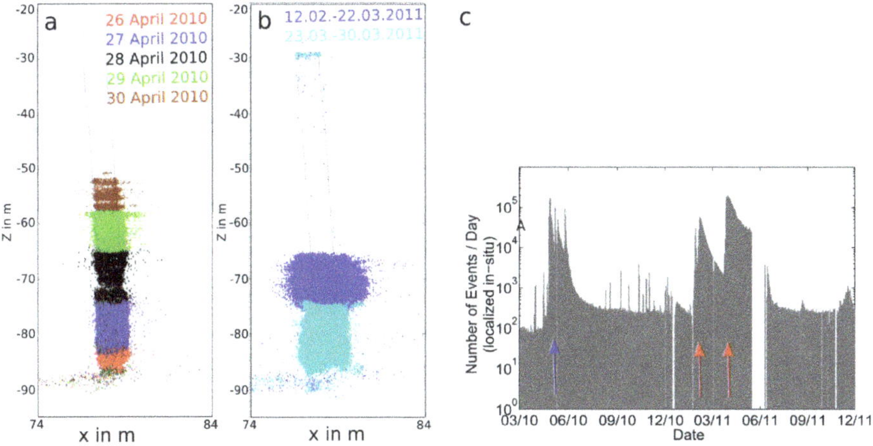

Figure 6. Temporal and spatial distribution of AE events in Merkers 1 experiment. The location of AE events during excavation (**a**) and cementation (**b**) are given in side view projection. The number of events per day of the whole monitoring period from March 2010 to December 2011 is shown in (**c**). The blue and red vertical arrows indicate the start of excavation and cementation, respectively [82].

It should be noted that from June 2011 on the wide-diameter borehole was loaded stepwise with compressed air. The pressurization took place in several steps. Each time, when pressure was increased, the daily activity rate increased by a few hundred events.

In January 2012 at a pressure of 60 bars for the first time AE events were observed that migrate as far as 20.2 m from the wide-diameter borehole. Those events were concentrated on a layer with minor vertical expansion and outline the migration of gas and brine. On 24 January 2012a gas and brine breakthrough was observed at two monitoring boreholes in combination with a pressure drop from 68 bars down to 56 bars. It was discussed that the AE events represent the break down of grain boundaries (source radius of a few centimeters) during percolation of the gas and brine mixture.

Figure 7 shows AE swarm activity during gas-brine break-through in top view (Figure 7a) and a lateral view (Figure 7b). For orientation, the neighboring pillars are shown in dark grey areas in (Figure 7a). Here color-coded lines mark the most outward extension of AE activity with time, which showed a migration of approximately 0.6 m per day [82]. In Figure 7b the cemented plug is shown in grey and the positions of AE sensors are shown by red triangles.

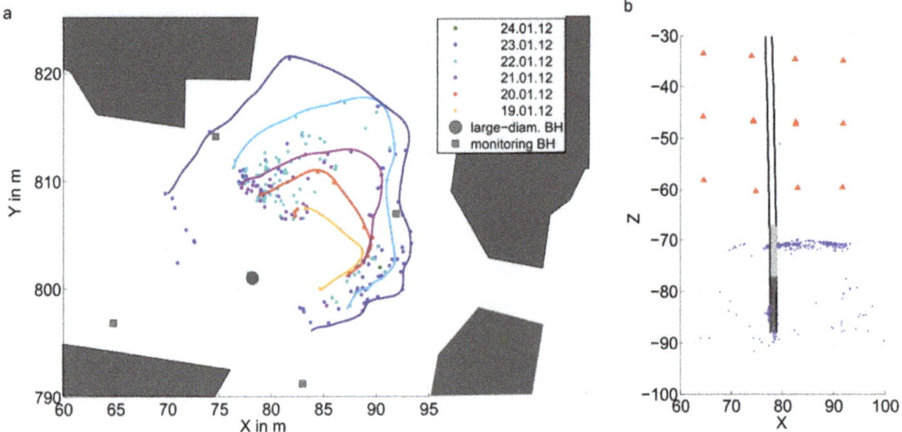

Figure 7. AE swarm activity during gas-brine break-through in map view (**a**) and side view (**b**). For orientation, the neighboring pillars are shown in dark grey in (**a**). Here color-coded lines mark the most outward extension of AE seismicity with time, which shows a migration of approximately 0.6 m per day. In (**b**) the cemented plug is shown in grey and the positions of AE sensors are shown by red triangles [82].

The experiment shows that quite different processes that influence the strength and the permeability of rock are successfully monitored using in situ AE monitoring technique. Not only is the influence of mechanical penetration (drilling) mapped by AE events (formation of the so-called excavation disturbed zone (EDZ)), but also the influence of environmental influences (humidity and temperature during cementation). Last but not least, the aseismical opening of a pathway for gas-brine percolation is accompanied by AE events.

3.3. Salt Mine Morsleben in Germany

The following example of in situ AE monitoring originated from the central section of the underground repository of Morsleben in Germany. Mining in these areas continued until the 1960s, but most of the rooms in the rock salt were mined more than 80 years ago. In this section, in situ AE monitoring has been performed since 1994. In the central section of the underground repository, the borehole sensors are distributed at three excavation levels and installed in 3 to 20 m deep boreholes. Originally, a network of 24 AE sensors monitored this section and covered an area of

150 m × 100 m × 120 m. This network was enlarged to 48 channels and covers a rock volume of about 250 m × 200 m × 120 m [37,39,69]. The average depth of the monitored volume is 400 m. This in situ AE monitoring provides a dataset of currently approximately 15 million located events per year [92]. For most events no waveforms are stored, but only the results of real-time processing.

The aim of in situ AE monitoring is to investigate micro- and macrocracking, which are important for the evaluation of the stability of cavities and the hydraulic integrity of the rock, which is of special interest in the case of an underground disposal of hazardous waste in salt rock [39]. Figure 8 shows a cross section (sketch) through the central part perpendicular to the average direction of strike, where cavities in rock salt were mined beneath thick anhydrite blocks. The actual geological situation and the arrangement of cavities is more complicated and strongly varying along the strike. Stress redistribution causes high AE activity at the walls of the cavities (in the EDZ) particularly, where cavities are close to one another and at the boundary between rock salt and anhydrite [39].

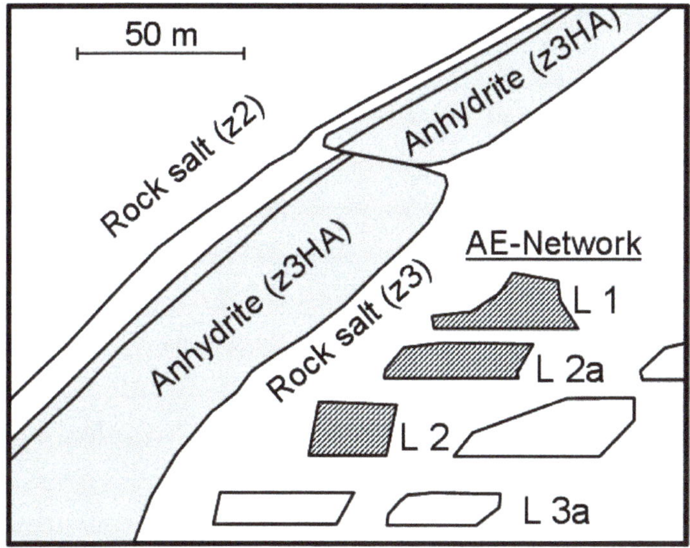

Figure 8. Sketch of geology and geometry of rooms in the central mine segment, where the AE network was installed (levels L1 to L3) [39]. Reproduced with permission.

However, the temporal and spatial occurrence of the events differs. Apart from seasonal fluctuations that may be explained by variations in humidity, the AE activity along the walls of the cavities does not vary with time. Outside the cavities near the anhydrite border, the AE events occurred in clusters. In some cases, such clusters were repeatedly located in the same volume; in other cases, significant emission occurred only within a limited time period [37].

To maintain the integrity of the barrier to the top of the salt deposit and the stability of the rooms for a long time, the rooms in the central part were backfilled with salt concrete from September 2003 to January 2011. During and after backfilling, the rock in the vicinity of these rooms were additionally loaded due to thermally induced stresses by released heat during hydration of the salted concrete for a period of several months to several years. Figure 9 shows the development of the located AE events per hour over a very long period, including the time before backfilling. One recognizes the strong increase of the microcracking activity up to approximately 1200 events per hour with beginning of the backfilling starting from September 2003 [92].

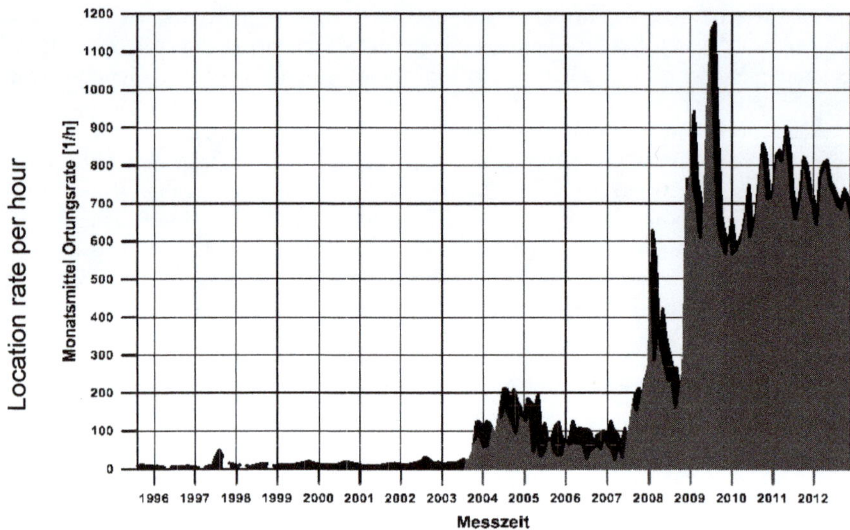

Figure 9. Development of the location rate in the time period from August 1995 to December 2012 in the area of the AE network in the central part of the salt mine Morsleben [92]. Reproduced with permission.

Figure 10 shows a perspective view of the central part of the salt mine Bartensleben with the rooms at Levels L1 to L3. The located AE events and the AE borehole sensors are marked by blue and red dots, respectively. The AE events were located during backfilling within a time period of two days in May 2010. It can be stated, that most of the activity took place near the cavities with especially high AE activity at the roof of the cavities [92].

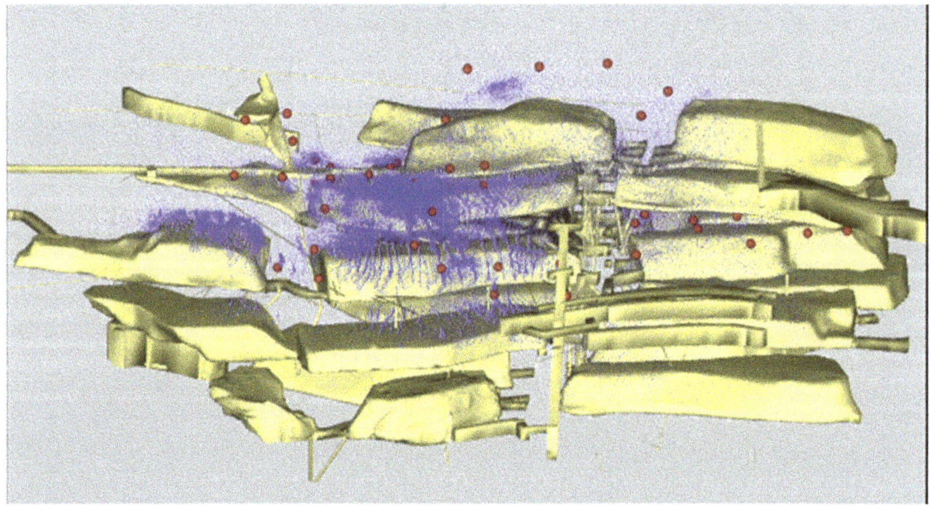

Figure 10. Perspective view of the central part of the salt mine Bartensleben (Levels L1 to L3) with location of AE events (blue dots) and AE borehole sensors (red dots). The area shown has an extension of about 300 m in the N–S direction and 240 m in height. The AE events were located within 2 days in May 2010 (see https://www.bgr.bund.de) [92]. Reproduced with permission.

Almost a year after backfilling of the cavities from 2003, AE events are distributed with distinctive stripe shapes above cavities at different depth levels (see Figure 11). The physical forces driving the creation of these stripes are still unknown. One possible explanation might be that these spatial patterns of the AE activity originated from the extensional stress developing in the cavities roofs. This strip-shaped pattern runs transversely to the longitudinal axis of the excavation in the ridge area of the room [92].

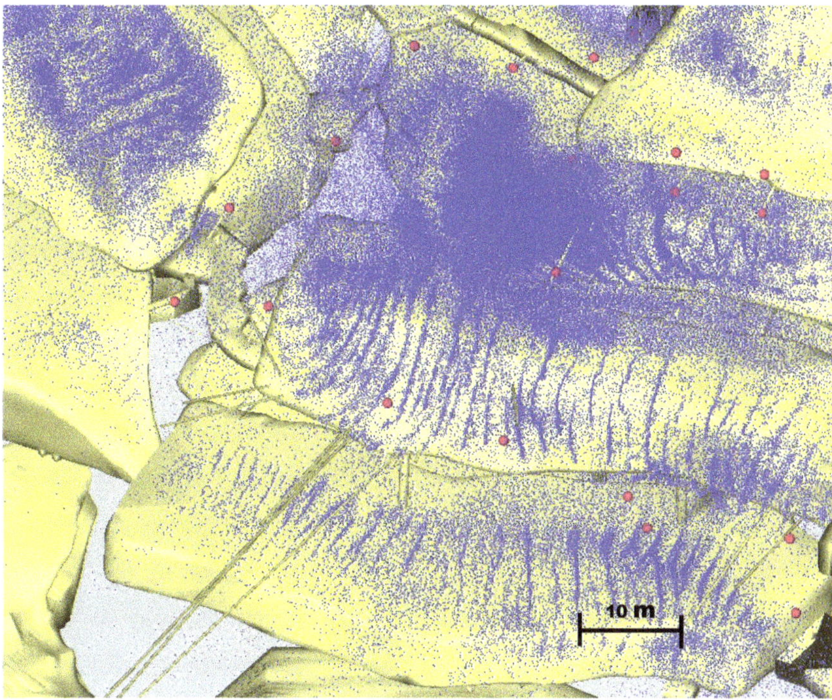

Figure 11. Patterns of approximately parallel stripes of AE activity over the ridges of backfilled excavation chambers. The view is downwards in diagonal direction [92]. Reproduced with permission.

In situ AE monitoring in the salt mine Morsleben provide a very large and unique dataset of approximately 100 million located AE events and it offers a wide range of options for evaluating fracture processes in a salt mine. In conclusion, the AE activity in salt rock is detected around open cavities and at the boundaries between different rock types. Creep processes cause high AE activity due to high deviatoric stresses at the walls of the cavities in the EDZ. This kind of AE activity is interpreted as ongoing deformation (convergence) in the vicinity of the open cavities and it is always present until convergence has been stopped, e.g., by backfilling of the open cavities. Apart from seasonal fluctuations due to the variation of humidity, the AE activity does not vary with time.

Because of its ductile behavior, rock salt is usually capable of performing creep deformation without occurrence of microcracking at stresses below the so-called dilatancy boundary. Above the dilatancy boundary, microcracking occurs. Most of the microcracks occur on grain boundaries and form no continuous macroscopic fractures. Deviatoric stresses above the dilatancy boundary result in the growth and opening of these microcracks, which are mainly responsible for dilatancy and the increase of permeability for fluids [37].

Although these microcracks have small dimensions in the order of the grain size (millimeter or centimeter) of the rock salt, in the course of time in dilatant zones microcracks may join and form macroscopic fractures especially in pillars between open cavities and in locations of high stress due to the edges of rooms, which are superposed in different levels. In such highly stressed zones, local instabilities may arise as spalling from the walls and roof falls. In this case, the EDZ migrates into the intact rock salt. AE activity in the EDZ will be stopped not before the open cavities are closed by convergence or backfilling. Closed cavities stabilize the environmental rock and microcrack formation will be stopped [92].

Figure 12 displays the evolution of AE location rates per hour (average of one day) and the evolution of temperature in the open space beneath the roof of a room. Backfilling started in October 2003 and the room was filled completely by the end of March 2004. Clearly, a very good correlation of AE location rates and temperature could be stated in the first three month. The general increase of temperature resulted from the heating due to the concrete setting of hydration. The peaks of the curve are caused by the fact that the fresh concrete cools down the surface of the concrete body during the working week. At the weekend backfilling is interrupted and temperature rises quickly until the beginning of the next week. The quick temperature increase is accompanied by a very sharp and high increase in the location rate of AE events characterizing the intensity of microcrack processes taking place. When temperatures decreased after the weekend, the location rates also decreased instantaneously indicating that the microcrack activity was generated by thermoelastic effects due to the heating of the rock mass by the setting of the concrete.

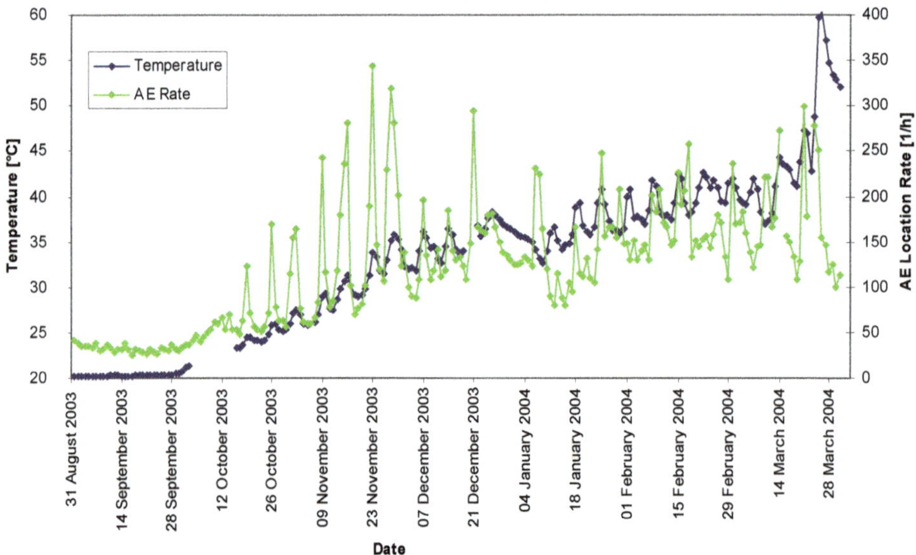

Figure 12. AE location rate and temperature in a room in the central part of the salt mine Bartensleben during backfilling [38].

Becker et al. [72] compare in a study the spatio-temporal evolution of the AE event distribution (Figure 12) with results from a 2D-finite element study of the evolving stress field well constrained by the known geometry of the structure and the material properties. They found out that for the first five thermal loading cycles a pronounced Kaiser effect can be observed. The so called Felicity Ratio is the ratio of the AE onset stress value to the previous peak stress [93]. It is a measure of the quality of the Kaiser effect with a value of 1 indicating a perfect stress memory effect. The observed Felicity Ratio of greater than 0.96 for the first five loading cycles indicates a very pronounced Kaiser effect. The deviation from the Kaiser effect during later loading cycles seems to be caused by the initiation of a planar macroscopic crack, which is subsequently reactivated. AE activity tends to concentrate along this macrocrack.

4. In Situ AE Monitoring in Gold Mines

4.1. Gold Mine Cook 4 in South Africa

Performing in situ AE monitoring inside the deep mines in South Africa is challenging, because these mines are active working mines that come with an especially harsh environment. What is beneficial for AE monitoring is the hard rock environment that allows to observe seismic waves with frequencies far above 25 kHz from significant distances from more than 100 m [36]. For SHM, in situ AE monitoring was shown to be a valuable tool, because damage zones prone to failure can be identified beforehand and rock bursts can be analyzed in great detail after.

The SATREPS project [33], that took place in Cooke 4 Mine, near Westonaria in South Africa at about 1 km depth demonstrated how to monitor a mining front in quartzite from afar, using a network of 24 field AE sensors and 6 triaxial accelerometers installed in development tunnels approximately 20 m to 50 m below the stope (network dimension 95 m × 50 m × 30 m, monitoring area 100 m × 180 m × 50 m). Figure 13 shows configuration of the in situ AE monitoring system installed 1 km beneath the ground in the Cooke 4 shaft. This figure shows the top view of the stope and the positions of sensors. The black squares denote the positions of the borehole sensors. The hatched area represents the excavated area. The mining faces are typically 30 m across and 1 to 2 m high, and the stope is sub-horizontal orientated. By daily blasting, the mining face advances to the north by about 10 m per month (red arrows). The mining face was located about 20 m above the AE network.

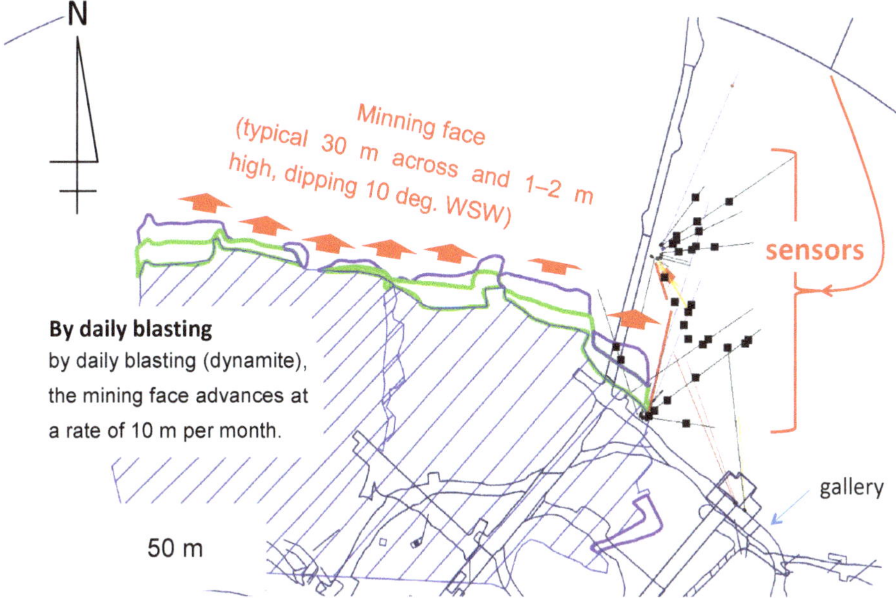

Figure 13. Top view of the stope and the positions of the borehole sensors (black squares). The shaded area represents the excavated area. The mining faces are typically 30 m across and 1 to 2 m high and the stope is sub-horizontal orientated. By daily blasting, the mining face advances to the north by about 10 m per month (red arrows) [34].

Over a time period from 11 July 2011 to 24 August 2011 (approximately 50 days), about 290,000 AE events were recorded in trigger mode recording and localized using the joint hypocenter determination

(JHD) approach and double-difference re-localization. The moment magnitudes M_W of events ranged from −3.7 to 1.0 [53]. Figure 14 shows the JHD result of 289,015 events, that had RMS residuals smaller than 1.0 ms in a top view (a) at an elevation of 690 m above sea level (about 1 km beneath ground), and (b) NNE–SSW vertical projection viewed from the west.

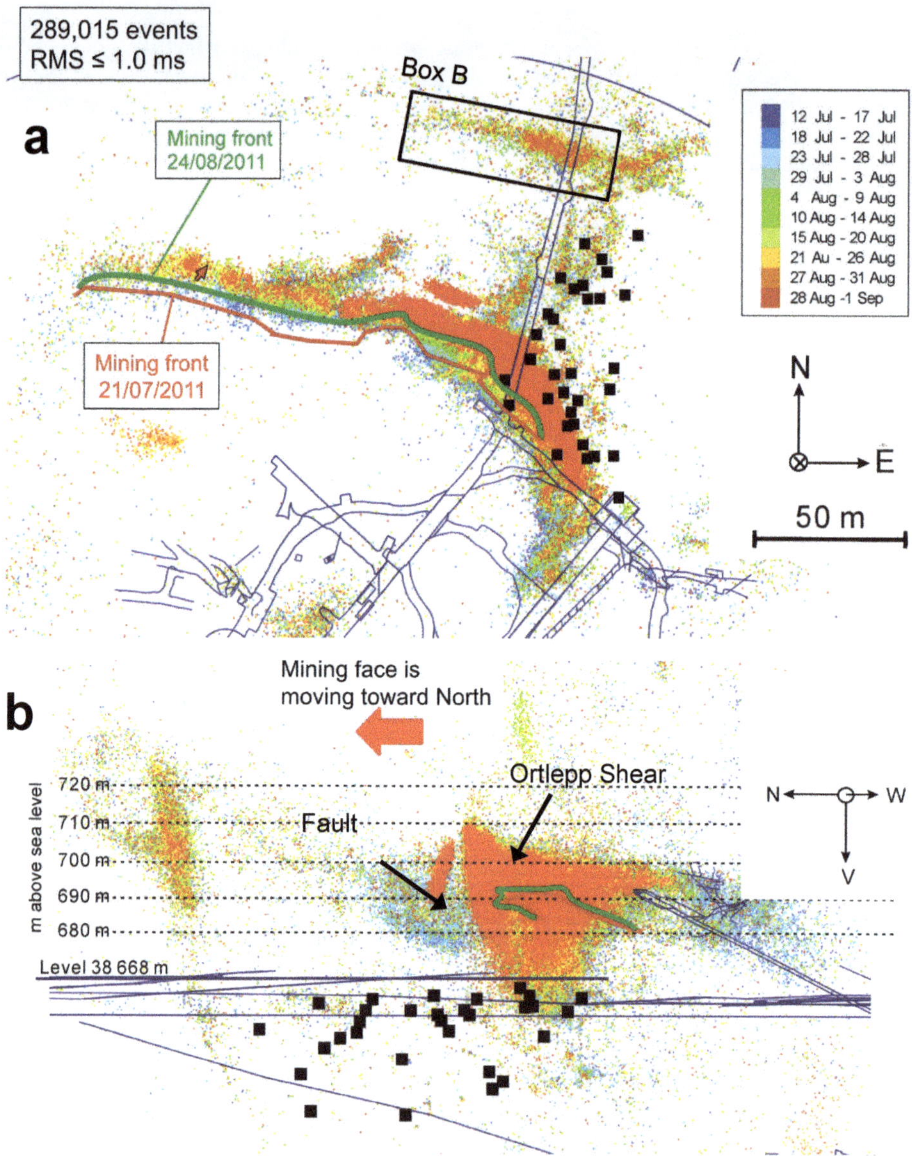

Figure 14. Source locations of AE events determined by JHD in a top view (**a**) at an elevation of 690 m above sea level (about 1 km beneath ground), and (**b**) NNE–SSW vertical projection viewed from the west. Black squares indicate AE sensor locations. Thick cyan and red lines represent positions of the mining face on 21 July and 24 August 2011, respectively. Blue lines are outlines of the mining infrastructure. Box B outlines the area that is shown in Figure 15 in detail. A total of 289,015 AE events were located, with their dates of occurrence indicated by the color scale [34].

Black squares indicate AE sensor locations. Thick cyan and red lines represent positions of the mining face on 21st July and 24th August 2011 respectively. Blue lines are outlines of the mining infrastructure. The 289,015 AE events were located, with their dates of occurrence indicated by the color scale. Nearly all (90%) of the AE events occurred in a zone extending about 20 m ahead (north) of the active mining face, forming what is referred to as the stope-front cloud. AE events clearly clustered on structures' outlining planes of localized damage, which is why the authors referred to the monitoring as an advanced AE mapping technique.

Naoi et al. [84–86] relocated clustered AE events by using the double difference technique [94]. After a pre-selection of events, the cross-correlation technique for all event pairs whose inter-event distance was smaller than 4 m was applied. The number of events in Box B in Figure 14a before relocation was about 10,688. The number of events successfully relocated as AE events was 10,337. Figure 15a,b show relocated AE events in projections onto the x-y plane and x-z plane, respectively. A local coordinate system aligned with the clear discernible plane is used. The x coordinate is along the strike direction (positive eastward), the y coordinate is in the normal direction to the plane (positive northward), and the z coordinate is along the dip direction (positive upward). Figure 15c shows projections onto the x-y plane for z coordinate between 13.5 m to 34.5 m in steps of 3 m thickness in the z direction. Blue dots are clustered AE events (approximately 4900) defined at the plane, which are situated in the red dashed frames in (a) and (c). Gray dots are other events, which belongs not to the cluster. The AE hypocenters projected onto x–y planes exhibit clear traces continuous over several meters or more. It was shown that the events of this cluster correspond to the Zebra fault, a local fault.

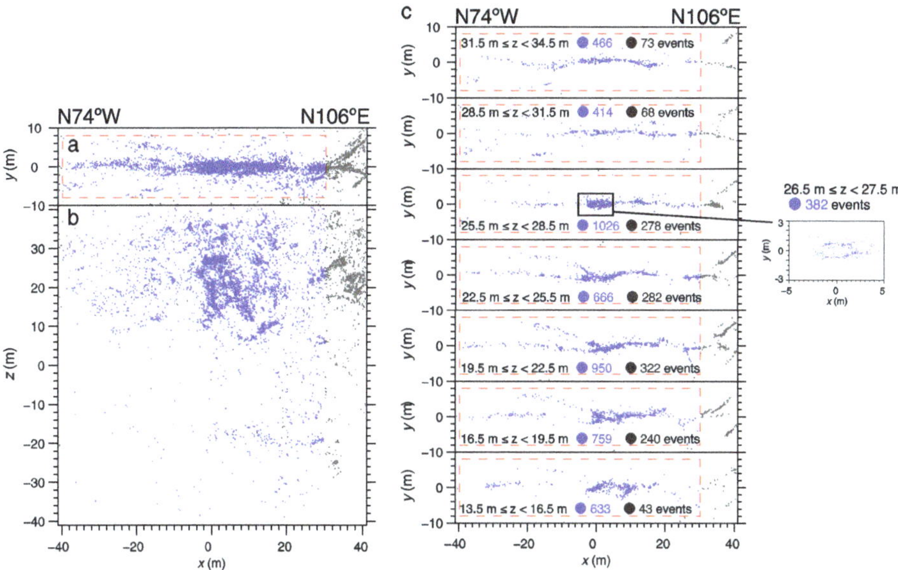

Figure 15. Close up of area marked by Box B in Figure 13. The relocated AE events in projections onto the x-y plane (**a**) and x-z plane (**b**) are shown. A local coordinate system aligned with clear discernible plane is used. x coordinate is along strike direction (positive eastward), y coordinate is in normal direction to the plane (positive northward), and z coordinate is along dip direction (positive upward). (**c**) Shows projections onto the x–y plane for z coordinate between 13.5 m to 34.5 m in steps of 3 m thickness in the z direction. Blue dots are clustered, which are located in the red dashed frames in (**a**,**c**). Gray dots are other events [84].

The information gained from studying the characteristics of AE event on such structures revealed in-depth information useful for risk assessment and stope planning. For example, the formation of

Ortlepp shear fractures [95,96] due to the approaching stope front could be observed (see Figure 14b); AE activity picking up on the newly created formation already, when the mining front was still more than 20 m away [34].

At the same time, pre-existing discontinuities such as faults or dikes that get seismically activated by the stress re-distribution due to the excavation were also identified [34,84]. Higher-order features such as branches and step-overs could be observed owing to the high localization certainty. Detailed analysis of AE event properties including a b-value study revealed that both quasi-static slip [85] as well as dynamic slip events is observed on faults [84].

Interestingly the frequent observation of repeating AE events on faults monitored during the SATRAPS project suggests that AE events on faults loaded by stress-redistribution due to mining undergo a similar process as tectonic faults subject to tectonic stresses [86]. Note that both Ortlepp shear fractures as well as the re-activation of faults cause on a regular basis violent rock burst events (M1 to M4), which are a significant threat to people working underground [96–98].

4.2. Gold Mine Mponeng in South Africa

The joint Japanese-German Underground Acoustic Emission Research project (JAGUARS) (see Table 3) in South Africa measures AE events in the frequency range from 700 Hz to 200 kHz. In the JAGUARS project [73] conducted in Mponeng Gold Mine in Carltonville in South Africa in 3.3 km depth the full evolution of a rock burst with moment magnitude $M_W = 1.9$ could be monitored using an in situ AE monitoring network. The JAGUARS network was installed in spring 2007.

Table 3. Group members of the joint Japanese-German Underground Acoustic Emission Research in South Africa (JAGUARS) project [36,73].

Institute	Main Contributors
Japan	-
Earthquake Research Institute of the University of Tokyo	M. Nakatani, M. Naoi;
Tohoku University	Y. Yabe
Ritsumeikan University	H. Ogasawara
Germany	-
GFZ: German Research Centre for Geosciences Potsdam	K. Plenkers, G. Kwiatek, G. Dresen, S. Stanchits
GMuG mbH, Bad Nauheim	J. Philipp
South Africa	-
Seismogen CC, Carletonville	
AngloGoldAshanti Ltd.	
Institute of Mine Seismology (IMS), Stellenbosch	G. Morema, T. Ward
Council for Scientific and Industrial Research (CSIR),	C. Miller, T. Nortje, R. Carstens
Johannesburg	E. Pinder, G. van Aswegen
	S. Spottiswoode

The gold is mined from the Witwatersrand formation. The gold-carrying sedimentary layer (Ventersdorp Contact Reef) is embedded in a thick series of quartzite, dipping with 26.5° toward the south-east, and reaches a thickness of 0.5 to 1 m. The JAGUARS network is located approximately 90 m below the reef, next to a gabbroic dike (Pink-Green (PG) dike, Figure 16). This dominant geological feature is 30 m wide and dips nearly vertically. It cuts through the reef and serves as a support pillar for the exploitation.

Naoi et al. [99] estimated the seismic velocities of rock types in the vicinity of the network from velocity tomography using AE transmission measurements. For the quartzite host rock, Naoi et al. [99] estimated seismic velocities of $v_P = 6.2$ km/s for P waves and $v_S = 3.8$ km/s for S waves. The velocities within the pink-green dike were found to be slightly higher, with $v_P = 6.9$ km/s and $v_S = 3.9$ km/s.

Mining in the vicinity of the JAGUARS network (Figure 16) started in early 2007. The network focuses on the dike-host rock contact close to the mining front, where larger events with magnitudes up to $M_W = 3.0$ were expected due to stress concentration induced by mining (personal conversation with S. Spottiswoode, 2009). The seismic network is consisting of 8 borehole AE sensors and one triaxial accelerometer. All sensors were installed in short boreholes of 6 m to 15 m length. The data were recorded in trigger-mode recording.

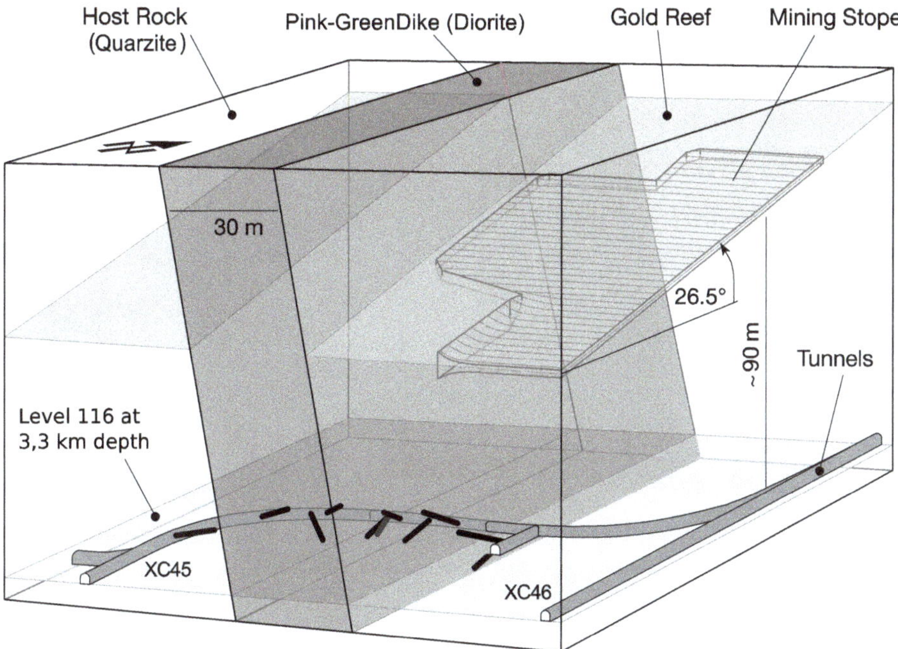

Figure 16. Sketch of the location of the JAGUARS network in the Mponeng gold mine and its surrounding geologic formations. The JAGUARS network is located at 3540 m depth. The sensors are located in boreholes, that are shown as bold black lines. The Pink-Green (PG) dike is shown in gray. The gold reef with active mining located above the JAGUARS network is shown in light gray. The mining stope width is 0.5 m to 1 m. The development tunnels XC45 and XC46 are shown in dark gray [52]. Reproduced with permission.

The stress changes owing to the approaching mining front on the PG dyke are modeled by Ziegler [77]. A main shock with magnitude of $M_W = 1.9$ occurred on 27 December 2007 in the center of the AE monitoring network [33,36,74]. Naoi et al. [74] manually picked P- and S-wave arrival times to locate more than 20,000 AE events, that occurred within 150 hours, following the main shock. The location error for events within a radius of about 40 m of the center of the AE network was less than 1 m. Most of the AE events from this period occurred within 50 m to 100 m of the network, where the spatial coverage of the network is best. The events contain signals with a broad range of high frequencies, which allowed the sensitivity of the network toward very high frequencies above 25 kHz to be analysed in greater detail [36,52]. Owing to the resulting excellent recording of the aftershock AE event sequence with most AE events after the main shock, the rupture process and the rupture plane could be studied in great detail.

The spatial distribution of 9,444 aftershocks is shown in map view in Figure 17a. The main shock hypocenter is shown by a grey star. The positions of cross sections I to IV are shown by magenta rectangulars. In Figure 17b the cross sections are presented the location of manual re-located AE events in a side-view. The PG dyke is shown by two grey lines. The re-located events reveal that the rupture plane started within the PG dyke, but reached the geological boundary, where branching and bending occurred. Secondary features (marked with VII and IX) are observed that likely correspond to the rupture of major aftershocks.

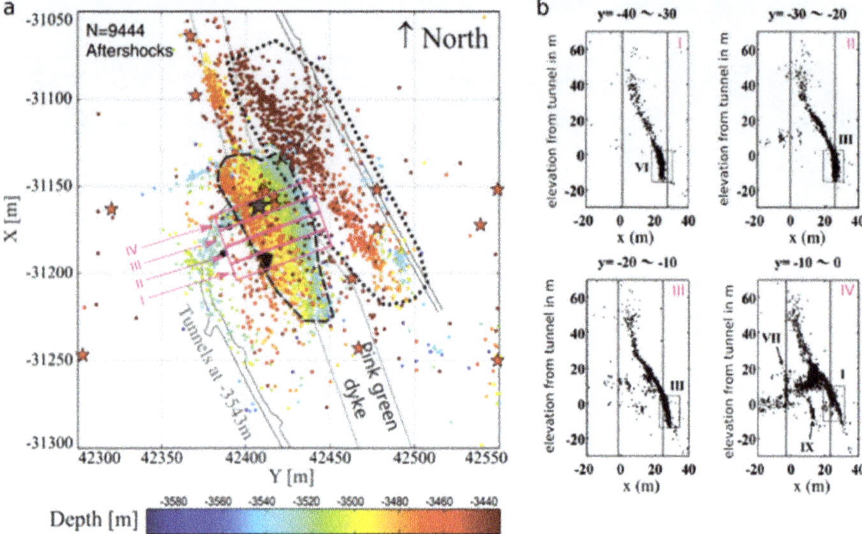

Figure 17. Mainshock-Aftershock sequence from 27 December 2007 recorded by the JAGUARS network. The spatial distribution of 9,444 aftershocks is shown in map view in (**a**). These aftershocks comprise a subset of the 25,000 aftershocks recorded which come with small residuals of the automatic locations. The mainshock location is shown by a grey star. Red stars show the location of all aftershocks recorded by the in-mine geophone network. The event depth (shaft depth) is color-coded, which shows that aftershocks were triggered not only on rupture plane, but also at the stope face above the mainshock. The positions of cross sections I to IV are shown by magenta rectangulars. In (**b**) cross sections are presented that show the location of manual re-located AE events in side-view. The PG dyke is shown by two grey lines.

Yabe et al. [78] showed that the aftershock AE events clearly delineated a plane in the PG dike with a strike of N22W and a dip of 68° toward N68E (Figure 17). Because waveforms of the main shock recorded by the AE network were saturated in AE recordings, the main shock was analyzed using waveform data of the in-mine geophone network. Naoi et al. [74] were able to resolve the complexity of the rupture plane (approximately extension 100 m × 80 m), which underwent branching and bending according to geological heterogeneities present. Naoi et al. [74] applied the master-event location technique to locate the hypocenter of the main shock relative to the aftershock AE events. The main shock hypocenter obtained was about 30 m above the AE network and on the aftershock plane. The focal mechanism solution of the normal fault for the main shock using seismic waveforms recorded by the seismic network operated by the mine pointed out that one of the nodal planes agreed well with the aftershock plane. Therefore, the aftershock plane is considered to correspond to the rupture plane of the main shock plane, which demonstrates clearly that the AE aftershock activity outlines the main shock's rupture plane. Kozlowska [79] show that the aftershocks occur in areas of positive Coulomb stress change as determined using rate and state based stress modeling.

Source parameter analysis demonstrated that the aftershock AE event sequence has the same characteristics as tectonic aftershock events, i.e., they follow the Omori law [36], the Gutenberg–Richter distribution [55] and a static stress drop [16]. The magnitude ranges between −5.0 and −0.8. The magnitude of completeness varied strongly in space, but was estimated to $M_C = -4.8$ in the network center [52].

The area of rupture initiation was subject to foreshock activity that is interpreted as the breakdown of asperities [78]. Interestingly, there are indications that AE events announce the main shock in advance. Figure 18a,b display the results of manual and automatic source location, respectively, applied to the same AE events, which are within 5 m of the aftershock surface. The events occurred during the periods before the main shock and within 150 h following the main shock. The black dashed contour represents the area of significant aftershock activity, as defined by Naoi et al. [74]. As this area can be presumed to represent the rupture area of the main shock, AE events that took place in the aftershock area and within 5 m of the aftershock surface before the main shock are hereafter referred to as "foreshocks". Four foreshock Clusters F1 to F4, and two aftershock Clusters A1 and A2 were identified from a concentration of manually located AE events. Both the foreshock and the aftershock clusters barely overlap one another. For comparison, the automatically located AE events, which are much more numerous than manually located ones are shown in Figure 18b as a density plot.

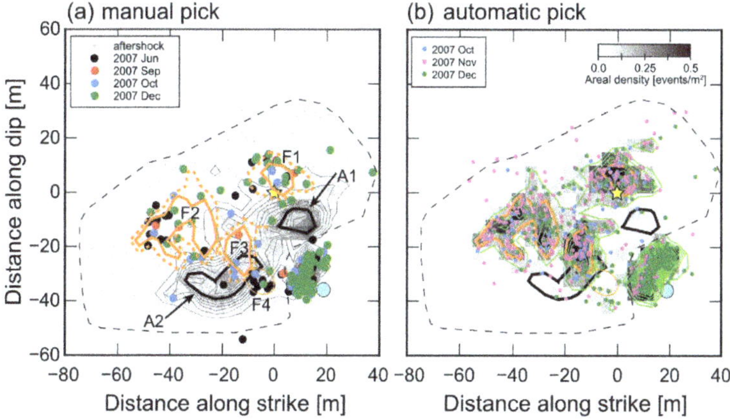

Figure 18. Foreshock activity compared to aftershock activity. (**a**) Distribution of the manually located AE events that occurred within 5 m of the aftershock surface during the period from 13 June 2007 to 150 h after the main shock. The yellow star indicates the main shock hypocenter. Black, red, blue, and green solid circles denote the events in June, September, October, and December, respectively. Orange and black thick solid lines enclose the foreshock clusters (F1, F2, and F3) and aftershock clusters (A1 and A2), respectively. Gray dots denote the aftershocks. Gray contours show the areal density of the aftershocks drawn by feeding areal densities in 5 m × 5m cells. The thin black dashed contour indicates the aftershock area defined by Naoi et al. [78]. The light blue solid circle is the access tunnel along which the observation network was deployed. (**b**) Distribution of the automatically located events that occurred before the main shock and within 5 m of the aftershock surface. Their densities in 5 m × 5 m cells are shown by gray scale. Blue, purple, and green solid circles denote events in October, November, and December, respectively. Light green contours show the areal density of the automatically located foreshocks.

5. In Situ AE Monitoring During Hydraulic Fracturing in Mines

A slightly different aspect of SHM concerns the monitoring of hydraulic fracturing (HF) using in situ AE monitoring. Engineered fractures generated underground by packer probes in boreholes are facilitated in a broad variety of contexts, many of which require detailed knowledge on the damage process actively initiated. HF is a common tool for underground stress determination and provides important input for designing the stope layout and for risk assessment [100–102]. HF has become a widely used engineering tool in reservoir enhancement of geothermal systems, shale gas, or conventional oil and gas extraction as it effectively increases the permeability [103–106]. In addition,

HF is successful in increasing the productivity in ore production e.g., HF is used in fragmenting ore bodies [107].

Several research projects have addressed HF using in situ AE monitoring or strain cells in order to increase the understanding of the rock response to HF, to study the evolution of fracture generating and predict the stimulation of existing fractures [67,108,109]. Two recent research projects in crystalline rock did not only record very interesting and rich data, but have pushed the limits in highly sensitive monitoring and advanced signal processing. Zang et al. [43] report on HF at the hard rock underground laboratory Äspö in Sweden (Figure 19a).

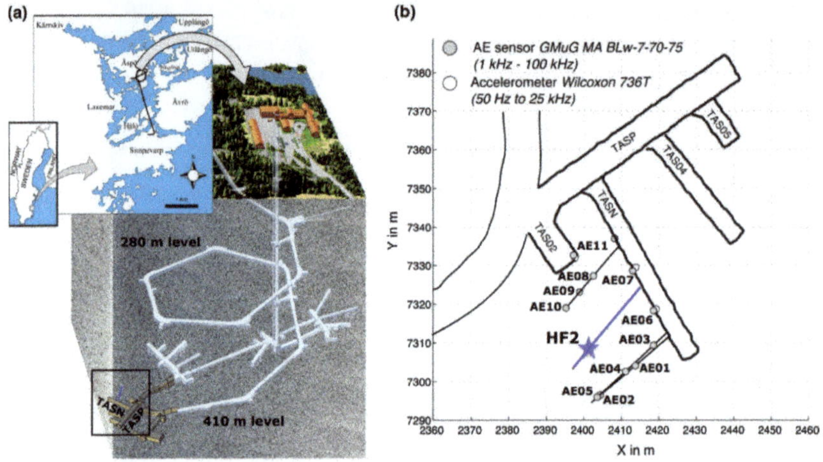

Figure 19. (**a**) Test site of hydraulic fracturing tests in the underground Äspö Hard Rock Laboratory in Sweden (see http://www.skb.se/upload/publications/pdf/Aspo_Laboratory.pdf). (**b**) Location of the AE borehole sensors together with the central injection well (blue line). The blue star identifies the fluid injection segment corresponding to the hydraulic fracturing (HF2) experiment [43,87].

Zang et al. [43] utilize three different monitoring networks, namely in situ AE monitoring, microseismic monitoring, and electromagnetic monitoring. Figure 19b shows a top view of the AE monitoring network, which consists of 11 AE sensors of GMuG type MA BLw-7-70-75 and four Wilcoxon 736T accelerometers.

Data is recorded both in trigger mode and continuous mode (1 MHz sampling frequencies). AE sensors are installed in monitoring boreholes of 22 m to 30 m length parallel to the injection borehole and in short boreholes along the tunnels.

Overall, six HF stimulations are performed using three different injection schemes (continuous, progressive and pulse pressurization), from which four produced significant AE activity outlining the fracture orientation, extension and temporal evolution. Within 20 days, about 69,400 triggers were recorded in situ, from which many correspond to noise events due to dripping water or anthropogenic activities. The strongest AE events recorded in situ with best signal-to-noise ratio and most reliable location certainty (maximum location residual 0.3 m) formed a catalog of 196 seismic events, of which all correlated in time and space directly to the HF periods. All relocated AE events are shown in a perspective view (Figure 20a) and a lateral view in a rotated coordinate system (Figure 20b) for seismically active HF stimulations (HF1 to HF4 and HF6). AE events recorded during the different fracture experiments clearly delineate the fractures and display differences between the different hydraulic fractures generated.

A faster and further expansion of AE events away from the stimulation point is observed with each subsequent re-stimulation stage outlining the damage extension. In-depth source analysis of the largest AE events including energy estimation, moment tensor inversion, source parameter estimation, and stress inversion by Kwiatek et al. [27] estimated the moment magnitude M_W of the AE events ranges from −4.2 to −3.5. The source analysis clearly reveals that most events correspond to shear slip events on pre-existing fractures. AE event activity starts as soon as a certain pressure level is reached. First optimum oriented fault planes fail but, overall, differently oriented fault planes are observed. Stress inversion reveals stress rotation during and after the stimulation.

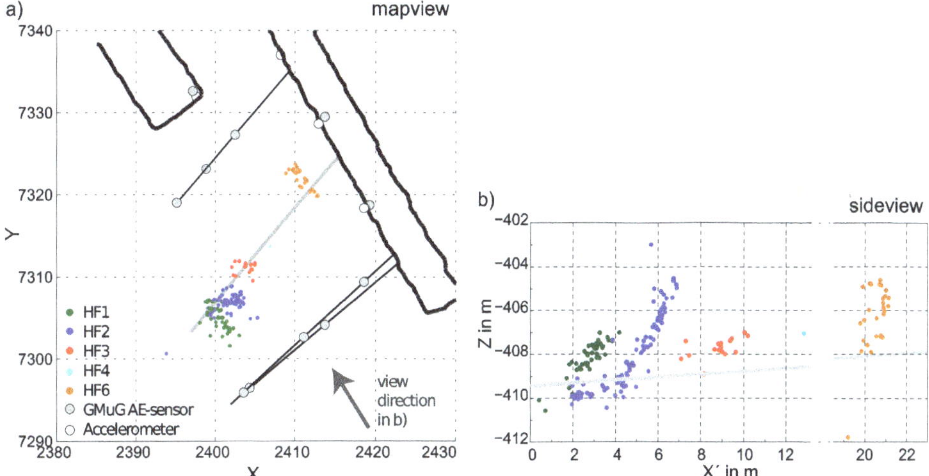

Figure 20. Location of AE events of six HF stimulations (HF1 to HF4 and HF6) presented in a top view (**a**) and a lateral view (**b**). For lateral view the coordinate system was rotated in that way to indicate the preferred fracture traces. The solid grey line outlines injection well and the observation boreholes (modified from [43]). All dimensions are given in meter.

Using the continuous data recordings from the Äspö experiment, López-Comino et al. [87] demonstrated that using automated full waveform detection algorithm during post-processing could significantly increase the amount of triggered AE events. No seismic fracture event was recorded by the microseismic monitoring network or the Wilcoxon accelerometers, although the latter is capable of measuring the frequency range of the observed AE events. Owing to the small nature of all AE events recorded, not only seismic monitoring in the kHz range was essential, but also the significantly higher sensitivity of the in situ AE sensors.

A different experiment was conducted at the Grimsel Test Site (GTS) in Switzerland (Figure 21a,b) operated by the Swiss National Cooperative for the Disposal of Radioactive Waste (Nagra). The GTS is located at 1,733 m above the sea level and has an overburden of 400 to 500 m. Gischig et al. [54] implement in situ AE monitoring for stress determination by HF in the Grimsel site. The so-called in situ stimulation and circulation (ISC) [110] was performed between two tunnels i.e., the VE and the AU tunnel (see Figure 21c), and the injection and monitoring boreholes were mostly drilled from the AU cavern at the southern end of the AU tunnel (Figure 21c). The host rock is the so-called Grimsel granodiorite, which changes into the Central Aar granite about 50 m north of the experiment volume [111]. The rock mass in the experiment volume is exceptionally intact. The Grimsel test site was monitored using 28 AE sensors (type GMuG MA-Bls-7-70) and four Wilcoxon accelerometers. Most AE sensors were installed on the tunnel wall on polished rock face, while eight AE sensors were installed in a water-filled vertical borehole (Borehole SBH1 in Figure 21c).

For stress determination, a series of hydraulic fracturing tests and overcoring were performed. During hydraulic fracturing, nearly 2000 AE events were recorded with a source-receiver distances smaller than 30 m that outlined, similar to the Äspö experiment, clearly the fracture plane that extended up to 5 m from the injection point (see Figure 22). Events occurred mostly during the refracturing cycles once a critical injection volume of 0.5 to 1 liter was exceeded and less during the initial fracturing cycle. A comparison of the fracture plane outlined by AE events and stress measurements using an imprint packer and overcoring testing, revealed significant deviations. The imprint packer revealed that fractures initiated at the borehole wall within the foliation plane, but the fracture growth than rotated, as outlined by AE events, in such way that it extends normal to the minimum principal stress. The deviation of the overcoring stress measurement result to the actual fracture plane observed could be explained by using a transversely isotropic elasticity model.

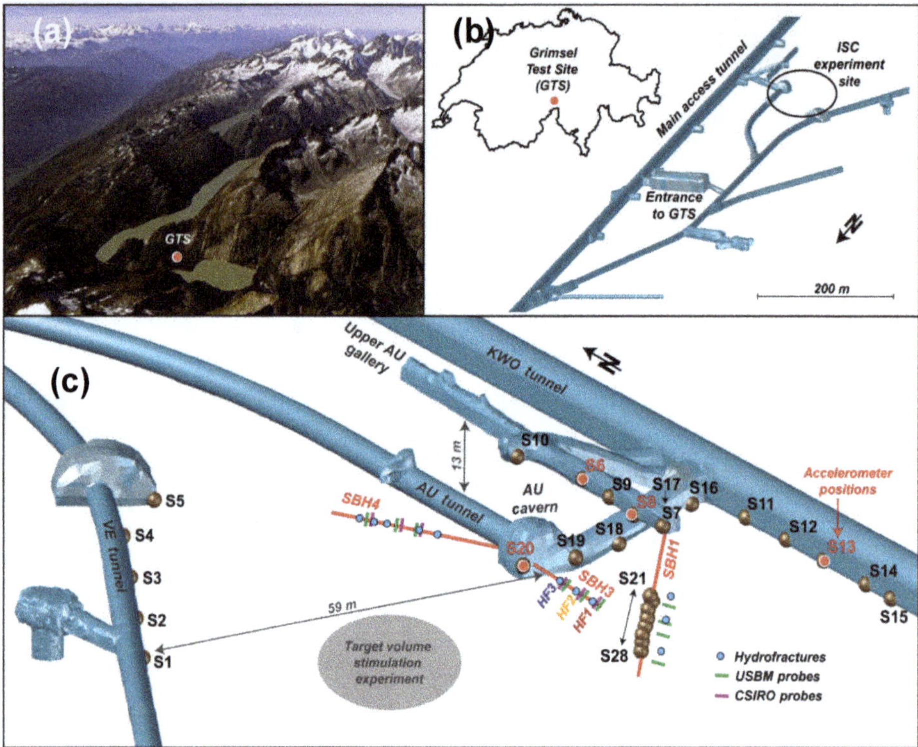

Figure 21. (**a,b**) Grimsel test site is located in the Bernese Alps in southern Switzerland (see www.grimselstrom.ch). (**c**) In situ AE monitoring was performed during hydraulic fracturing tests HF1 to HF3 in Borehole SBH3. Positions S1 to S28 mark the location of the AE sensors and accelerometers [88].

The authors conclude that AE monitoring was crucial for the combined interpretation of the stress characterization results and to maintain meaningful stress estimation.

We summarize that HF routine monitoring using in situ AE monitoring systems becomes feasible for underground production, if sensitive AE sensors for in situ operation are used. Two upcoming projects performing underground medium-scale HF testing will implement AE monitoring accordingly: the enhanced geothermal system (EGS) Collab project's stimulation experiment in Sanford Underground Research Facility in the former Homestake Gold mine, USA [112] and the STIMTEC stimulation experiment in the underground laboratory in the silver mine "Reiche Zeche" in Germany (personal conversation with J. Renner and G. Dresen, 2018) (more details at http://stimtec.rub.de).

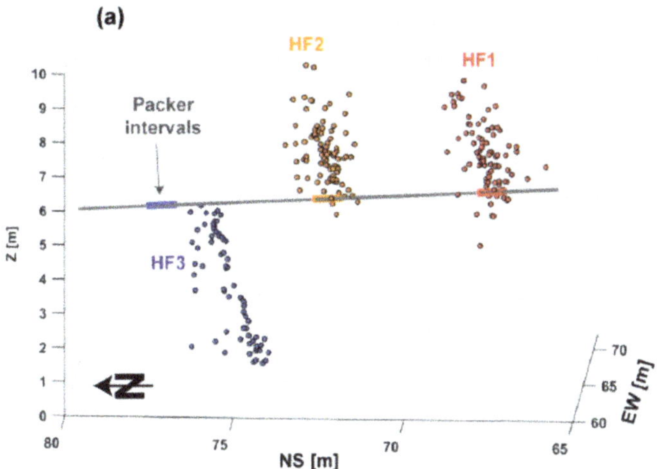

Figure 22. Perspective view of AE events detected during hydraulic fracturing tests HF1 to HF3. The continuous line indicates the injection well with the position of the injection intervals [54].

6. Concluding Remarks

In conclusion, this article summarizes the capability of in situ AE monitoring in the context of SHM based on the results of monitoring projects in mines. The in situ AE method is capable of detecting microcracking, in high resolution and sensitivity, which is caused by very small deformation processes at high deviatoric stresses. This means that in situ AE monitoring provides detailed insights into the ongoing deformation processes.

In contrast to in situ AE monitoring, microseismic monitoring is used to measure large-scale deformations in mines, which may cause rock bursts or roof falls. Due to limitations in frequency range and sensitivity, microseismic networks are not able to detect microcracks. Therefore, small AE events are very often not considered for stability assessment and interpretation of geomechanical conditions of the rock. This work clearly shows that in situ AE monitoring is able to detect very small AE events in zones of weakness related to dynamic processes like dilatation. Therefore, in situ AE monitoring is a useful tool to monitor the geomechanical conditions of the host rock.

Real-time processing gives direct information on the location of AE events as well as on clustering, migration of AE activity or aftershock sequences of microseismic events. Recent advances in computer storage capacity allow recording of continuous data streams with 1 MHz sampling in addition to trigger mode recording, which makes advanced post-processing techniques possible.

The results shown here also demonstrate that monitoring of larger rock volumes with in situ AE measurements is possible in various rock types. Detections of AE events from distances much greater than 100 m is possible in rock with low wave attenuation like salt rock or hard rock. In this case, rock volumes far away from the AE network can be monitored. On the other hand, the in situ AE monitoring method is able to identify "aseismic" zones because AE activity is expected during significant damage processes. Finally, in situ AE monitoring is capable of detecting zones in mines where instability may appears long before macroscopic damage becomes visible, which is the objective of SHM.

Author Contributions: G.M. and K.P. wrote the article.

Acknowledgments: The authors thank K. Ono and two anonymous reviewers for their comments and suggestions to improve this article. The authors are grateful to T. Spies and D. Kaiser (BGR in Hannover), G. Kwiatek (GFZ in Potsdam), and H. Moriya (Tohuko University in Sendai) for providing updated figures for this review.

Conflicts of Interest: The authors declare no conflicts of interest.

References

1. *Condition Monitoring and Diagnostics of Machines—General Guidelines*; ISO 17359:2011; Beuth Verlag GmbH: Berlin, Germany, 2011.
2. Cawley, P. Structural health monitoring: Closing the gap between research and industrial deployment. *Int. J. Struct. Health Monit.* **2018**. [CrossRef]
3. Chang, P.C.; Liu, S.C. Recent research in nondestructive evaluation of civil infrastructures. *J. Mater. Civil. Eng.* **2003**, *15*, 298–304. [CrossRef]
4. McCrea, A.; Chamberlain, D.; Navon, R. Automated inspection and restoration of steel bridges—A critical review of methods and enabling technologies. *Automat. Constr.* **2002**, *11*, 351–373. [CrossRef]
5. Rens, K.L.; Wipf, T.J.; Klaiber, F.W. Review of non-destructive evaluation techniques of civil infrastructure. *J. Perform. Constr. Fac.* **1997**, *11*, 152–160. [CrossRef]
6. Chang, P.C.; Flatau, A.; Liu, S.C. Review paper: Health monitoring of civil infrastructure. *Struct. Hlth. Monit.* **2003**, *2*, 257–267. [CrossRef]
7. Farrar, C.R.; Doebling, S.W.; Nix, D.A. Vibration-based structural damage identification. *Philos. Trans. R. Soc. A* **2001**, *359*, 131–149. [CrossRef]
8. Shih, H.W.; Thambiratnam, D.P.; Chan, T.H.T. Vibration based structural damage detection in flexural members using multi-criteria approach. *J. Sound Vib.* **2009**, *323*, 645–661. [CrossRef]
9. Ono, K. Structural Integrity Evaluation Using Acoustic Emission. *J. Acoust. Emiss.* **2007**, *25*, 1–20.
10. Ono, K. Review on Structural Health Evaluation with Acoustic Emission. *Appl. Sci.* **2018**, *8*, 958. [CrossRef]
11. Eisenblätter, J. The origin of acoustic emission—Mechanisms and models. In *Acoustic Emission*; Conference in Bad Nauheim, Deutsche Gesellschaft für Metallkunde e.V.: Berlin, Germany, 1980; pp. 189–204.
12. Manthei, G.; Eisenblätter, J. Acoustic Emission in Study of Rock Stability. In *Acoustic Emission Testing*; Grosse, C.U., Ohtsu, M., Eds.; Springer: Berlin/Heidelberg, Germany, 2008; pp. 239–310.
13. Kwiateka, G.; Plenkers, K.; Martínez-Garzóna, P.; Leonhardt, M.; Zang, A.; Dresen, G. New Insights into Fracture Process through In-Situ Acoustic Emission Monitoring during Fatigue Hydraulic Fracture Experiment in Äspö Hard Rock Laboratory. *Procedia Eng.* **2017**, *191*, 618–622. [CrossRef]
14. Stork, A.L.; Butcher, A.C.; Verdon, J.P.; Koe, V.A.; Kendall, J.-M. Geophysical investigations to assess the extent of disturbance in excavated rock faces at hinkley point C. In Proceedings of the 23rd European Meeting of Environmental and Engineering Geophysics, Malmö, Sweden, 3–7 September 2017.
15. Girard, L.; Beutel, J.; Gruber, S.; Hunziker, J.; Lim, R.; Weber, S. A custom acoustic emission monitoring system for harsh environments: Application to freezing-induced damage in alpine rock walls. *Geosci. Instrum. Meth.* **2012**, *1*, 155–167. [CrossRef]
16. Kwiatek, G.; Plenkers, K.; Dresen, G.; JAGUARS Research Group. Source Parameters of Picoseismicity Recorded at Mponeng Deep Gold Mine, South Africa: Implications for Scaling Relations. *Bull. Seismol. Soc. Am.* **2011**, *101*, 2592–2608. [CrossRef]
17. Bohnhoff, M.; Kwiatek, G.; Dresen, G. Von der Gesteinsprobe bis zur Plattengrenze: Skalenübergreifende Analyse von Bruchprozessen. *Syst. Erde* **2016**, *6*, 50–55. (In German)
18. Kwiatek, G.; Bulut, F.; Bohnhoff, M.; Dresen, G. High-resolution analysis of seismicity induced at Berlín geothermal field, El Salvador. *Geothermics* **2014**, *52*, 98–111. [CrossRef]
19. Kwiatek, G.; Bohnhoff, M.; Dresen, G.; Schulze, A.; Schulte, T.; Zimmermann, G.; Huenges, E. Microseismicity induced during fluid-injection: A case study from the geothermal site at Groß Schönebeck, North German Basin. *Acta Geophys.* **2010**, *58*, 995–1020. [CrossRef]
20. Harrington, R.M.; Kwiatek, G.; Moran, S.C. Self-similar rupture implied by scaling properties of volcanic earthquakes occurring during the 2004-2008 eruption of Mount St. Helens, Washington. *J. Geophys. Res.* **2015**, *120*, 4966–4982. [CrossRef]
21. Kwiatek, G.; Martinez Garzon, P.; Dresen, G.; Bohnhoff, M.; Sone, H.; Hartline, C. Effects of long-term fluid injection on induced seismicity parameters and maximum magnitude in northwestern part of The Geysers geothermal field. *J. Geophys. Res.* **2015**, *120*, 7085–7101. [CrossRef]
22. Kwiatek, G.; Ben-Zion, Y. Theoretical limits on detection and analysis of small earthquakes. *J. Geophys. Res.* **2016**, *21*, 5898–5916. [CrossRef]
23. Madariaga, R. Dynamics of an expanding circular fault. *Bull. Seismol. Soc. Am.* **1976**, *66*, 639–666.

24. Brune, J.N. Tectonic stress and the spectra of seismic shear waves from earthquakes. *J. Geophys. Res.* **1970**, *75*, 4997–5009. [CrossRef]

25. Hanks, T.; Kanamori, H. A moment magnitude scale. *J. Geophys. Res.* **1979**, *84*, 2348–2350. [CrossRef]

26. Manthei, G.; Eisenblätter, J.; Dahm, T. Moment Tensor Evaluation of Acoustic Emission Sources in Salt Rock. *Constr. Build. Mater.* **2001**, *15*, 297–309. [CrossRef]

27. Kwiatek, G.; Marinez-Garzon, P.; Plenkers, K.; Leonhardt, M.; Zang, A.; Specht, S.; Dresen, G.; Bohnhoff, M. Insights into complex sub-decimeter fracturing processes occurring during a water-injection experiment at depth in Äspö Hard Rock Laboratory, Sweden. *J. Geophys. Res. Sol. Earth* **2018**. [CrossRef]

28. Manthei, G.; Eisenblätter, J.; Spies, T. Source Mechanisms of Acoustic Emission Events between huge Underground Cavities in Rock Salt. In Proceedings of the Advances in Acoustic Emission 2007—Proceedings of 6th International Conference on Acoustic Emission, Lake Tahoe, NV, USA, 28 October–2 November 2007; pp. 288–293.

29. Manthei, G. Moment tensor evaluation of acoustic emission sources in rock. In *Rock Mechanics and Engineering*; Feng, X.-T., Ed.; CRC Press/Balkema: Leiden, The Netherlands, 2017; Volume 4, pp. 478–527, ISBN 9781138027640.

30. Kwiatek, G.; Ben-Zion, Y. Assessment of P and S wave energy radiated from very small shear-tensile seismic events in a deep South African mine. *J. Geophys. Res.* **2013**, *118*, 3630–3641. [CrossRef]

31. Kwiatek, G.; Goebel, T.; Dresen, G. Seismic moment tensor and b value variations over successive seismic cycles in laboratory stick-slip experiments. *Geophys. Res. Lett.* **2014**, *41*, 5838–5846. [CrossRef]

32. Lockner, D.A. The Role of Acoustic Emission in the Study of Rock. *Int. J. Rock Mech. Min. Sci.* **1993**, *30*, 883–899. [CrossRef]

33. Yabe, Y.; Philipp, J.; Nakatani, M.; Morema, G.; Naoi, M.; Kawakata, H.; Igarashi, T.; Dresen, G.; Ogasawara, H.; JAGUARS Research Group. Observation of numerous aftershocks of an 1.9 earthquake with an AE network installed in a deep gold mine in South Africa. *Earth Planets Space* **2009**, *61*, e49–e52. [CrossRef]

34. Moriya, H.; Naoi, M.; Nakatani, M.; van Aswegen, G.; Murakami, O.; Kgarume, T.; Ward, A.K.; Durrheim, R.J.; Philipp, J.; Yabe, Y.; et al. Delineation of large localized damage structures forming ahead of an active mining front by using advanced acoustic emission mapping techniques. *Int. J. Rock Mech. Min.* **2015**, *79*, 157–165. [CrossRef]

35. Manthei, G.; Eisenblätter, J.; Spies, T. Determination of Wave Attenuation in Rock Salt in the Frequency Range 1–100 kHz using Located Acoustic Emission Events. *J. Acoust. Emiss.* **2006**, *24*, 179–186.

36. Plenkers, K.; Kwiatek, G.; Nakatani, M.; Dresen, G.; JAGUARS Research Group. Observation of Seismic Events with Frequencies f > 25 kHz at Mponeng Gold Mine, South Africa. *Seism. Res. Lett.* **2010**, *81*, 467–479. [CrossRef]

37. Spies, T.; Eisenblätter, J. Acoustic emission investigation of microcrack generation at geological boundaries. *Eng. Geol.* **2001**, *61*, 181–188. [CrossRef]

38. Philipp, J.; Plenkers, K.; Gärtner, G.; Teichmann, L. On the potential of In-Situ Acoustic Emission (AE) technology for the monitoring of dynamic processes in salt mines. In *Proceedings of the Conference on Mechanical Behavior of Salt, South Dakota School of Mines and Technology, Mechanical Behavior of Salt VIII, Rapid City, SD, USA, 26–28 May 2015*; Lance, R., Mellegard, K., Hansen, F., Eds.; CRC Press/Balkema: Leiden, The Netherlands, 2015; pp. 89–98, ISBN 9781138028401.

39. Spies, T.; Hesser, J.; Eisenblätter, J.; Eilers, J. Measurements of acoustic emission during backfilling of large excavations. In Proceedings of the 6th International Symposium on Rockbursts and Seismicity Mines (RaSiM6), Perth, Australia, 9–11 March 2005; pp. 379–384.

40. Becker, D.; Cailleau, B.; Dahm, T.; Shapiro, S.; Kaiser, D. Stress triggering and stress memory observed from acoustic emission records in a salt mine. *Geophys. J. Int.* **2010**, *182*, 933–948. [CrossRef]

41. Eisenblätter, J.; Manthei, G.; Meister, D. Monitoring of Microcrack Formation around Galleries in Salt Rock. In *Proceedings of the Sixth Conference on Acoustic Emission/Microseismic Activity in Geologic Structures and Materials, Pennsylvania State University, University Park, PA, USA, 9–11 June 1998*; Hardy, H.R., Jr., Ed.; Trans Tech Publications: Clausthal-Zellerfeld, Germany, 1998; pp. 227–243.

42. Manthei, G.; Philipp, J.; Dörner, D. Acoustic Emission monitoring around gas-pressure loaded boreholes in rock salt. In *Mechanical Behavior of Salt VII*; Berest, P.B., Ghoreychi, M., Hadj-Hassen, F., Tijani, M., Eds.; Taylor & Francis (Balkema): London, UK, 2012; pp. 185–192, ISBN 9780415621229.

43. Zang, A.; Stephansson, O.; Stenberg, L.; Plenkers, K.; Specht, S.; Milkereit, K.; Schill, E.; Kwiatek, G.; Dresen, G.; Zimmermann, G.; et al. Hydraulic fracture monitoring in hard rock at 410 m depth with an advanced fluid-injection protocol and extensive sensor array. *Geophys. J. Int.* **2017**, *208*, 790–813. [CrossRef]
44. Hsu, N.N.; Breckenridge, F. Characterization of acoustic emission sensors. *Mater. Eval.* **1981**, *39*, 60–68.
45. Theobald, P.D.; Esward, T.J.; Dowson, S.P.; Preston, R.C. Acoustic emission transducers development of a facility for traceable out-of-plane displacement calibration. *Ultrasonics* **2005**, *43*, 343–350. [CrossRef] [PubMed]
46. Kim, K.Y.; Castagnede, B.; Sachse, W. Miniaturized capacitive transducer for detection of broadband ultrasonic signals. *Rev. Sci. Instrum.* **1989**, *60*, 2785–2788. [CrossRef]
47. Manthei, G. Characterization of Acoustic Emission Sources in a Rock Salt Specimen under Triaxial Compression. *Bull. Seismol. Soc. Am.* **2005**, *95*, 1674–1700. [CrossRef]
48. McLaskey, G.; Glaser, S. Acoustic emission sensor calibration for absolute source measurements. *J. Nondest. Eval.* **2012**, *31*, 157–168. [CrossRef]
49. Ono, K. Critical examination of ultrasonic transducer characteristics and calibration methods. *Res. Nondestruct. Eval.* **2017**. [CrossRef]
50. Ono, K. Calibration Methods of Acoustic Emission Sensors. *Materials* **2016**, *9*, 508. [CrossRef] [PubMed]
51. Manthei, G. Characterization of AE Sensors for Moment Tensor Analysis. In Proceedings of the International Acoustic Emission Conference, Kumamoto University, Kumamoto, Japan, 12–15 September 2010; pp. 19–24.
52. Plenkers, K.; Schorlemmer, D.; Kwiatek, G.; JAGUARS Research Group. On the Probability of Detecting Picoseismicity. *Bull. Seismol. Soc. Am.* **2011**, *101*, 2579–2591. [CrossRef]
53. Naoi, M.; Nakatani, M.; Horiuchi, S.; Yabe, Y.; Philipp, J.; Kgarume, T.; Morema, G.; Khambule, S.; Masakale, T.; Ribeiro, L.; et al. Frequency-magnitude distribution of $-3.7 \leq$ Mw ≤ 1 mining-induced earthquakes around a mining front and b-value invariance with post-blast time. *Pure Appl. Geophys.* **2013**, *171*, 2665–2684. [CrossRef]
54. Gischig, V.S.; Doetsch, J.; Maurer, H.; Krietsch, H.; Amann, F.; Evans, K.F.; Nejati, M.; Jalali, M.; Valley, B.; Obermann, A.C.; et al. On the link between stress field and small-scale hydraulic fracture growth in anisotropic rock derived from microseismicity. *Solid Earth* **2018**, *9*, 39–61. [CrossRef]
55. Kwiatek, G.; Plenkers, K.; Nakatani, M.; Yabe, Y.; Dresen, G.; JAGUARS Research Group. Frequency-magnitude characteristics down to magnitude –4.4 for induced seismicity recorded at Mponeng gold mine, South Africa. *Bull. Seismol. Soc. Am.* **2010**, *100*, 1167–1173. [CrossRef]
56. Eisenblätter, J.; Spies, T. Ein Magnitudenmaß für Schallemissionsanalyse und Mikroakustik. In *Deutsche Gesellschaft für zerstörungsfreie Prüfung; 12. Kolloquium Schallemission*; DGZfP Berichtsband: Jena, Germany, 2000; pp. 29–41. (In German)
57. Sasaki, S.; Ishida, T.; Kanagawa, T. Source location and focal mechanisms of AE events during the hydraulic fracturing. In *CRIEPI Rep. U86032*; Central Research Institute of Electric Power Industry: Abiko, Japan, 1987.
58. Ohtsu, M. Simplified Moment Tensor Analysis and Unified Decomposition of Acoustic Emission Sources: Application to In Situ Hydrofracturing Test. *J. Geophys. Res.* **1991**, *96*, 6211–6221. [CrossRef]
59. Young, R.P.; Collins, D.S. Monitoring an experimental tunnel seal in granite using acoustic emission and ultrasonic velocity. In *Rock Mechanics for Industry*; Amadei, B., Kranz, R.L., Scott, G.A., Smeallie, P.H., Eds.; Balkema: Leiden, The Netherlands, 1999; pp. 869–876.
60. Young, R.P.; Hazzard, J.F.; Pettitt, W.S. Seismic and micromechanical studies of rock fracture. *Geophys. Res. Lett.* **2000**, 1767–1770. [CrossRef]
61. Young, R.P.; Collins, D.S. Seismic studies of rock fracture at the Underground Research Laboratory, Canada. *Int. J. Rock Mech.* **2001**, *38*, 787–799. [CrossRef]
62. Collins, D.S.; Pettitt, W.S.; Young, R.P. High-resolution mechanics of a micro-earthquake sequence. *Pure Appl. Geophys.* **2002**, *159*, 197–219. [CrossRef]
63. Young, R.P.; Collins, D.S.; Reyes-Montes, J.M.; Baker, C. Quantification and interpretation of seismicity. *Int. J. Rock Mech.* **2004**, *41*, 1317–1327. [CrossRef]
64. Dahm, T.; Manthei, G.; Eisenblätter, J. Relative Moment Tensors of Thermally Induced Microcracks in Salt Rock. *Tectonophysics* **1998**, *289*, 61–74. [CrossRef]

65. Manthei, G.; Eisenblätter, J.; Salzer, K. Acoustic Emission Studies on Thermally and Mechanically Induced Cracking in Salt Rock. In *Proceedings of the Sixth Conference on Acoustic Emission/Microseismic Activity in Geologic Structures and Materials, Pennsylvania State University, University Park, PA, USA, 9–11 June 1998*; Hardy, H.R., Jr., Ed.; Trans Tech Publications: Clausthal-Zellerfeld, Germany, 1998; pp. 245–265.
66. Dahm, T.; Manthei, G.; Eisenblätter, J. Automated Moment Tensor Inversion to Estimate Source Mechanism of Hydraulically Induced Micro-Seismicity in Salt Rock. *Tectonophysics* **1999**, *306*, 1–17. [CrossRef]
67. Manthei, G.; Eisenblätter, J.; Kamlot, P. Stress measurements in salt mines using a special hydraulic fracturing borehole tool. In Proceedings of the International Symposium on Geotechnical Measurements and Modelling, Karlsruhe, Germany, 23–26 September 2003; pp. 355–360.
68. Pettitt, W.S.; Baker, C.; Young, R.P. Using acoustic emission and ultrasonic techniques for assessment of damage around critical engineering structures. In *Proceedings of 5th North American Rock Mechanics Symposium*; University of Toronto Press: Toronto, ON, Canada, 2002; pp. 1161–1170.
69. Spies, T.; Hesser, J.; Eisenblätter, J.; Eilers, G. Monitoring of the rockmass in the final repository Morsleben: Experiences with acoustic emission measurements and conclusions. In Proceedings of the DisTec 2004, Berlin, Germany, 26–28 April 2004; pp. 303–311.
70. Manthei, G.; Eisenblätter, J.; Spies, T.; Eilers, G. Source Parameters of Acoustic Emission Events in Salt Rock. *J. Acoust. Emiss.* **2001**, *19*, 100–108.
71. Köhler, D.; Spies, T.; Dahm, T. Seismicity patterns and variation of the frequency-magnitude distribution of microcracks in salt. *Geophys. J. Int.* **2009**, *179*, 489–499. [CrossRef]
72. Becker, D.; Cailleau, B.; Kaiser, D.; Dahm, T. Macroscopic Failure Processes at Mines Revealed by Acoustic Emission (AE) Monitoring. *Bull. Seismol. Soc. Am.* **2014**, *104*, 1785–1801. [CrossRef]
73. Nakatani, M.; Yabe, Y.; Philipp, J.; Morema, G.; Stanchits, S.; Dresen, G. Acoustic emission measurements in a deep gold mine in South Africa: project overview and some typical waveforms. *Seismol. Res. Lett.* **2008**, *79*, 311.
74. Naoi, M.; Nakatani, M.; Yabe, Y.; Kwiatek, G.; Igarashi, T.; Plenkers, K. Twenty thousand aftershocks of a very small (M 2) earthquake and their relation to the mainshock rupture and geological structures. *Bull. Seismol. Soc. Am.* **2011**, *101*, 2399–2407. [CrossRef]
75. Davidsen, J.; Kwiatek, G. Earthquake Interevent Time Distribution for Induced Micro-, Nano-, and Picoseismicity. *Phys. Rev. Lett.* **2013**, *110*. [CrossRef] [PubMed]
76. Davidsen, J.; Kwiatek, G.; Dresen, G. No Evidence of Magnitude Clustering in an Aftershock Sequence of Nano- and Picoseismicity. *Phys. Rev. Lett.* **2012**, *108*. [CrossRef] [PubMed]
77. Ziegler, M.; Reiter, K.; Heidbach, O.; Zang, A.; Kwiatek, G.; Stromeyer, D.; Dahm, T.; Dresen, G.; Hofmann, G. Mining-Induced Stress Transfer and Its Relation to a Mw 1.9 Seismic Event in an Ultra-deep South African Gold Mine. *Pure Appl. Geophys.* **2015**, *172*, 2557–2570. [CrossRef]
78. Yabe, Y.; Nakatani, M.; Naoi, M.; Philipp, J.; Janssen, C.; Watanabe, T.; Katsura, T.; Kawakata, H.; Dresen, G.; Ogasawara, H. Nucleation process of an M2 earthquake in a deep gold mine in South Africa inferred from on-fault foreshock activity. *J. Geophys. Res. Sol. Earth* **2015**, *120*, 5574–5594. [CrossRef]
79. Kozłowska, M.; Orlecka-Sikora, B.; Kwiatek, G.; Boettcher, M.S.; Dresen, G. Nanoseismicity and picoseismicity rate changes from static stress triggering caused by a Mw 2.2 earthquake in Mponeng gold mine, South Africa. *J. Geophys. Res. Sol. Earth* **2015**, *120*, 290–307. [CrossRef]
80. Dörner, D.; Philipp, J.; Manthei, G.; Popp, T. Monitoring of AE activity around a large-diameter borehole in rock salt. In Proceedings of the International Acoustic Emission Conference, Progress in Acoustic Emission XVI, Okinawa, Japan, 21–24 October 1986; pp. 187–192.
81. Popp, T.; Minkley, W.; Wiedemann, M.; Salzer, K.; Dörner, D. Gas pressure effects on salt—the large scale in-situ test Merkers. In *Proceedings of the Conference on Mechanical Behavior of Salt, South Dakota School of Mines and Technology, Mechanical Behavior of Salt VIII, Rapid City, SD, USA, 26–28 May 2015*; Lance, R., Mellegard, K., Hansen, F., Eds.; CRC Press/Balkema: Leiden, The Netherlands, 2015; pp. 127–136.
82. Plenkers, K.; Philipp, P.; Dörner, D.; Minkley, W.; Popp, T.; Wiedemann, M. Observation of seismic and aseismic rock behavior during large-scale loading experiment. In Proceedings of the Mechanical Behavior of Salt IX, Hannover, Germany, 12–14 September 2018.
83. Le Gonidec, Y.; Schubnel, A.; Wassermann, J.; Gibert, D.; Nussbaum, C.; Kergosien, B.; Sarout, J.; Maineult, A.; Guéguen, Y. Field-scale acoustic investigation of a damaged anisotropic shale during a gallery excavation. *Int. J. Rock Mech. Min.* **2012**, *51*, 136–148. [CrossRef]

84. Naoi, M.; Nakatani, M.; Otsuki, K.; Yabe, Y.; Kgarume, T.; Murakami, O.; Masakale, T.; Ribeiro, L.; Ward, A.; Moriya, H.; et al. Steady activity of microfractures on geological faults loaded by mining stress. *Tectonophysics* **2015**, 100–114. [CrossRef]

85. Naoi, M.; Nakatani, M.; Kgarume, T.; Khambule, S.; Masakale, T.; Ribeiro, L.; Philipp, J.; Horiuchi, S.; Otsuki, K.; Miyakawa, K.; et al. Quasi-static slip patch growth to 20 m on a geological fault inferred from acoustic emissions in a South African gold mine. *J. Geophys. Res.* **2015**, *120*, 1692–1707. [CrossRef]

86. Naoi, M.; Nakatani, M.; Igarashi, T.; Otsuki, K.; Yabe, Y.; Kgarume, T.; Murakami, O.; Masakale, T.; Ribeiro, L.; Ward, A.; et al. Unexpectedly frequent occurrence of very small repeating earthquakes ($-5.1 \leq$ Mw ≤ -3.6) in a South African gold mine: Implications for monitoring intraplate faults. *J. Geophys. Res. Sol. Earth* **2015**, *120*, 8478–8493. [CrossRef]

87. López-Comino, J.A.; Heimann, S.; Cesca, S.; Milkereit, C.; Dahm, T.; Zang, A. Automated full waveform detection and location algorithm of acoustic emissions from hydraulic fracturing experiment. *Proc. Eng.* **2017**, *191*, 697–702. [CrossRef]

88. Jalali, M.; Gischig, V.; Doetsch, J.; Näf, R.; Krietsch, H.; Klepikova, M.; Amann, F.; Giardini, D. Transmissivity changes and microseismicity induced by small-scale hydraulic fracturing tests in crystalline rock. *Geophys. Res. Lett.* **2018**, *45*, 1–9. [CrossRef]

89. Allen, R. Automatic phase pickers: their present use and future prospects. *Bull. Seismol. Soc. Am.* **1982**, *72*, 225–242.

90. Kamlot, P.; Günther, R.-M.; Stockmann, N.; Gärtner, G. Modeling of strain softening and dilatancy in the mining system of the southern flank of the Asse II mine. In *Mechanical Behavior of Salt VII*; Berest, P.B., Ghoreychi, M., Hadj-Hassen, F., Tijani, M., Eds.; Taylor & Francis (Balkema): London, UK, 2012; pp. 327–336.

91. Kamlot, P.; Weise, D.; Gärtner, G.; Teichmann, L. Drift sealing in the Asse II mine as a component of the emergency concept—Assessment of the hydro-mechanical functionality. In *Mechanical Behavior of Salt VII*; Berest, P.B., Ghoreychi, M., Hadj-Hassen, F., Tijani, M., Eds.; Taylor & Francis (Balkema): London, UK, 2012; pp. 479–489.

92. Kaiser, D.; Spies, T.; Schmitz, H. Mikroakustisches Monitoring in Bergwerken zur Bewertung aktueller Rissprozesse. In Proceedings of the GeoMonitoring 2013, Hannover, Germany, 14–15 March 2013; Sörgel, U., Schack, L., Eds.; 2013; pp. 39–55. (In German)

93. Lavrov, A.; Vervoort, A.; Filimonov, Y.; Wevers, M.; Mertens, J. Acoustic emission in host-rock material for radioactive waste disposal: comparison between clay and rock salt. *Bull. Eng. Geol. Environ.* **2002**, *61*, 379–387. [CrossRef]

94. Waldhauser, F.; Ellsworth, W.L. A double-difference earthquake location algorithm: method and application to the Northern Hayward Fault, California. *Bull. Seismol. Soc. Am.* **2000**, *90*, 1353–1368. [CrossRef]

95. Ortlepp, W.D. Observation of mining-induced faults in an intact rock mass at depth. *Int. J. Rock Mech. Min.* **2012**, *37*, 423–436. [CrossRef]

96. van Aswegen, G. Forensic rock mechanics, Ortlepp shears and other mining induced structures. In Proceedings of the 8th International Symposium on Rockbursts and Seismicity in Mines (RaSiM8), St. Petersburg and Moscow, Russia, 1–8 September 2013; Balkema: Rotterdam, The Netherlands, 2013; pp. 1–19.

97. Heesakkers, V.; Murphy, S.K.; Reches, Z. Earthquake Rupture at Focal Depth, Part I: Structure and Rupture of the Pretorius Fault, TauTona Mine, South Africa. *Pure Appl. Geophys.* **2011**, *168*, 2395–2425. [CrossRef]

98. McGarr, A. Violent deformation of rock near deep-level, tabular excavation-seismic events. *Bull. Seismol. Soc. Am.* **1971**, *61*, 1453–1466.

99. Naoi, M.; Nakatani, M.; Yabe, Y.; Philipp, J.; JAGUARS Research Group. Very high frequency AE (<200 kHz) and micro seismicity observation in a deep South African gold mine—Evaluation of the acoustic properties of the site by in-situ transmission test. *Seismol. Res. Lett.* **2008**, *79*, 330.

100. Stacey, T.R.; Wesseloo, J. In situ stresses in mining areas in South Africa. *J. S. Afr. Inst. Min. Metall.* **1998**, 365–368.

101. Haimson, B.C.; Cornet, F.H. ISRM Suggested Methods for rock stress estimation—Part 3: hydraulic fracturing (HF) and/or hydraulic testing of pre-existing fractures (HTPF). *Int. J. Rock Mech. Min.* **2003**, *40*, 1011–1020. [CrossRef]

102. Kaiser, P.K.; Valley, B.; Dusseault, M.B.; Duff, D. Hydraulic fracturing mine back trials—Design rationale and project status. In *Proceedings ISRM International Conference for Effective and Sustainable Hydraulic Fracturing*; International Society for Rock Mechanics, IntechOpen Limited: London, UK, 2013. [CrossRef]

103. Warpinski, N.R.; Mayerhofer, M.; Agarwal, K.; Du, J. Hydraulic-fracture geomechanics and microseismic-source mechanisms. *SPE J.* **1992**, *18*, 766–780. [CrossRef]

104. Economides, M.J.; Nolte, K.G.; Ahmed, U.; Schlumberger, D. *Reservoir Stimulation*; Wiley: Chichester, UK, 2000.

105. Häring, M.O.; Schanz, U.; Ladner, F.; Dyer, B.C. Characterisation of the Basel 1 enhanced geothermal system. *Geothermics* **2008**, *37*, 469–495. [CrossRef]

106. Schindler, M.; Nami, P.; Schellschmidt, R.; Teza, D.; Tischner, T. Summary of hydraulic stimulation operations in the 5 km deep crystalline HDR/EGS reservoir at Soultz-sous-Forêts. In Proceedings of the 33rd Workshop on Geothermal Reservoir Engineering, Stanford, CA, USA, 28–30 January 2008; pp. 325–333.

107. Jeffrey, R.G. Hydraulic Fracturing of Ore Bodies. U.S. Patent No. 6,123,394, 2000.

108. Niitsuma, H.; Nagano, K.; Hisamatsu, K. Analysis of acoustic emission from hydraulically induced tensile fracture of rock. *J. Acoust. Emiss.* **1993**, *11*, S1–S18.

109. Guglielmi, Y.; Cappa, F.; Avouac, J.P.; Henry, P.; Elsworth, D. Seismicity triggered by fluid injection-induced aseismic slip. *Science* **2015**, *348*, 1224–1226. [CrossRef] [PubMed]

110. Amann, F.; Gischig, V.; Evans, K.; Doetsch, J.; Jalali, R.; Valley, B.; Krietsch, H.; Dutler, N.; Villiger, L.; Brixel, B.; et al. The seismo-hydro-mechanical behaviour during deep geothermal reservoir stimulations: Open questions tackled in a decameterscale in-situ stimulation experimen. *Solid Earth* **2018**, *9*, 115–137. [CrossRef]

111. Keusen, H.; Ganguin, J.; Schuler, P.; Buletti, M. *Grimsel Test Site: Geology, Report*; Nationale Genossenschaft für die Lagerung Radioaktiver Abfälle (NAGRA): Wettingen, Switzerland, 1989.

112. Morris, J.P.; Dobson, P.; Knox, H.; Ajo-Franklin, J.; White, M.D.; Fu, P.; Burghardt, J.; Kneafsey, T.J.; Blankenship, D.; EGS Collab Team. Experimental Design for Hydrofracturing and Fluid Flow at the DOE Collab Testbed. In Proceedings of the 43rd Workshop on Geothermal Reservoir Engineering Stanford University, Stanford, CA, USA, 12–14 February 2018.

 *applied sciences*

Article

Acoustic Emission/Seismicity at Depth Beneath an Artificial Lake after the 2011 Tohoku Earthquake

Hirokazu Moriya

Graduate School of Engineering, Tohoku University, 6-6-04, Aramaki Aza Aoba, Aoba-ku, Sendai 980-8579, Japan; hirokazu.moriya.e1@tohoku.ac.jp; Tel.: +81-22-795-7996

Received: 20 July 2018; Accepted: 16 August 2018; Published: 20 August 2018

Abstract: Acoustic emission (AE)/seismicity activity increased near the city of Sendai, Japan, after the 11 March 2011 Tohoku earthquake in a newly seismically active region near the Nagamachi-Rifu fault, which caused a magnitude 5.0 earthquake in 1998. The source of this activity was around 12 km beneath an artificial lake. At the same time, activity on the Nagamachi-Rifu fault nearly ceased. More than 1550 micro-earthquakes were observed between 11 March 2011 and 1 August 2012, of which 63% exhibited similar waveforms and defined 64 multiplets. It appears that crustal extension of about 2 m during the Tohoku earthquake and additional extension of about 1 m during the following year changed the stress field in this region, thus generating micro-earthquakes and controlling their frequency. However, it has been presumed that crustal movement during the Tohoku earthquake did not affect the direction of principal stress, and that these events induced repeated quasi-static slips at asperities and the resultant micro-earthquakes.

Keywords: acoustic emission swarm; 2011 Tohoku earthquake; repeating earthquake; multiplet; crustal movement

---

1. Introduction

Acoustic emission (AE)/seismicity is important for evaluating the stability of geological structures; for instance, the delineation of fracture systems from AE activity in geothermal reservoirs is indispensable for thermal energy extraction [1]. The stability evaluation of fractures during hydraulic stimulation in geothermal reservoirs is necessary to avoid inducing large AE events, where the relationship between large AE events and the pore-pressure distribution in reservoirs is studied [2]. The stability of geological structures is also important from the viewpoint of disaster prevention. The dynamic behavior of geological structures has been studied by many seismologists who have examined repeating earthquakes and slow slip at seismic zones [3–5].

The 11 March 2011 Tohoku earthquake and its many aftershocks brought widespread destruction to eastern Japan [6–8]. This seismic activity occurred on both offshore and inland faults. Although thrust events dominated seismicity before and during the Tohoku earthquake, eastward movement of the shallow crust of the overriding tectonic plate caused extension that relaxed the subduction-related horizontal compressional stress locally, resulting in normal movement on some inland faults after the Tohoku earthquake [9].

Close observations of seismic activity are indispensable for the monitoring of the stability of geological structures. This paper describes a seismic swarm that occurred beneath an artificial lake near the plane of an active fault in Sendai city, Japan, after the Tohoku earthquake. It has long been known that changes in the water level of reservoirs can trigger substantial earthquakes, and it is shown that the seismic activity was sensitive to the small change of tectonic stress caused by the water level of the artificial lake.

2. Study Area

The study area is in the western part of Sendai city, Miyagi Prefecture, Japan, about 150 km west of the Tohoku earthquake epicenter (Figure 1a). Here, a river terrace is present on both sides of the Hirose River. A double arch dam, built across a tributary to the Hirose River in 1961, created a reservoir named Lake Okura, with a 1.6 km^2 surface area and a storage capacity of 2.8×10^7 m^3. The water level is normally maintained between 240 and 270 m above sea level (masl), which corresponds to water depths of 0 to 30 m (Miyagi Prefecture, http://www.pref.miyagi.jp/snd-dam/data/okdam-cyosui.htm). The Nagamachi-Rifu fault extends approximately 15 km under Sendai city, striking northeast–southwest and dipping about 45° to the northwest [10,11]. Estimates of the stress field made by the multiple inverse method [12] indicate that the direction of minimum principal stress is vertical and the direction of maximum principal stress varies from east–west to northwest–southeast.

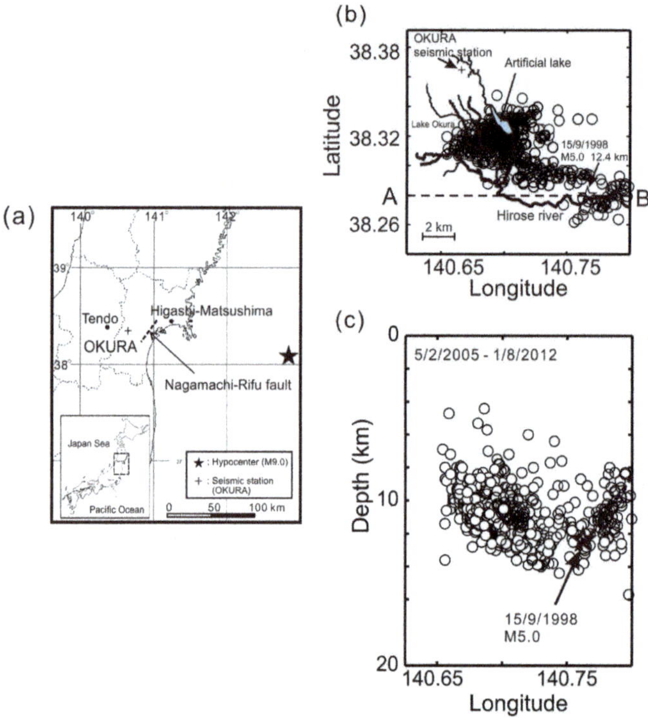

Figure 1. (**a**) Map of northeastern Honshu, Japan, with inset showing location of study area. The solid star marks the epicenter of the Tohoku earthquake; (**b**) Map showing epicenters of micro-earthquakes from 5 February 2005 to 1 August 2012; (**c**) Vertical cross section showing hypocenters of micro-earthquakes projected onto line A–B. Open stars show the location of the magnitude 5.0 earthquake of 15 September 1998.

At Sendai city, very high seismic intensities (greater than 6 on the Japan Meteorological Agency scale) were recorded during the Tohoku earthquake. GPS (Global Positioning System) data documented crustal movements of more than 5 m at the coast and about 2.5 m in Sendai city. The Nagamachi-Rifu fault is known to be active in the Sendai area [10,13,14]. Seismic activity increased near this fault after the Tohoku earthquake, and more than 1550 events were observed between 11 March 2011 and 1 August 2012. A micro-earthquake is physically the same to as an AE and the term "AE" is generally used for much smaller seismic events with smaller magnitude, and this paper discusses the stability of geological structures using micro-earthquakes in the same manner.

3. Tectonic Conditions before and after the Tohoku Earthquake

On 15 September 1998, an earthquake of magnitude (M) 5.0 occurred at a 12.4 km depth, on the deepest part of the Nagamachi-Rifu fault [14]. The mechanism of this earthquake and its aftershocks was reverse faulting, and the P-axes of the events described a line oriented east–west to northwest–southeast; the maximum principal stress direction in the deep part of the active fault was northwest–southeast.

On the basis of GPS data from an observation point about 6 km southeast of the dam, the Geospatial Information Authority of Japan (GSI) reported crustal movement of 2.55 m eastward and 16 cm ground subsidence during the Tohoku earthquake (e.g., [15]). The GSI also measured eastward crustal movements of 1.85 m in Tendo city, 30 km to the west, and 3.98 m in Higashi-Matsushima city, 70 km to the east (GSI, http://www.gsi.go.jp/index.html; Figure 1a). From these data plus the assumption of an elastic rock mass in the area, the strain change resulting from the Tohoku earthquake was estimated to be 2.13×10^{-5} in the study area. Eastward crustal movement around Sendai continued after the earthquake, reaching about 1 m during the following year (GSI; http://www.gsi.go.jp/cais/chikakuhendo40012.html).

It appears that the direction of maximum principal stress in the study area was roughly east–west both before and after the Tohoku earthquake [7]. However, at the Kamaishi mine, $\approx$170 km northeast of Sendai, horizontal displacement of 3.3 m was observed during the Tohoku earthquake [16]. In situ measurements of tectonic stress at the mine documented a large change in the stress field in the shallow crust in the two years after the Tohoku earthquake, with the principal stress decreasing in the east–west direction and increasing in the north–south direction [16].

The evidence suggests that although the Tohoku earthquake caused large changes of the stress field in some parts of the shallow crust, changes in the orientation of the stress field at seismogenic depths were small in the study area, which implies that the stress field in the study area still favors reverse faulting.

4. AE Activity before and after the Tohoku Earthquake

A large number of micro-earthquakes were recorded in western Sendai city from 5 February 2005 to 1 August 2012, a period spanning the time of the Tohoku earthquake (Figure 1). The distribution of hypocenters defines two areas of AE activity, largely corresponding to seismic events before and after the Tohoku earthquake, respectively (Figures 2 and 3).

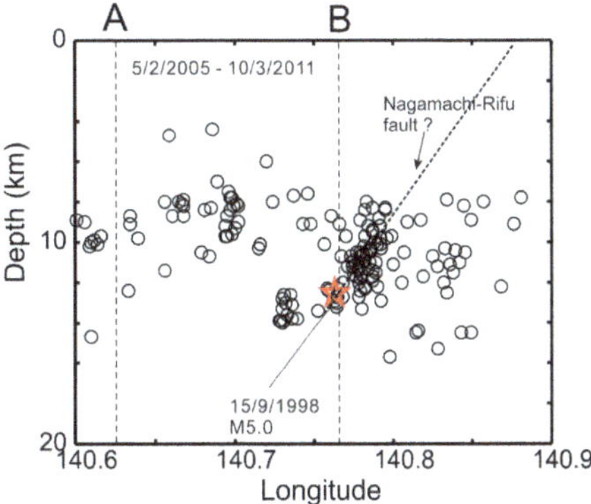

Figure 2. Vertical cross section showing hypocenters of micro-earthquakes before the Tohoku earthquake projected onto the vertical plane defined by line A–B (location in Figure 1b). The dashed line shows the downward projection of the Nagamachi-Rifu fault. The red star represents the source location of the earthquake of magnitude (M) 5.0 occurred on 15 September 1998.

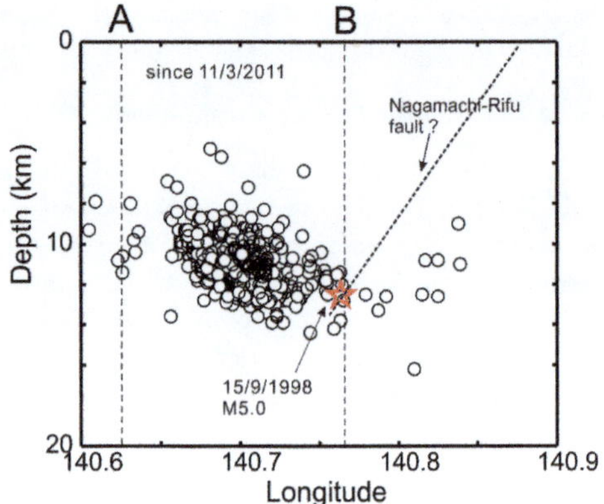

Figure 3. Vertical cross section showing hypocenters of micro-earthquakes after the Tohoku earthquake projected onto the vertical plane defined by line A–B. The red star represents the source location of the earthquake of magnitude (M) 5.0 occurred on 15 September 1998.

The AE events before the Tohoku earthquake occurred at depths between 8 and 14 km on the Nagamachi-Rifu fault. However, after the earthquake, this activity ceased and a new active zone appeared to the west, generating hundreds of events between 11 March 2011 and 1 August 2012 at depths of 8 to 14 km. Activity in the study area did not increase immediately after the Tohoku earthquake, but began a gradual rise on 3 April and totaled 11 events (maximum magnitude 2.3) by 21 April. After 24 April, AE activity increased further, and by 1 August 2012, more than 1550 AE events had occurred. It is known that smaller earthquakes (e.g., aftershocks) occur after a previous large earthquake in the same area of the main shock, and the frequency of aftershocks decreases with the reciprocal of time after the main shock (Omori's law). However, clear mainshock and aftershocks are not identified after the Tohoku earthquake in Figure 4; thus, the typical decay of earthquake swarms is difficult to discern. The relationship between the event activity and the seasonal variations in water level in other years were investigated. However, the clear relationship as seen in Figure 4 was not observed before the Tohoku earthquake and after the studied period, and the seismic activity was very low in 2017. Figure 5 shows the Gutenberg-Richter relation before and after 11 March 2011. The b-value was 1.0 before 11 March and 0.92 after 11 March, indicating that larger events became more frequent after the Tohoku earthquake.

Among the micro-earthquakes after 11 March 2011, we identified events that shared similar waveforms (Figure 6), known as repeating earthquakes or multiplets [17–19]. By calculating coherences for all possible pairs of events, and using coherence greater than 0.98 as the criterion for similarity, we identified 985 similar events (63% of those observed), which defined 64 multiplets. Travel-time differences of P- and S-waves were determined by cross-correlating the P- and S-components with their master events.

The mean and standard deviation of the travel-time differences between P- and S-waves were 0.024 and 0.04 s, respectively. Source locations were not corrected according to travel-time differences because the number of seismic stations was insufficient to properly constrain them. However, the similarity of P- and S-wave travel times suggests that the multiplet sources were close together on adjacent fractures.

During the period following the Tohoku earthquake, eight earthquakes larger than M 3.0 were observed at more than 40 stations, making it possible to estimate their source mechanisms by using P-wave polarity distributions. Figure 7a shows the fault plane solution of the largest of these events (M 3.4 on 3 June 2011), for which only P-wave polarities that were clearly detected by visual inspection were used to determine the nodal planes. The other seven of these events had the same mechanism, indicating that the fault planes strike north-northeast–south-southwest and dip at about 65°, and that the slip included both left-lateral strike-slip and reverse components. The accuracy of the source locations was not enough to identify the orientation of the fault plane which caused the micro-earthquakes. If the source locations were approximated to a plane, the plane was dipping around 45 degrees to the east and striking nearly north-south, and this orientation was similar to one of the nodal planes in Figure 7a. The orientation of nodal planes for the mainshock of the 1998 M 5.0 earthquake (Figure 7b) [14] was similar to those of the eight M > 3 events (Figure 7a), but included a right-lateral strike-slip component.

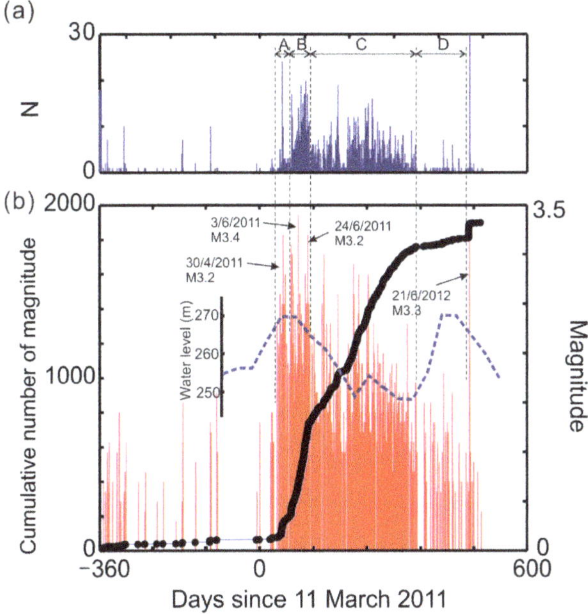

Figure 4. (**a**) Number of micro-earthquakes per day and (**b**) magnitudes and cumulative number of seismic events (N) during the 360 days before and 600 days after the Tohoku earthquake along with the water level of Lake Okura. The record ends on 1 August 2012.

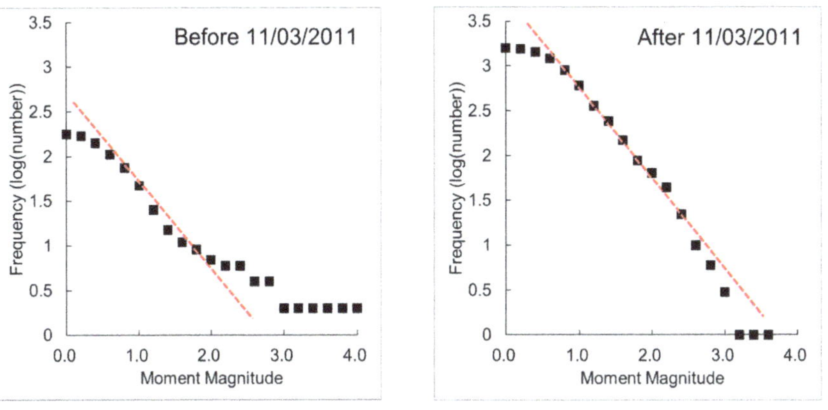

Figure 5. Histograms showing the Gutenberg-Richter relation.

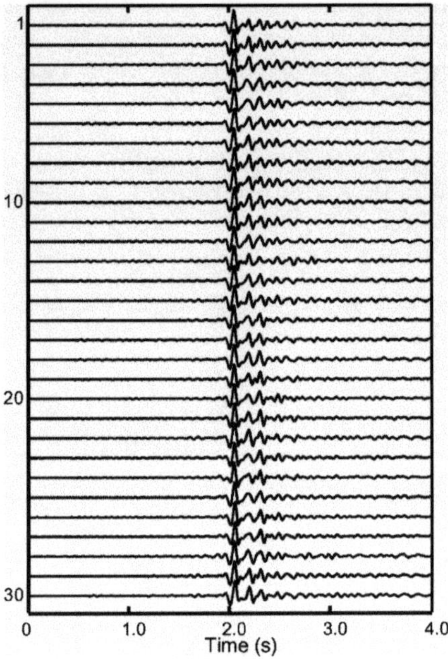

Figure 6. Example of a set of waveforms constituting a multiplet. Amplitudes are normalized to the maximum amplitude of each waveform. The vertical axis represents the number of events.

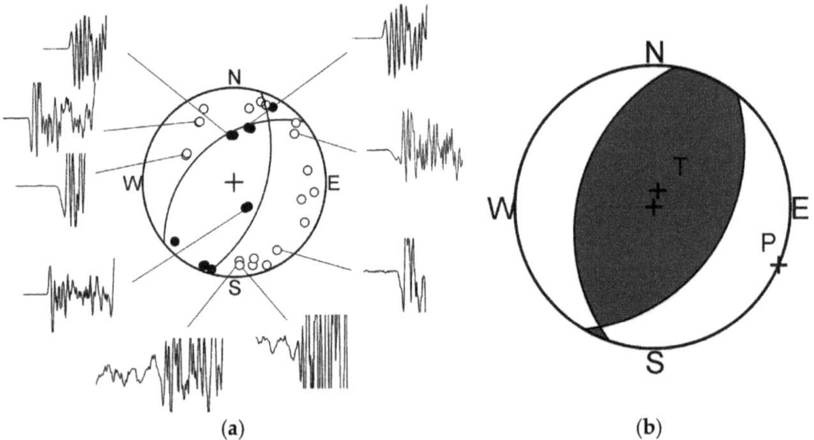

(a)   (b)

Figure 7. (a) Waveforms of eight M > 3 earthquakes after the Tohoku earthquake and fault plane solution for the largest event (M 3.4, 3 June 2011). Open circle and closed circle represent negative and positive P-wave polarities, respectively; (b) Fault plane solution for the 1998 M 5.0 earthquake on the Nagamachi-Rifu fault, which was reproduced following the result by Umino et al. [14].

5. Discussion

East–west crustal extension during the Tohoku earthquake relaxed the subduction-related horizontal compressional stress in the study area. The extension decreased normal stresses on east- and west-dipping fault planes and, because shear stress on the reverse faults decreased, they were dynamically stabilized against shear slip. Thus, the number of normal faulting events increased after the Tohoku earthquake [7], and earthquakes on the Nagamachi-Rifu fault nearly ceased after 11 March 2011, as horizontal compressional stress on the fault surface decreased.

After the Tohoku earthquake, seismic activity increased in a region 6 km west of the source of the 1998 earthquake (Figure 3), apparently on newly activated fractures. Fault plane solutions (Figure 7a) indicate that the newly activated fractures dip either east or west and slipped with left-lateral strike-slip and reverse mechanisms. Seismic activity on these fractures, which were previously under critical conditions for reverse faulting, would be expected to decrease when horizontal compressional stress was relaxed during the Tohoku earthquake and lowered the shear stress on the fault plane. However, seismic activity increased after the earthquake, so we must seek other contributing factors.

One possible explanation is a relative increase of principal stress in the north–south direction [16]. This would have created conditions favoring left-lateral strike-slip faulting. Fault plane solutions after the Tohoku earthquake (Figure 7a) include both left-lateral and reverse components, whereas for the 1998 earthquake (Figure 7b), although a left-lateral component was present, reverse faulting was dominant. This result suggests that a relative increase of principal stress in the north–south direction may have also increased the left-lateral component, but the Tohoku earthquake did not dramatically change the contrast between the principal east–west and north–south stresses. This observation implies that the differential stress was greater before the Tohoku earthquake, and that the Tohoku earthquake did not significantly affect the stress field in the study area.

The triggering of earthquakes is sensitive to changes in stress when conditions are critical for shear slip. For example, tidal changes are a likely factor in triggering certain seismic events [20] even though the stress change due to earth tides is no greater than about 10^3 Pa [21]. On the other hand, the water level of Lake Okura changes by about 20 m, which corresponds to a change in overburden pressure of 200×10^3 Pa. This increase in the overburden stress would increase the pore pressure in the fractures under the lake. AE events can be triggered by increased pore pressure due to fluid diffusion (or pressure diffusion) associated with increases of water level in a lake or reservoir, and it is known that small changes of pore pressure can trigger earthquakes [1,22]. Earthquakes triggered by this mechanism have been reported at several kilometers depth beneath a reservoir in the Koyna district of India [23]. Here, micro-earthquakes started when the water level of Lake Okura approached its maximum level of 270 masl, and the rate of events peaked at 25/day just after that water level was reached. Although those facts do not establish a correlation, fluctuations of the lake level should not be precluded as a possibility in explaining the swarm.

The increase of seismic activity after the Tohoku earthquake can be explained as follows. Before the earthquake, strain energy was accumulating in fractures such that they reached a critical condition for shear slip before 11 March 2011, but cohesion on the fracture surfaces kept the fault locked. Given that crustal movement at this time was a few centimeters per year, cohesive strengthening would have been aseismic. In the process that increases frictional resistance to sliding, generally referred to as healing, micromechanical processes (e.g., increases of asperity contact areas) result in cohesive strengthening [24]. Crustal extension during the Tohoku earthquake changed the direction of shear stress on fractures and weakened the contacts across fracture planes, thus activating shear slip at asperities that were under critical conditions for shear slip. It appears that quasi-static slips occurred repeatedly in the newly activated fracture zone after the Tohoku earthquake, causing AE events as great as M 3.4. This interpretation is supported by the observation of many AE events with similar waveforms (accounting for 63% of located micro-earthquakes) and the decrease after 11 March of the gradient of the high-magnitude tail of earthquake magnitude distributions (Figure 5).

6. Conclusions

About 40 days after the 2011 Tohoku earthquake, a micro-earthquake swarm began west of Sendai, Japan, centered about 8–14 km deep beneath an artificial lake near the down-dip projection of an active fault that caused an M 5 earthquake in 1998. Observations suggest that the new fractures were seismically activated, even though fracturing should have been stabilized by a decrease of principal stress in the east–west direction that was caused by the Tohoku earthquake. It is typically presumed that crustal movement during the Tohoku earthquake did not affect the direction of principal stress; however, a relative increase of the principal stress magnitude in the north–south direction in the study area may have caused a change in the direction of shear stress on fracture surfaces, thus weakening the contacts at asperities. These events induced repeated quasi-static slips at asperities and the resultant AE events. The magnitudes and frequency of the micro-earthquakes indicate that these fractures were capable of causing micro-earthquakes with magnitudes of about 4. The results of this study indicate that ongoing monitoring of the earthquake swarm is important for mitigation of damage caused by micro-earthquakes at these newly activated inland faults. The findings of this study may be helpful as a case history to understand the stability of geological structures in at the kilometer scale.

Author Contributions: H.M. has performed the works of conceptualization, methodology, writing–original draft preparation, writing–review and editing.

Funding: This research received no external funding.

Acknowledgments: The earthquake waveform data and source location data used in this study were acquired from the Hi-net website of the National Research Institute for Earth Science and Disaster Prevention (NIED), Japan.

Conflicts of Interest: The authors declare no conflicts of interest.

References

1. Mukuhira, Y.; Moriya, H.; Ito, T.; Asanuma, H.; Häring, M. Pore pressure migration during hydraulic stimulation due to permeability enhancement by low-pressure subcritical fracture slip. *Geophys. Res. Lett.* **2017**, *44*, 3109–3118. [CrossRef]
2. Moriya, H.; Naoi, M.; Nakatani, M.; van Aswegen, G.; Murakami, O.; Kgarume, T.; Ward, A.K.; Durrheim, R.J.; Philipp, J.; Yabe, Y.; et al. Delineation of large localized damage structures forming ahead of an active mining front by using advanced acoustic emission mapping techniques. *Int. J. Rock Mech. Min. Sci.* **2015**, *79*, 157–165. [CrossRef]
3. Talwani, P. On the nature of reservoir-induced seismicity. *Pure Appl. Geophys.* **1977**, *150*, 473–492. [CrossRef]
4. Nakajima, J.; Uchida, N. Repeated drainage from megathrusts during episodic slow slip. *Nat. Geosci.* **2018**, *11*, 351–356. [CrossRef]
5. Ariyoshi, K.; Uchida, N.; Matsuzawa, T.; Hino, R.; Hasegawa, A.; Hori, T.; Kaneda, Y. A trial estimation of frictional properties, focusing on aperiodicity off Kamaishi just after the 2011 Tohoku earthquake. *Geophys. Res. Lett.* **2014**, *41*, 8325–8334. [CrossRef]
6. Hirose, F.; Miyaoka, K.; Hayashimoto, N.; Yamazaki, T.; Nakamura, M. Outline of the 2011 off the Pacific coast of Tohoku Earthquake (Mw 9.0)—Seismicity: Foreshocks, mainshock, aftershocks, and induced activity. *Earth Planets Space* **2011**, *63*, 513–518. [CrossRef]
7. Ide, S.; Baltay, A.; Beroza, G.C. Shallow dynamic overshoot and energetic deep rupture in the 2011 Mw 9.0 Tohoku-Oki Earthquake. *Science* **2011**, *332*, 1426–1429. [CrossRef] [PubMed]
8. Sato, M.; Ishikawa, T.; Ujihara, N.; Yoshida, S.; Fujita, M.; Mochizuki, M.; Asada, A. Displacement above the hypocenter of the 2011 Tohoku-Oki earthquake. *Science* **2011**, *332*, 1395. [CrossRef] [PubMed]
9. Yoshida, K.; Hasegawa, A.; Okada, T.; Iinuma, T.; Ito, Y.; Asano, Y. Stress before and after the 2011 great Tohoku-oki earthquake and induced earthquakes in inland areas of eastern Japan. *Geophys. Res. Lett.* **2012**, *39*, L03302. [CrossRef]
10. Nakamura, A.; Asano, Y.; Hasegawa, A. Estimation of deep fault geometry of the Nagamachi-Rifu fault from seismic array observations. *Earth Planets Space* **2002**, *54*, 1027–1031. [CrossRef]
11. Nakajima, J.; Hasegawa, A.; Horiuchi, S.; Yoshimoto, K.; Yoshida, T.; Umino, N. Crustal heterogeneity around the Nagamachi-Rifu fault, northeastern Japan, as inferred from travel-time tomography. *Earth Planets Space* **2006**, *58*, 843–853. [CrossRef]
12. Yamaji, A. The multiple inverse method: A new technique to separate stresses from heterogeneous fault-slip data. *J. Struct. Geol.* **2000**, *22*, 441–452. [CrossRef]
13. Umino, N.; Ujikawa, H.; Hori, S.; Hasegawa, A. Distinct S-wave reflectors (bright spots) detected beneath the Nagamachi-Rifu fault, NE Japan. *Earth Planets Space* **2002**, *54*, 1021–1026. [CrossRef]

14. Umino, N.; Okada, T.; Hasegawa, A. Foreshock and aftershock sequence of 1998 M5.0 Sendai, northeastern Japan, earthquake and its implications for earthquake nucleation. *Bull. Seismol. Soc. Am.* **2002**, *92*, 2465–2477. [CrossRef]

15. Ozawa, S.; Nishimura, T.; Suito, H.; Kobayashi, T.; Tobita, M.; Imakiire, T. Coseismic and postseismic slip of the 2011 magnitude-9 Tohoku-Oki earthquake. *Nature* **2011**, *475*, 373–376. [CrossRef] [PubMed]

16. Sakaguchi, K.; Koyano, T.; Yokoyama, T. Change in stress field in the Kamaishi Mine, associated with the 2011 Tohoku-oki earthquake. In Proceedings of the 13th Japan Symposium on Rock Mechanics, Okinawa, Japan, 9–11 January 2013; pp. 513–517. (In Japanese)

17. Moriya, H.; Niitsuma, H.; Baria, R. Multiplet-clustering analysis reveals structural details within the seismic cloud at the Soultz geothermal field, France. *Bull. Seismol. Soc. Am.* **2003**, *93*, 1606–1620. [CrossRef]

18. Nadeau, R.M.; McEvilly, T.V. Periodic pulsing characteristic microearthquakes on the San Andreas fault. *Science* **2004**, *303*, 220–222. [CrossRef] [PubMed]

19. Igarashi, T. Spatial changes of inter-plate coupling inferred from sequences of small repeating earthquakes in Japan. *Geophys. Res. Lett.* **2010**, *37*, L20304. [CrossRef]

20. Chen, H.J.; Chen, C.C.; Tseng, C.Y.; Wang, J.H. Effect of tidal triggering on seismicity in Taiwan revealed by the empirical mode decomposition method. *Nat. Hazards Earth Syst. Sci.* **2012**, *12*, 2193–2202. [CrossRef]

21. Tsuruoka, H.; Ohtake, M.; Sato, H. Statistical test of the tidal triggering of earthquakes: Contribution of the ocean tide loading effect. *Geophys. J. Int.* **1995**, *122*, 183–194. [CrossRef]

22. Talwani, P.; Cobb, J.S.; Schaeffer, M.F. In situ measurements of hydraulic properties of a shear zone in northwestern South Carolina. *J. Geophys. Res.* **1999**, *104*, 14993–15003. [CrossRef]

23. Talwani, P. Seismogenic properties of the crust inferred from recent studies of reservoir-induced seismicity–Application to Koyna. *Curr. Sci.* **2000**, *79*, 1327–1333.

24. Scholz, C.H.; Engelder, T. Role of asperity indentation and ploughing in rock friction. *Int. J. Rock Mech. Min. Sci.* **1976**, *13*, 149–154. [CrossRef]

Review

Evaluation of Low-Temperature Cracking Performance of Asphalt Pavements Using Acoustic Emission: A Review

Behzad Behnia [1], William Buttlar [2] and Henrique Reis [3,*

[1] Department of Civil and Environmental Engineering, Clarkson University, Potsdam, NY 13699, USA; bbehnia@clarkson.edu

[2] Department of Civil and Environmental Engineering, University of Missouri, Columbia, MO 65211, USA; buttlarw@missouri.edu

[3] Department of Industrial and Enterprise Systems Engineering, University of Illinois, Urbana, IL 61801, USA

* Correspondence: h-reis@illinois.edu; Tel.: +1-217-333-1228 or +1-217-898-7063

Received: 11 January 2018; Accepted: 14 February 2018; Published: 21 February 2018

Abstract: Low-temperature cracking is a major form of distress that can compromise the structural integrity of asphalt pavements located in cold regions. A review of an Acoustic Emission (AE)-based approach is presented that is capable of assessing the low-temperature cracking performance of asphalt binders and asphalt pavement materials through determining their embrittlement temperatures. A review of the background and fundamental aspects of the AE-based approach with a brief overview of its application to estimate low-temperature performance of unaged, short-term, and long-term aged binders, as well as asphalt materials, is presented. The application of asphalt pavements containing recycled asphalt pavement (RAP) and recycled asphalt shingles (RAS) materials to thermal cracking assessment is also presented and discussed. Using the Felicity effect, the approach is capable of evaluating the self-healing characteristics of asphalt pavements and the effect of cooling cycles upon their fracture behavior. Using an iterative AE source location technique, the approach is also used to evaluate the efficiency of rejuvenators, which can restore aged asphalt pavements to their original crack-resistant state. Results indicate that AE allows for relatively rapid and inexpensive characterization of pavement materials and can be used towards enhancing pavement sustainability and resiliency to thermal loading.

Keywords: acoustic emission; thermal cracking; asphalt pavements; embrittlement temperatures; recycled asphalt pavements; recycled asphalt shingles; cooling cycles

1. Introduction

The majority of the roads in United States are surfaced with asphalt material, indicating the importance of this material in U.S. transportation infrastructure. In asphalt roads located in cold regions or in milder climate regions with large daily temperature fluctuations, a major form of deterioration is associated with the formation of low-temperature cracks, a.k.a., thermal cracks, in the pavement. Thermal cracks are categorized into two groups based on their formation mechanism: (1) "single-event thermal cracks", which occur in cold climates due to fast cooling rates; and (2) "thermal fatigue cracks", which develop after several cooling cycles in regions with a milder climate and large daily temperature fluctuations [1–4]. Thermal cracking phenomenon demonstrates itself as a group of parallel, evenly-spaced, transversely-oriented, surface-initiated cracks in pavements. A typical thermal cracking pattern in asphalt pavements is depicted in Figure 1a. As the temperature decreases, thermal stresses build up in the pavement due to the tendency of the "restrained" continuous layer of asphalt pavement to contract. The distribution of thermal stresses through the pavement thickness is non-uniform, with the highest stresses occurring at pavement surface, see Figure 1b. Thermal cracks happen when the thermally-induced stresses exceed the material fracture strength.

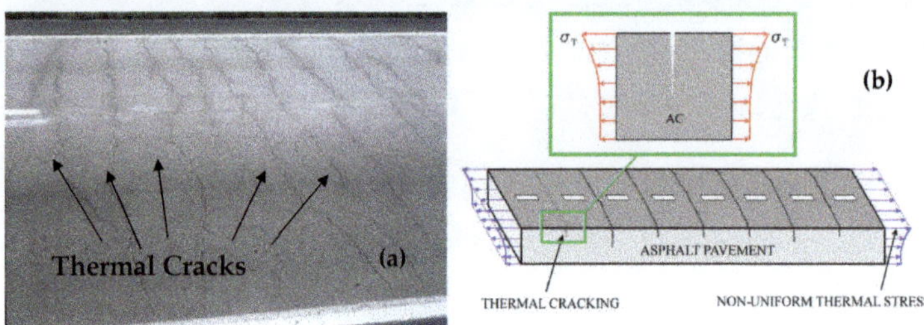

Figure 1. Thermal cracks as a result of oxidation; (**a**) typical thermal cracking pattern in asphalt pavement, (**b**) thermal cracking formation and non-uniform distribution of thermal stresses through the pavement thickness.

On the local scale, another source of thermal stresses in asphalt pavements is the thermal contraction mismatch between aggregates, i.e., crashed stone, and the surrounding asphalt mastic material. As the pavement temperature drops, asphalt mastic tends to contract more (often more than ten times) than the aggregate particles, causing increasing thermally-induced stresses in the pavement structure. At the same time as the temperature drops, the asphalt mastic becomes increasingly more brittle with less resistance against fracture. As a result, microcracks develop within the asphalt mastic when the thermally-induced tensile stresses overcome the fracture resistance of mastic material. In addition to mastic cracking, another type of damage, i.e., debondings, also occurs at low temperatures at the interfaces between asphalt mastic and aggregates, see Figure 2. Repair and rehabilitation of low-temperature cracks in pavements costs millions of dollars every year. In addition, damaged pavements also develop higher surface roughness, which leads to vehicle damage. A study by Islam and Buttlar showed that the presence of cracks in pavements would add over $300 per vehicle per 19,000 km (12,000 mi) driven in user costs [5].

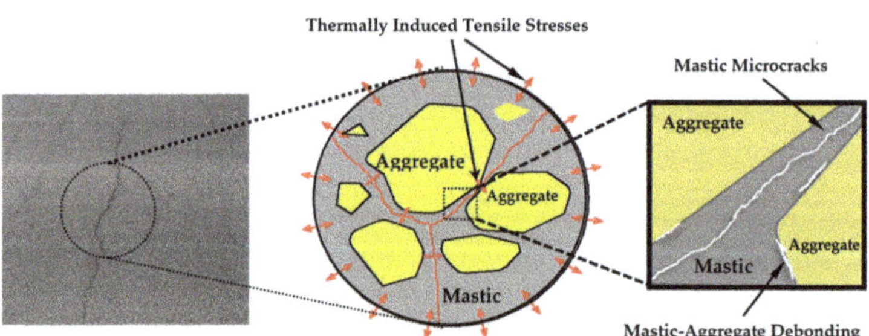

Figure 2. Schematic depiction of thermally-induced stresses within the asphalt mastic causing microcracks in the mastic and mastic-aggregate debondings.

Asphalt roads also suffer from oxidative aging [6,7]. Figure 3a shows two images of the same pavement section: one taken right after the construction and the other 19 months later when the pavement has already encountered some level of oxidative aging. Figure 3b shows a schematic diagram illustrating the typical steep gradation of material properties (e.g., complex modulus) caused by oxidative aging at the pavement top material layer. Computer models have already been developed to estimate the change in material properties due to oxidative aging for given climatic conditions [7]. In addition to changing the asphaltenes to maltenes ratio [6], oxidative aging also increases the stiffness of the top material layers, see Figure 3b, which increases the pavement vulnerability to cracking in cold environments. Oxidative aging is an important issue in asphalt pavements that negatively affects low temperature cracking of asphalt materials. The higher the oxidative aging level of asphalt pavements, the higher the extent of thermal cracking damage in the material.

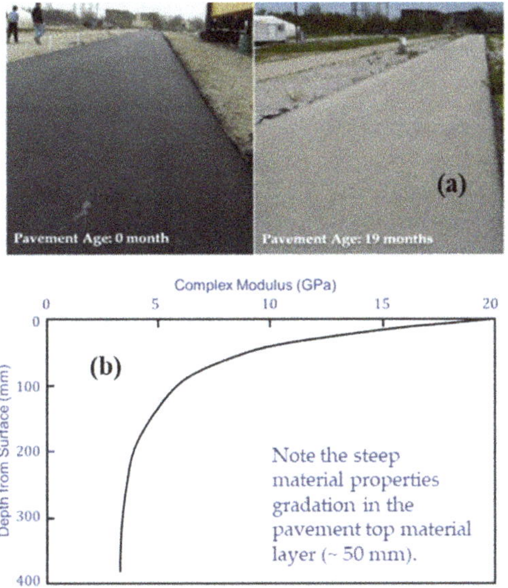

Figure 3. Oxidation of asphalt concrete pavements: (**a**) pavement section right after the construction (left), and the same pavement 19 months after construction (right); and (**b**) Schematic diagram illustrating the steep gradation of material properties (e.g., complex modulus) caused by oxidative aging at the pavement top material layer.

A great deal of research efforts has been directed towards the characterization and prevention of thermal cracks in pavement. The Superpave tests developed under the Strategic Highway Research Program (SHRP) in the 1990s have significantly improved the performance tests to assess the behavior of asphalt pavements by providing fundamental material tests over a broad range of production and service temperatures. However, the Superpave tests were not developed for (1) the characterization of highly modified binders, (2) the characterization of asphalt pavements containing recycled materials such as Reclaimed Asphalt Pavement (RAP) or Reclaimed Asphalt Shingles (RAS), and (3) the characterization of warm-mix materials. In addition to Superpave performance tests [8–12], studies conducted by several researchers [7,13–29] also provide valuable and significant insight into the estimation of asphalt materials low-temperature performance.

Acoustic Emission (AE) has been used extensively for damage detection and assessment of several materials including concrete, steel, wood, and rock. However, there has been limited application of AE for evaluating damage mechanisms in asphalt materials. Khosla and Goetz [21] used AE techniques to detect crack initiation and propagation in indirect tensile (IDT) specimens at −23 °C. The study found that failure by fracture is indicated by a sharp increase in total AE counts, and that significant AE counts occur at about 80% of the peak load. Valkering and Jongeneel [22] used AE to monitor temperature cycling tests with restrained asphalt concrete specimens at low temperatures (−10 °C to −40 °C). They observed that the repeatability of AE measurements is good, that the AE activity (i.e., number of events) correlates with the thermal fracture temperatures, and that the AE activity in restrained specimens at low temperatures is caused by defect initiation in the binder. Hesp et al. [23] used AE measurements to detect crack initiation and propagation in restrained specimens at low temperatures (−32 °C to −20 °C). They concluded that the Styrene-Butadiene-Styrene (SBS)-modified mixes produced less AE activity than unmodified mixes. Li et al. [24–28] used AE techniques to characterize fracture in semi-circular bend asphalt specimens at low temperatures (−20 °C). They also concluded that large amounts of accumulated AE events occur at 70% of material strength, that the maximum intensity of AE peaks correlates with the development of macro-cracks, and that the location of AE events suggests that a several centimeter-sized process zone forms before the peak load. Nesvijski and Marasteanu [29,30] used an AE spectral analysis approach to characterize fracture in semi-circular bend asphalt specimens at low-temperatures, and concluded that an AE approach could be used for evaluation of asphalt pavements. All of this research work led to the development of standards by the American Association of State and Highway Transportation Officials (AASHTO MP1, 1998; AASHTO TP1, 1999; AASHTO MP1A, 2001) [8–10] that allowed the estimation of the low-temperature performance of binders based upon their rheological properties.

The Bending Beam Rheometer (BBR) Method

For unmodified binders, the determination of cracking temperature is based on results from the bending beam rheometer (BBR) test in accordance with the standard test methods AASHTO TP1 [8]. Two asphalt binder parameters, i.e., the stiffness and the *m*-value, are determined based on the bending beam rheometer (BBR) test results at a loading time of 60 s. The cracking temperature is defined as the temperature at which the stiffness reaches the value of 300 MPa or the m-value reaches the value of 0.3, or, moreover, the temperature at which one of the specification thresholds is reached as temperature is decreased. The cracking temperatures determined using AASHTO TP1 have been found to be predictive of low-temperature cracking in asphalt pavements constructed using unmodified binders. However, AASHTO TP1 was found to grossly over predict low-temperature cracking performance for modified asphalt binders. As a result, for modified asphalt binders, additional testing using the direct tension test [3] (DTT) was proposed to address the case of binders with a stiffness higher than 300 MPa and with an m-value greater than 0.3. Bouldin et al. [13] presented methods for predicting cracking temperatures using both the BBR and the DTT test data in the so-called dual instrument method (DIM).

These methods involve the use of sophisticated computer software for computing thermal stress in asphalt binders using the bending beam rheometer (BBR) data and estimating binder strength using the direct tension test (DTT). The dual instrument method (DIM) has been proposed as a standard method (MP1A) [10] for evaluating both modified and unmodified asphalt binders. To be able to use the MP1A, sophisticated equipment (DTT and BBR) and analysis software are required. Concerns have already been raised about the expensive nature of the MP1A approach, because its use has become prohibitive to many practitioners. Different approaches to alleviate these issues have also been proposed by Kim [15], and by Kim et al. [16], by Roy and Hesp [17], and by Shenoy [18], with some but not enough success to evaluate the low-temperature performance of asphalt materials.

For the practical low-temperature evaluation of binders, binder blends and mixtures for the purpose of formulation, design, control, and forensics, there is still a need for a test which is rapid, simple, compact, portable, and applicable to all modern binder types. Acoustic emission (AE) is a promising technique for evaluating embrittlement in asphalt binders, since it has been used successfully in other materials.

In this review paper, an acoustic emission (AE) approach is used to accurately and rapidly evaluate the thermal cracking performance of asphalt concrete materials including virgin, short-term, and long-term aged asphalt binders, as well as different asphalt concrete mixtures. A review of this technique along with applications of this approach are presented and discussed. For more details regarding experimental setups, etc., the readers are encouraged to refer to the appropriate citations.

2. Acoustic Emission Based Evaluation of Embrittlement Temperatures

The AE-based technique was implemented for low-temperature fracture assessment of asphalt materials. During the course of developing this testing approach, a wide range of asphalt binders and asphalt concrete materials with significantly different low-temperature cracking characteristics, as well as different oxidative aging levels (i.e., unaged, short-term aged, and long-term aged), were evaluated.

Figure 4a schematically shows the geometry of the developed test specimen for virgin (i.e., unaged) or aged asphalt binders, which consists of a thin layer of asphalt binder (with nominal dimensions of 125 mm long, 12 mm wide, and thickness of 6 mm) bonded to a granite substrate. These asphalt binder specimens are prepared using aluminum molds, and they have the same dimensions as the standard Bending Beam Rheometer (BBR) specimens [9]. To make the binder specimen, the asphalt binder is poured into the aluminum mold wrapped in Teflon tape and placed on the granite slab, which is heated to the temperature of 135 °C, see Figure 4a. Prepared binder samples are allowed to cool down to room temperature for two hours before conducting the corresponding AE tests [31–37]. Figure 4b shows the specimen after being tested using AE. Figure 5a shows the binder test specimen along with the position of the AE sensors and the thermocouple. The geometry of AE specimens utilized for testing compacted asphalt concrete mixtures is shown in Figure 5b, in which the position of the AE sensor and the thermocouple is also shown. These specimens are typically semicircular, measuring 150 mm in diameter and 50 mm in thickness. This geometry was selected after examining several shapes and different sample thicknesses. In addition, the semicircular geometry is very practical and easy to make from cylindrical extracted field cores or laboratory gyratory compacted samples [36,37].

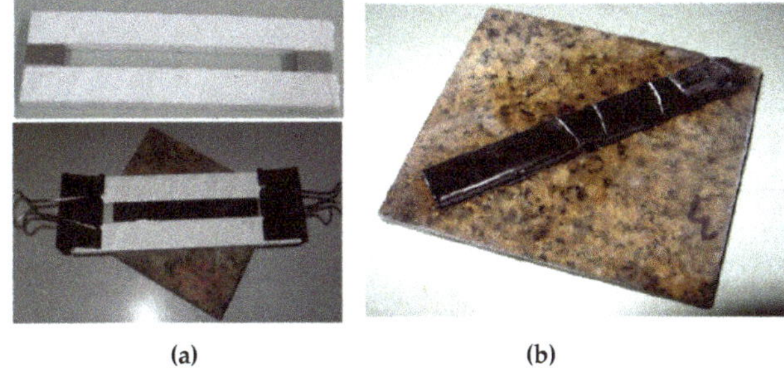

(a)   (b)

Figure 4. Fabrication of binder test samples: (**a**) top: molds for fabricating binder test samples; bottom: completed binder test sample in the mold; (**b**) typical visible crack patterns in asphalt binder test sample after AE testing. The binder samples have the same nominal dimensions as the specimens in the Bending Beam Rheometer (BBR) test (125 mm long, 12 mm wide, and 6 mm thick). The binder is poured into the mold seating on the granite (which has been heated to 135 °C) to assure good adhesion between the binder sample and the granite substrate.

Apart from different test sample geometries, the same testing set up and procedures are used for testing both asphalt binders and asphalt concrete specimens. To conduct the AE test, prepared specimens are placed inside the freezer and are cooled down from 20 °C to about −35 °C, or even to −50 °C, if necessary. A K-type thermocouple is placed on the specimens' surface to record the temperature of the samples. Because of the finite size of the test sample, there is a thermal lag at the beginning of the test, which becomes negligible (almost zero) at temperature lower than −10 °C. Considering that the embrittlement temperature of almost all binders is below −10 °C, measuring the temperature at the specimen's surface appears to be a proper place [3,32,33]. Figure 5c shows an AE test system, while Figure 5d shows a temperature versus time plot while cooling the sample in a freezer.

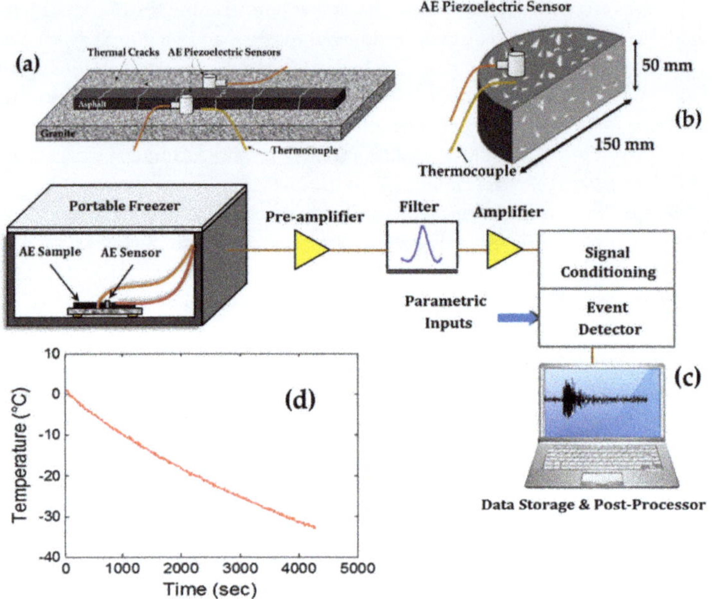

Figure 5. AE testing of asphalt materials; (**a**) asphalt binder test sample after AE testing showing two AE sensors and one thermocouple; (**b**) test sample for asphalt concrete mixtures; (**c**) AE test system for evaluating binders and asphalt concrete mixtures test samples; (**d**) typical temperature vs. time cooling plot for asphalt concrete mixture test samples.

Wideband AE piezoelectric sensors (Model B1025, Digital Wave Co., Denver, CO, USA) with a relatively flat sensitivity in the frequency band of 50 kHz to 1.5 MHz were utilized to monitor and record the acoustic activities of the samples during the test. AE stress waves were detected using the AE piezoelectric sensors, amplified, filtered, and then recorded. High-vacuum grease was used to couple the AE sensors to the test sample. AE signals were pre-amplified 20 dB using broad-band pre-amplifiers to reduce extraneous noise. The signals were then further amplified 21 dB (for a total of 41 dB) and filtered using a 20 kHz high-pass double-pole filter using the Fracture Wave Detector (FWD, Digital Wave Co., Denver, CO, USA signal condition unit. The signals were then digitized using a 16-bit analog to digital converter (ICS 645B-8) using a sampling frequency of 2 MHz and a length of 2048 points per channel per acquisition trigger. The outputs were stored for later processing using Digital Wave software (Wave-Explorer TM V7.2.6) [9,31–37]. The Embrittlement Temperature (T_{EMB}) of asphalt materials is defined as the temperature corresponding to the first high-energy AE event above a chosen energy threshold.

2.1. Testing Asphalt Binders

The performance of asphalt binders is extremely susceptible to changes in temperature. Binders are performance Graded (PG) with high- and low-temperature performance designations, which are based on the pavement's expected operating temperatures [11,12]. This grading designation allows for appropriate binders to be chosen to prevent temperature-driven pavement failures, such as rutting at high temperatures and cracking at low temperatures. Performance graded binders are designated as *PG XX-YY*, in which *XX* is the high-temperature designation, which corresponds to the average temperature the binder is expected to encounter over a seven day period, and *YY* is the low-temperature designation, which corresponds to the lowest expected pavement temperature [11,12].

In asphalt binder test samples, the thermal contraction coefficient of asphalt binders is several times greater than that of granite substrate. Therefore, as the binder specimen cools down, due to the differential thermal contraction between asphalt and granite substrate, thermal stresses build up in the restrained asphalt binder layer. At the same time, the tensile strength of the asphalt binder reduces with temperature. Eventually, when the thermal stress exceeds the strength of the binder material, thermal cracks develop, which are accompanied by the release of elastic strain energy in the form of transient mechanical stress waves, i.e., acoustic emissions. Both Figures 4b and 5a show the asphalt binder test sample after the AE test, illustrating several thermal cracks [2,4,31–34].

The typical time domain response and the corresponding power spectral density curve of an AE event is shown in Figure 6. Analyses of AE activities of samples are performed on recorded AE signals and are associated testing temperature. The energy of the AE events was computed using Equation (1), in which E_{AE} is the AE energy of an event (V^2-μs) with duration of time t (μs) and recorded voltage of V(t) [31–34].

$$E_{AE} = \int_0^t V^2(t)dt \tag{1}$$

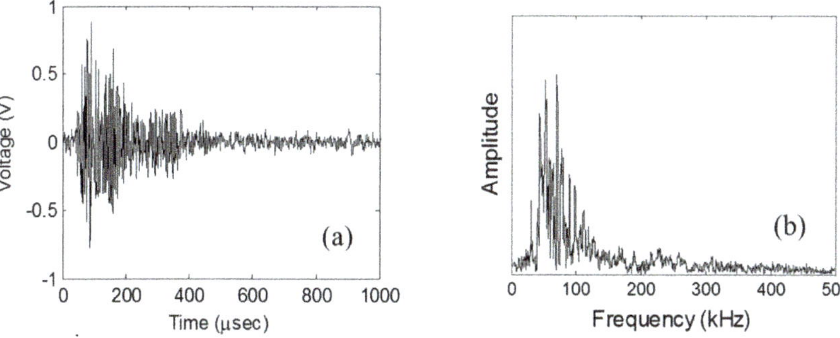

Figure 6. Typical acoustic emission event waveform (**a**) and corresponding spectral content (**b**).

A typical plot of AE cumulative events versus temperature for asphalt binders is shown in Figure 7, in which the energy of events versus temperature is also shown. In Figure 7, no events are recorded in the pre-cracking region, mainly because of the energy filtering process used. For each AE system gain, and using a binder of known cracking temperature, the energy threshold is selected (i.e., calibrated) to assure that the AE-obtained embrittlement temperature equals the known binder cracking temperature value grade, which is currently obtained using the binder rheological properties, i.e., the BBR methods. The AE-based embrittlement temperatures along the corresponding BBR-based critical cracking temperatures for various asphalt binders are provided in Table 1, in which, for comparison, the coefficient of variation (COV%) is also shown for both methods. Please note that while the coefficient of variation is temperature scale-dependent, for the present application in which the embrittlement temperatures are in a relatively narrow range and sufficiently below zero, the COV was deemed to be a useful statistical parameter to describe the repeatability of the measurements obtained via the two approaches. Results from both approaches are also presented in Figure 8. Comparison of results also indicate that $T_{cracking}$ (TANK) < $T_{cracking}$ (RFTO) < $T_{cracking}$ (PAV), in which RFTO and PAV stand for short-term and long-term aging, respectively, and TANK denotes virgin, i.e., unaged, binder. In Table 1 and in Figure 8, the term Rolling Thin-Film Oven (RTFO) indicates the binder was submitted to short-term aging, the Term Pressure Aging Vessel (PAV) indicates the binder was submitted to long-term aging (or a 7 to 10 year period), and TANK indicates virgin, i.e., unaged binder.

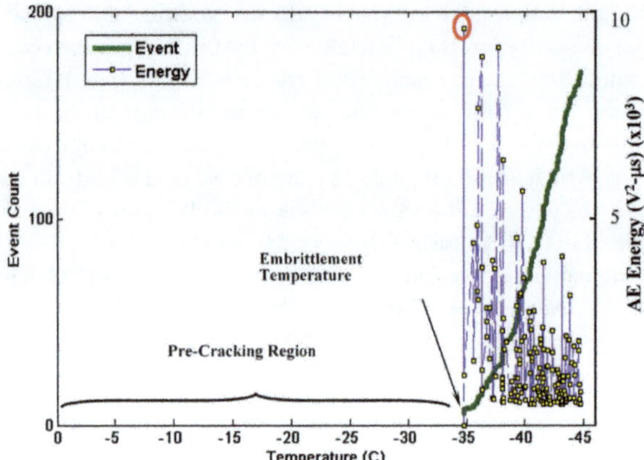

Figure 7. Typical experimental plot of cumulative events and energy versus temperature for asphalt binder test samples.

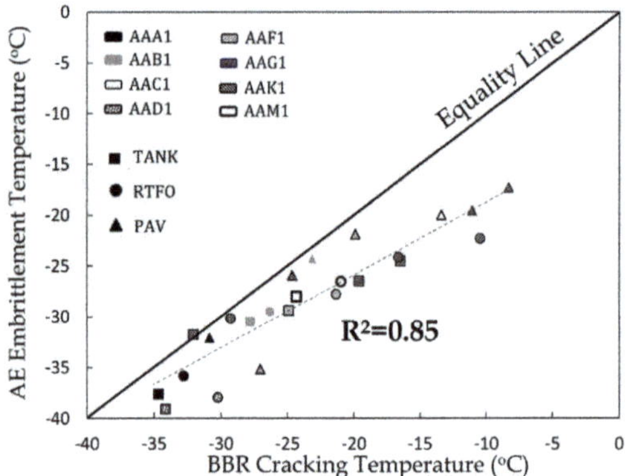

Figure 8. Correlation between AE embrittlement temperature and BBR-based cracking temperature, illustrating the conservative nature of the BBR-based cracking temperatures, see Table 1.

Table 1. AE-based embrittlement temperature and BBR cracking temperatures of several different binders, each with three aging levels.

Asphalt Binder *	AE-Based Embrittlement Temperatures (°C)		BBR Cracking Temperatures (°C)		
	T_{ENB} (°C)	COV [#] (%)	T_{BBR} (°C)	COV [#] (%)	T_{BBR}-T_{EMB} (°C)
TANK AAA1 (PG58-28)	−37.58	2.98	−34.69	7.54	2.89
RFFO AAA1 (PG58-28)	−35.78	3.68	−32.83	5.34	2.95
PAV AAA1 (PG58-28)	−32.09	3.68	−32.83	8.07	1.20
TANK AAB1 (PG58-22)	−30.46	3.56	−27.81	14.34	2.65
RFFO AAB1 (PG58-22)	−29.45	2.21	−26.32	3.27	3.13
PAV AAB1 (PG58-22)	−24.33	3.25	−23.13	7.92	1.20
TANK AAC1 (PG58-16)	−29.36	3.77	−24.90	8.99	4.46
RFFO AAC1 (PG58-16)	−27.77	1.26	−21.37	11.86	6.40
PAV AAC1 (PG58-16)	−21.86	5.30	−19.90	11.66	1.96
TANK AAD1 (PG58-28)	−39.09	2.66	−34.17	9.87	4.92
RFFO AAD1 (PG58-28)	−37.93	2.07	−30.24	5.13	7.69
PAV AAD1 (PG58-28)	−35.19	2.79	−27.03	3..35	8.16
TANK AAF1 (PG64-10)	−24.47	4.74	−16.49	7.35	7.99
RTFO AAF1 (PG64-10)	−22.29	3.35	−10.47	14.33	11.82
PAV AAF1 (PG64-10)	−17.30	5.15	−8.31	10.63	8.98
TANK AAG1 (PG58-10)	−26.48	3.56	−19.63	17.05	6.84
RTFO AAG1 (PG58-10)	−24.12	2.55	−16.63	10.45	7.49
PAV AAG1 (PG58-10)	−19.54	5.39	−11.08	5.77	8.46
TANK AAK1 (PG64-22)	−31.76	3.24	−32.12	9.00	−0.36
RTFO AAK1 (PG64-22)	−30.15	2.76	−29.29	3.94	0.83
PAV AAK1 (PG64-22)	−25.95	5.03	−24.63	6.21	1.32
TANK AAM1 (PG64-16)	−28.01	2.67	−24.32	4.58	3.70
RTFO AAM1 (PG64-16)	−26.52	3.60	−21.01	7.32	5.52
PAV AAM1 (PG64-16)	−19.98	4.40	−13.40	9.81	6.58

* TANK, Rolling Thin-Film Oven (RTFO), and Pressure Aging Vessel (PAV) stand for unaged, i.e., virgin, short-term, and long-term aging, respectively. [#] A minimum of four replicas was used to estimate average values and corresponding Coefficients of Variation (COVs).

Results also indicate that AE-based embrittlement temperatures are lower than the corresponding BBR-based critical cracking temperatures. This is not surprising, mainly because the AE-based embrittlement temperatures denote the measured values, while the BBR-base critical temperatures are based upon the binder's rheological material properties and include an inherent factor of safety to avoid low-temperature pavement cracking. For additional information, the readers are referred to References [3,4,36]. The developed AE-based approach was successfully employed to estimate the low-temperature performance grade of virgin, short-term, and long term-aged asphalt binders.

In addition to the initial transverse cracks shown in Figure 9, which divide the binder test sample into several blocks, Figure 9 also shows three-dimensional spiral cracks, which became visible only after removal of the top material from the test sample [38]. These spiral cracks are a result of the three-dimensional state-of-stress field that develops in each block (created by the transverse cracks) by the constraints imposed by the granite block. These spiral cracks were modeled as three-dimensional logarithm spirals using three parameters, (a spiral tightness parameter "b", an apparent length scale parameter "A", and the pitch angle "φ") that control how tightly and in which direction the spiral is wrapped. In addition to observing that the spiral pattern represents the crack trajectory with maximum energy release, it was also observed that the AE obtained embrittlement temperatures and fracture energy (obtained using indirect tension tests) are related to the spiral crack tightness parameter [38]. For additional information regarding modeling spiral cracks as the mode of failure in asphalt materials, the readers are referred to Behnia et al. [38].

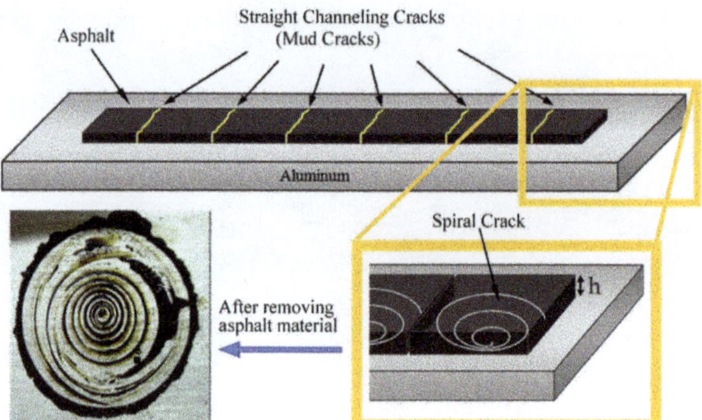

Figure 9. Asphalt binder showing the visible straight channeling cracks, i.e., mud cracks, and the spiral cracks that are only observed after the asphalt material is removed. The observed spiral cracks develop in each block due to the three-dimension state-of-stress induced by the thermal mismatch between the asphalt blocks and the granite substrate.

The promising AE results for low-temperature cracking performance of asphalt materials suggests that AE could be considered as a viable alternative for the ASHTO protocols, which estimate the low-temperature performance based upon the rheological binder properties. Furthermore, the AE approach has the advantage of being faster, having less variability, and capable of being used for all types of binders, including modified binders.

2.2. Testing Asphalt Concrete Materials

When AE asphalt concrete test samples are cooled down, thermally-induced stresses also develop in the test sample due to thermal contraction coefficient mismatch between the aggregates and the surrounding asphalt mastic. Different forms of thermal damage such as aggregate-asphalt mastic debondings, as well as thermal microcracks in the mastic, begin to occur in the sample when the local thermal stresses reach the local material strength. To visualize thermal damage in asphalt concrete test samples, X-ray computed micro-tomography (micro-CT) was employed using cubic asphalt concrete samples of one inch on-the-side, see Figure 10a. The X-ray micro-CT was conducted on asphalt concrete materials before and after two hours of conditioning the sample at $-50\ °C$. Figure 10b shows X-ray tomographic images of the thermal damage in the sample with damaged areas circled in red. The following two types of damage can be identified by (1) microcracks within the mastic, and (2) debondings at the interface between aggregates and mastic. These thermally induced microdamages in the sample act as sources of acoustic activity during the AE test.

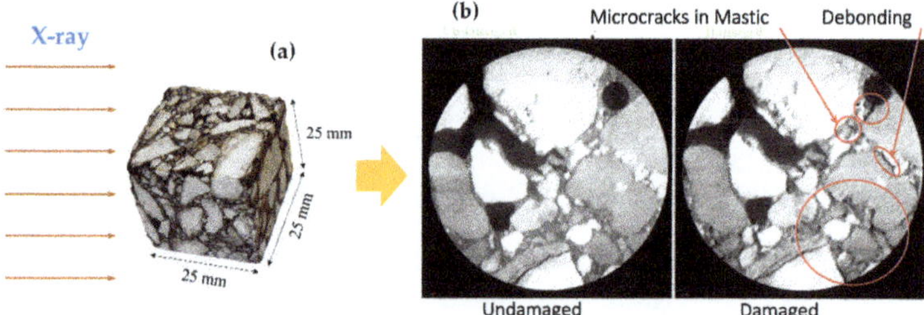

Figure 10. X-Ray micro-computer tomography imaging; (**a**) geometry and dimensions of used asphalt concrete sample; (**b**) images of undamaged (left) versus thermally damaged (right) asphalt concrete sample showing damaged regions.

Based upon many experimental observations, a typical schematic diagram of AE cumulative events versus temperature for asphalt concrete samples was observed as shown in Figure 11. Four distinct regions exist in the AE cumulative events vs. temperature plot, namely: pre-cracking, transition, stable cracking, and fully cracked regions. In the "pre-cracking region", thermal stresses begin to accumulate in the sample. However, thermal stresses are not yet high enough to cause any thermal damage in the material. As a result, no AE events are detected within this region. Progressively increasing thermal stresses in the specimen eventually result in the formation of thermal microcracks in the material, which are accompanied by the release of mechanical transient stress waves, i.e., AE events. The second region, i.e., the "transition region", is defined as the point in time when thermal micro-cracking in the specimen, as indicted by higher energy events, begins to occur. The temperature corresponding to the event with the energy level above a predetermined threshold has been termed the "*Embrittlement temperature (T_{EMB})*", as shown in Figure 12. For a given system amplification, the energy threshold is typically determined by calibrating the AE system by using samples made with a binder of a known cracking temperature, which is obtained by the traditional rheological-based methods (i.e., the BBR based methods).

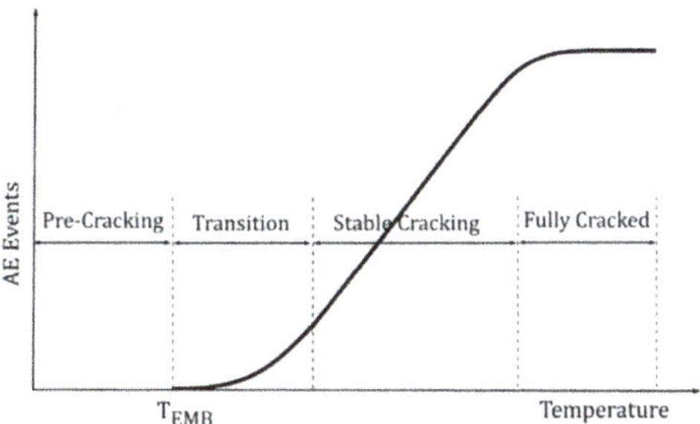

Figure 11. Schematic diagram of AE cumulative events vs. temperature showing four different regions for asphalt concrete materials.

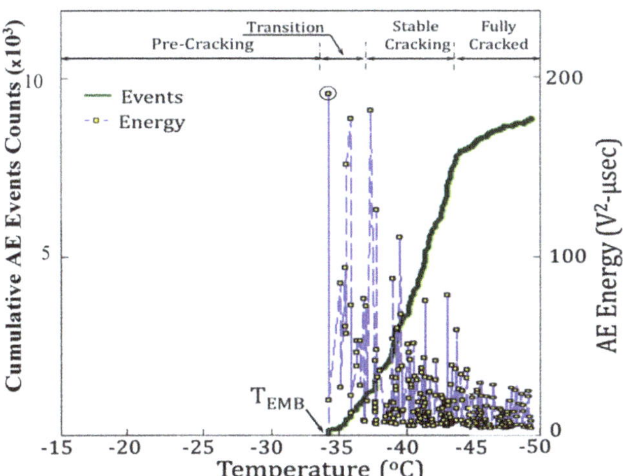

Figure 12. Typical plot of cumulative events and AE energy vs. temperature for asphalt mixtures.

The embrittlement temperature represents the onset of thermal damage in the asphalt concrete test sample. It is hypothesized that the embrittlement temperature represents a fundamental material property that is independent of material constraints, sample size (as long as a statistically representative volume or larger is used), and sample shape [3,32,33,38,39]. The "transition region" can be considered as the region where material behavior gradually changes from a quasi-brittle to a brittle state in which resistance to fracture is generally very low, allowing cracks to propagate readily. The "stable cracking region" usually initiates at a very low temperature, when the material is brittle and generates a significant amount of AE activity. The AE cumulative events versus temperature plot in this region usually has a steep slope that remains relatively constant. The "fully cracked region" starts right after the stable cracking region when the rate of AE activity of the samples begins to reduce until it reaches zero at the end of this region. Considering that the source of AE activities is the generation of new microdamage within the test sample, reduction in the rate of AE activity is an indication of the presence of plenty of microdamage in the sample. It should be noted that in AE-based tests, the fully cracked region is not usually observed unless the sample is cooled down to very cold temperatures, allowing all microdamage to fully develop within the test sample. Figure 12 shows a typical plot of cumulative event count and energy versus temperature for asphalt concrete test samples, where it is noted that the fully cracked region is not fully developed, mainly because only the embrittlement temperature is of interest.

3. Thermal Cracking Evaluation of Asphalt Roads Containing Recycled Materials

The use of AE-based approach to characterize pavements is now applied for asphalt concrete pavements containing recycled asphalt pavement (RAP) materials [36,37,40–42] and/or pavements containing materials from recycled asphalt shingles (RAS) [43]. The use of these recycled materials in asphalt roads has gained significant popularity in recent years, mainly because of environment and sustainability concerns. Additionally, using recycled materials can result in sustainable designs and cost savings by reducing the amount of virgin materials required in the production of new asphalt pavements. The AE-based technique was implemented to assess the effects of using recycled materials on thermal cracking performance of these asphalt pavements. Results showed that the AE-based technique could not only successfully evaluate the effect of recycled materials on low-temperature cracking performance of asphalt concrete, but also it could accurately detect the important phenomena of partial blending of recycled and virgin asphalt binders in asphalt mixtures.

Figure 13 presents the AE-obtained embrittlement temperature results of asphalt concrete samples of mixtures using the binder PG58-28 and containing the following different amounts of RAP: 0%, 10%, 30%, 40%, and 50%. Each data presented in the plot is the average of at least four test replicates for each mixture material. Two sets of results are demonstrated: Figure 13a shows the AE results of mixtures with partial blending of RAP and virgin binder, whereas Figure 13b presents AE results of asphalt mixtures with proper mixing of RAP and virgin materials at higher temperatures with a longer mixing time. In both cases, it is observed that adding RAP increases the embrittlement temperature of the material suggesting that RAP mixtures crack at warmer temperatures. In case of partial blending, surprisingly, there is not significant distinction between T_{EMB} of mixtures containing different amounts of RAP. This can be explained by noting that the AE testing procedure measures material properties on a local scale. If AE tests are conducted on a composite material system consisting of two types of materials, the T_{EMB} of the composite will be the temperature at which the weaker of the two materials starts to undergo damage and fracture. In the case of partial blending mixtures, regardless of the amount of RAP in the mixture, since the unblended RAP materials experience thermal cracking first, the measured T_{EMB} of those mixtures is in fact the embrittlement temperatures of the RAP material, which is the same for different mixtures with different amounts of RAP. However, in case of proper blending, it seems that the higher the amount of RAP content in the mixture, the warmer the cracking temperature of the mixture. Similar observations were noted for asphalt mixtures containing RAS materials [43]. Clearly, partial blending of recycled materials is an important issue. Application of AE-based test can significantly help reduce the partial blending in asphalt concrete and make the pavement more resilient to cooler temperature environments.

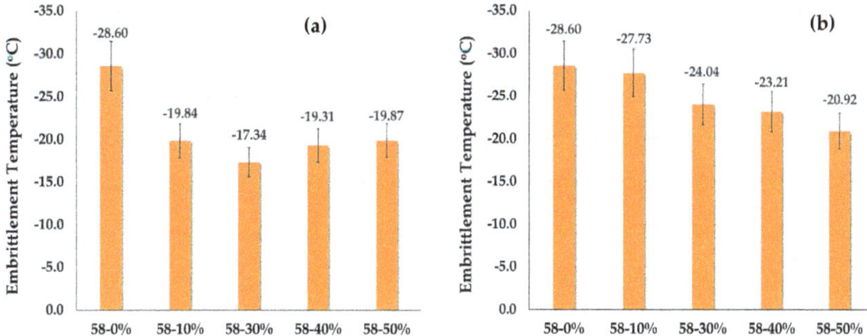

Figure 13. Embrittlement temperatures of PG58-28 mixtures containing different amounts of RAP: (**a**) mixtures with partial blending of RAP and virgin binders, (**b**) mixtures with proper blending of RAP and virgin binders.

4. Low-Temperature Characterization of Asphalt Mixtures Subjected to Cooling Cycles

In cold regions, cooling cycles cause thermally-induced micro-damages in the microstructure of asphalt concrete, which further reduces the asphalt pavements' resistance to fracture. Here, AE tests were conducted on field cores, as well as laboratory gyratory compacted samples, and the corresponding embrittlement temperature of samples were determined [44].

During thermal cycling of asphalt concrete, it was observed that asphalt concrete specimens generated AE activities before the previous minimum temperature (i.e., maximum thermal loading level) was reached. This phenomenon is known as the "Felicity effect", and is characterized by the presence of detectable AE activity during reloading of the material before the load level reaches the previous maximum applied load [45]. The Felicity effect observed in asphalt materials is illustrated in Figure 14, which shows the cumulative AE events versus applied thermal load. The loading path from B to C (unloading) and C to D to E (reloading) clearly indicates the absence of AE activity up to a loading level (T_D), which is below the level of the previous loading cycle (T_B). It should also be noted that another AE loading phenomenon, known as the Kaiser effect, may also take place if the mixture loses the ability of self-healing. The Kaiser effects occurs when no AE activity occurs until the previous maximum load level is reached, which makes the Kaiser effect a particular case of the Felicity effect (i.e., points B and D coincide). The Kaiser effect can be observed in highly oxidized mixtures and/or when no time is allowed between cycles for the self-healing to take place.

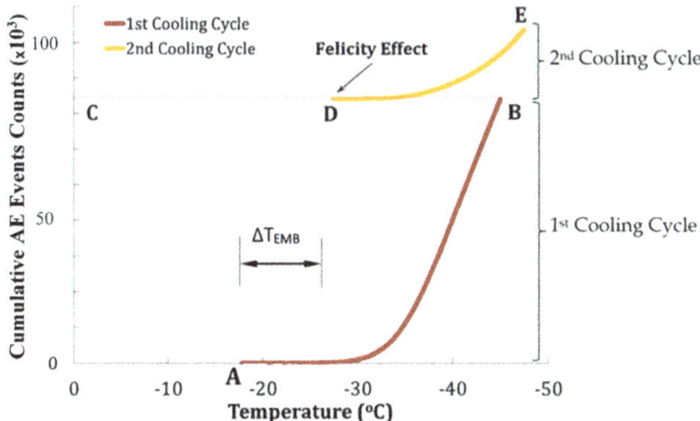

Figure 14. Typical AE test results of thermal cyclic loading of laboratory samples. Please note the presence of the Felicity effect as a result of self-healing.

Observation of the Felicity effect instead of the Kaiser effect in asphalt pavement materials suggests partial healing of microcracks in the material. Due to the adhesive nature of asphalt concrete materials, some level of micro-crack self-healing may take place within material during thermal unloading. The amount of micro-crack self-healing may vary depending on the binder chemical composition and upon the level of oxidative aging of the asphalt mixture, as well as on the ambient temperature and presence of moisture. This means that in thermal cyclic loading of adhesive materials such as asphalt mixtures with high potential of micro-crack healing, the Felicity effect is the one that would occur rather than the Kaiser effect. Here, the Felicity Ratio (FR) is defined as the ratio of temperature at point D to the temperature of point B, the lowest temperature of the first cooling cycle. The average Felicity ratio for tested asphalt concrete materials was around 0.648 [44,45]. To better quantify crack healing of asphalt material, a new parameter called the "Healing Index" was introduced and defined by the following equation.

$$Healing \ Index \ (\%) = 100(1 - FR) \tag{2}$$

The Healing Index was used as an indication of the amount of healing that may occur in asphalt concrete between the loading cycles over the period of thermal unloading. Depending on the sample temperature (i.e., thermal cycle), type of asphalt binder, and presence of moisture, the Healing Index of asphalt mixture ranges from 0% (no healing) to 100% (fully healed). For the mixtures tested, the average Healing Index of the tested samples was around 35%, which indicates partial healing of the asphalt concrete during the rest period between the first and second cooling cycles. Results also showed that the oxidative aging adversely affected the thermal cracking healing capability of asphalt pavements, as the asphalt materials with higher level of oxidative aging exhibited lower Healing Index. For additional information, the readers are referred to Reference [44].

5. Restoring Original Low-Temperature Performance to Aged Asphalt Pavements

Oxidative aging leads to an increase in stiffness, loss of ductility and cohesion of binders, and warmer embrittlement temperatures, see Table 1 and Figure 8. As a result, oxidative aging lowers the resistance to fracture of oxidized mixtures as compared to their virgin state [37,46–48]. To restore the crack-resistant state of oxidized asphalt concrete pavements, measures such as pavement surface milling and/or the application of rejuvenators are taken. Asphalt rejuvenators are asphalt additives and modifiers that are used to revitalize, provide sealing, and restore the physical and chemical properties of the aged asphalt concrete [49–51]. Rejuvenators address the issue of oxidative hardening by softening the aged asphalt binder through restoration of the asphaltenes to maltenes ratio [6]. After applying a thin layer of rejuvenator over the top surface of pavement, the rejuvenator penetrates the asphalt concrete using the pore and tortuosity structure via gravity and capillary action, and diffuses through the asphalt concrete to chemically react with the asphalt binder. The rejuvenator/binder reaction restores the binder's material properties to its original state, i.e., the material properties of the virgin material. As the asphalt binder is softened, it also increases its adhesive properties and reduces the susceptibility of the asphalt pavement to thermal cracking.

As described by Brown [49], with the exception of visual inspection, there is no standardized method to evaluate the performance of rejuvenators when applied in the field. Currently, the ability of rejuvenators to improve pavements' durability is typically evaluated by (1) estimating the penetration in samples at 25 °C using asphalt binder extracted from untreated and treated cores, (2) comparing the viscosity at 60 °C of asphalt binder samples extracted from untreated and treated cores, and (3) comparing the percentage of aggregate loss when untreated and treated samples are subjected to a pellet abrasion test. Mainly because these tests are cumbersome and time consuming, they are not often used. Here, AE source location is used to characterize and evaluate the rejuvenators' ability to restore mixtures to their original low-temperature performance grading, i.e., their original crack resistant state.

Because of the granular nature of asphalt concrete mixtures, the Geiger's iterative source location technique [51,52] was used to accurately locate the source of each event. The Geiger's method is the application of the Gauss-Newton algorithm, which requires data from at least four AE sensors to build the following arrival time function of the *i*th sensor:

$$f_i(x,y,z,t) = T_s + \frac{1}{v}\sqrt{(x_i - X_s)^2 + (y_i - Y_s)^2 + (z_i - Z_s)^2} \tag{3}$$

in which (X_s, Y_s, Z_s) represent the spatial coordinates of the source, (x_i, y_i, z_i) represent the coordinates of the ith sensor, v is the known wave velocity, and T_s and t_i are the unknown source event occurring time and the known receiving time by the *i*th sensor, respectively. Equation (3) can be expanded using Taylor series at a point (x_0, y_0, z_0), near the actual source, resulting in Equation (4):

$$f_i(x,y,z,t) = f_i(x_0, y_0, z_0, t_0) + \epsilon_i \tag{4}$$

in which ϵ_i, the residual term, a.k.a. the correction vector, is the difference between the calculated arrival time and the observed arrival time with respect to the *i*-th sensor. It can be calculated using the first order derivatives of the arrival time function. By going through several iterations of Equation (5), the Geiger's method tries to minimize the correction vector.

$$\epsilon_i = \frac{\partial f_i}{\partial x}\delta x + \frac{\partial f_i}{\partial y}\delta y + \frac{\partial f_i}{\partial z}\delta z + \frac{\partial f_i}{\partial t}\delta t \tag{5}$$

Gyratory compacted specimens with 5.6% of asphalt content were made using PG64-22 binder and nominal maximum aggregate size (NMAS) of 19 mm. Some specimens were short-term aged for two hours at 155 °C to simulate aging during plant production, and other specimens were aged for 36 h at 155 °C (in addition to the short term aging). The specimens' aging was performed on loose mixtures, which were hand-stirred every 12 h to ensure uniform aging. Each gyratory compacted specimen was then cut into two 5-cm tall cylindrical specimens with a diameter of 150 mm, for a total of eight test specimens. Figure 15 shows one of these specimens with four AE sensors coupled to the top and bottom plane surfaces of the cylindrical specimen. To avoid numerical instability, the sensors placed on the bottom of the specimens have a 45° offset angle with respect to the sensors coupled on the top surface.

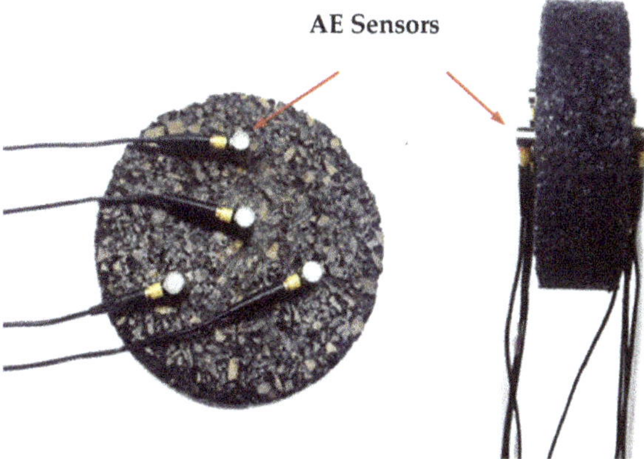

Figure 15. AE source location in asphalt mixture samples using eight piezoelectric sensors with four sensors mounted on each plane side of the specimen. To avoid numerical instabilities, the sensor pattern placed in one surface has a 45° offset angle with respect to the pattern of the opposite surface.

Out of the eight 5-cm tall specimens, two long term specimens (aged for 36 h) and two short term aged, i.e., virgin specimens, were tested using AE, and used as the control samples. The other four specimens were treated by spreading a thin layer of rejuvenator (e.g., Reclamite) on the top surface of the specimens in the amount of 10% by weight of the binder. The four specimens were then stored for a prescribed dwell time of two, four, six, and eight weeks before conducting AE tests. After each dwell time, each specimen was then tested using the same AE source location procedure used to test the 36 h and two hours aged specimens, which allowed the estimation of the embrittlement temperatures throughout the specimen thickness.

Figure 16 shows the combined embrittlement temperatures versus thickness results for all asphalt specimens using the Geiger's iterative source location method. Figure 16 shows that the embrittlement temperature of 36-h aged samples (−13 °C) is warmer than that of virgin, i.e., short-term aged sample (−22 °C) due to oxidative aging. Figure 16 clearly shows that after two weeks of dwell time, all of the specimen embrittlement temperature material properties have been recuperated. After six and eight weeks, the achieved embrittlement temperatures far exceeded the embrittlement temperatures of the virgin specimens. In addition, Figure 16 also shows that for the dwell times of two and four weeks, the method also captured the gradation the embrittlement temperature properties throughout the thickness, mainly because the rejuvenator has had additional time to act on the top material layers. This is an indication that the used AE approach can be used to evaluate the graded embrittlement temperature properties of aged pavements in the field, see Figure 3b, Figure 8, and Table 1. These findings shows that this AE approach may be used to intelligently select the best maintenance strategies by assessing and optimizing the relative amount of milling and surface replacement, or the levels of rejuvenation needed to restore pavement to the original crack-resistant state. The AE results using source location are consistent with the results obtained using non-collinear ultrasonic wave mixing [48,53,54].

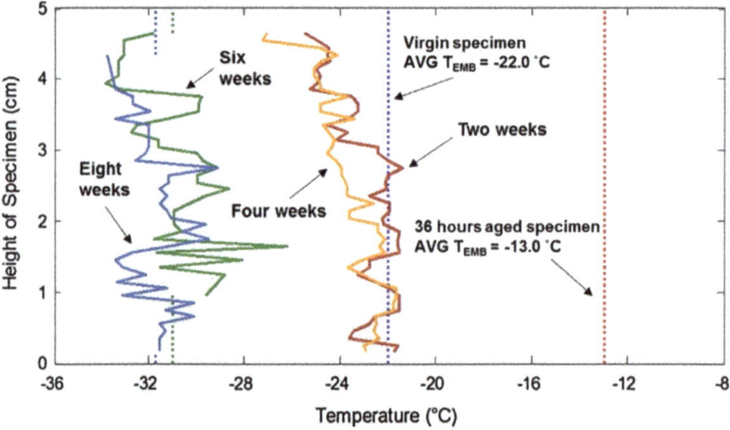

Figure 16. Average measured embrittlement temperatures of rejuvenator-treated oven-aged asphalt concrete samples (for 36 h at 135 °C) after dwell times of 2, 4, 6, and 8 weeks. For comparison, the embrittlement temperatures of the short-term aged (i.e., virgin) samples and of oven-aged samples for 36 h at 135 °C of −22.0 °C and −13 °C, respectively, are also noted (see dotted and dashed lines respectively). Note that after two weeks of dwell time, the unaged mixture embrittlement temperature was already restored. Also, at 2 and 4 weeks of dwell time, the top material layers have a cooler embrittlement temperature, because the rejuvenator had more time to act upon the binder.

6. Conclusions

An acoustic emission based testing approach was developed to address the shortage for accurate and reliable techniques to evaluate the low-temperature cracking performance of asphalt pavements. The developed AE approach was successfully implemented to evaluate and characterize both virgin, short-term, and long-term asphalt binders and asphalt concrete materials. The extension of AE-based method to asphalt concrete materials opened the door for the use of the technique for asphalt pavement condition assessment. The AE technique was also successfully employed in different areas such as evaluating asphalt pavements containing recycled materials such as RAP or RAS, assessing the effect of cooling cycles upon the structural integrity of pavements, and evaluating the thermal cracking performance of graded, i.e., aged asphalt pavements.

The developed acoustic emission-based testing technique appears to be a viable approach for the characterization of low-temperature cracking of asphalt pavements, and it could be a powerful tool for enhancing pavement sustainability when used for preventive maintenance and rehabilitation. This technique could yield a significant payoff to practice for both up-stream and down-stream suppliers and producers of asphalt concrete binders. Up-stream supplies of polymer, chemical, and other additives (warm-mix additives, antistrip agents) could use the proposed technology to rapidly assess the low-temperature characteristics of trial formulations, and could quickly assess the compatibility of blended additive systems. Asphalt mixture designers could also use the technology to verify binder grade selection, optimize the amount of recycled materials, and/or select the appropriate binder grade to use in RAP and RAS mixtures, assess and design warm-mixtures, or verify compatibility when multiple additives are used. Pavement owners may be able to use the AE approach for quality assurance of binders and mixtures, for the periodic assessment of pavement conditions, and for the scheduling of preventive maintenance and rehabilitation, when pavement cracking is of concern.

Acknowledgments: The authors are very grateful for the support provided by the NCHRP Innovations Deserving Exploratory Analysis (IDEA) Program (managed by Inam Jawed) under Project 144 and Project 170 [36,37], which led to several of the authors' articles cited in this paper. Without this support, the current review paper would not have been possible. The authors are also grateful for the partial support (Contract Number W912HZ-16-C-0006) of the US Airforce Civil Engineering Center (AFCEC), including the technical support of our technical contacts George Vansteenburg and Jeb S. Tingle. Their support and many technical discussions was invaluable. The authors are also very grateful to the Missouri Department of Transportation for their support of current work.

Author Contributions: The authors contributed equally to the manuscript.

Conflicts of Interest: The authors declare no conflicts of interest.

References

1. Kim, Y.R. *Modeling of Asphalt Concrete, McGraw-Hill Construction*; ASCE Press: New York, NY, USA, 2008; ISBN-13 978-0071464628.
2. Apeagyei, A.K.; Buttlar, W.G.; Reis, H. Estimation of low-temperature embrittlement for asphalt binders using an acoustic emission approach. *INSIGHT* **2009**, *51*, 129–136. [CrossRef]
3. Behnia, B. An Acoustic Emission-Based Test to Evaluate Low Temperature Behavior of Asphalt Materials. Ph.D. Thesis, University of Illinois at Urbana-Champaign, Urbana, IL, USA, 2013.
4. Behnia, B.; Buttlar, W.G.; Reis, H. Nondestructive evaluation of thermal damage in asphalt concrete materials. *J. Test. Eval. ASTM* **2017**, *45*, 1948–1958.
5. Islam, S.; Buttlar, W. Effect of pavement roughness on user costs. *Transp. Res. Rec. J. Transp. Res. Board* **2012**, *2285*, 47–55. [CrossRef]
6. Petersen, J. A review of the fundamentals of asphalt oxidation. In *Transportation Research Circular Number E-C140*; Transportation Research Board: Washington, DC, USA, 2009.
7. Mirza, M.W.; Witczak, M.W. Development of a global aging system for short term and long term aging of asphalt cements. *Assoc. Asph. Paving Technol. J.* **1995**, *64*, 393–430.

8.	AASHTO MP1. *Standard Specification for Performance- Graded Asphalt Binder*; American Association of State Highway and Transportation Officials (AASHTO): Washington, DC, USA, 1998.

9.	AASHTO TP1. *Standard Specification for Determining the Flexural Creep Stiffness of Asphalt Binder Using the Bending Beam Rheometer (BBR)*; American Association of State Highway and Transportation Officials (AASHTO): Washington, DC, USA, 1999.

10.	AASHTO MP1A. *Standard Specification for Determining Low-Temperature Performance Grade of Asphalt Binders*; American Association of State Highway and Transportation Officials (AASHTO): Washington, DC, USA, 2001.

11.	FHWA-SA-95-003. *Background of Superpave Asphalt Mixture Design and Analysis*; US Department of Transportation, Federal Highway Administration: Washington, DC, USA, 1995.

12.	Asphalt Institute Superpave Series No 2 (SP-2). *Superpave Mix Design*; Asphalt Institute: Lexington, KY, USA, 1996.

13.	Bouldin, M.G.; Dongre, R.; Rowe, G.M.; Sharrock, M.J.; Anderson, D.A. Predicting thermal cracking of pavements from binder properties: Theoretical basis and field validation. *Assoc. Asph. Pavement Technol.* **2000**, *69*, 455–496.

14.	Dongre, R.; Bouldin, M.G.; Anderson, D.A.; Reinke, G.; D'Angelo, J.; Kluttz, R.O.; Zanzotto, L. *Overview of the Development of the New Low-Temperature Binder Specification*; Technical Report; Federal Highway Administration Binder Expert Task Group: Washington, DC, USA, 1999.

15.	Kim, S.S. Direct measurement of asphalt binder thermal cracking. *J. Mater. Civ. Eng.* **2000**, *17*, 632–639. [CrossRef]

16.	Kim, S.S.; Wysong, Z.D.; Kovach, J. Low-temperature thermal cracking of asphalt binder by asphalt binder cracking device. *Transp. Res. Record* **2006**, *1962*, 28–35. [CrossRef]

17.	Roy, D.; Hesp, S.A.M. Low-temperature binder specification development: Thermal stress restrained specimen testing of asphalt binders and mixtures. *Transp. Res. Record* **2001**, *1766*, 7–14. [CrossRef]

18.	Shenoy, A. Single-event cracking temperature of asphalt pavements directly from bending beam rheometer data. *J. Transp. Eng.* **2000**, *128*, 465–471. [CrossRef]

19.	Marasteanu, M.; Zofka, A.; Turos, M.; Li, X.; Velasquez, R.; Li, X.; Williams, C.; Bausano, J.; Buttlar, W.; Paulino, G.; et al. *Investigation of Low Temperature Cracking in Asphalt Pavements*; Report No. 776; Minnesota Department of Transportation, Research Services MS 330: St. Paul, MN, USA, 2007.

20.	Marasteanu, M.; Buttlar, W.; Bahia, H.; Williams, C.; Moon, K.H.; Teshale, E.Z.; Falchetto, A.C.; Turos, M.; Dave, E.; Paulino, G.; et al. Investigation of Low Temperature Cracking in Asphalt Pavements National Pooled Fund Study–Phase II. 2012. Available online: https://www.dot.state.mn.us/research/TS/2012/2012-23.pdf (accessed on 20 February 2018).

21.	Khosla, N.P.; Goetz, W.H. *Tensile Characteristics of Bituminous Mixtures as Affected by Modified Binders*; Joint Highway Research Project; Purdue University and the Indiana State Highway Commission: West Lafayette, Indiana, 1979.

22.	Valkering, C.P.; Jongeneel, D.J. Acoustic emission for evaluating the relative performance of asphalt mixes under thermal loading conditions (With Discussion). *J. Assoc. Asph. Paving Technol.* **1991**, *60*, 160–187.

23.	Hesp, S.; Terlouw, T.; Vonk, W. Low temperature performance of SBS-modified asphalt mixes. *Asph. Paving Technol.* **2000**, *69*, 540–573.

24.	Li, X.; Marasteanu, M. Evaluation of the low temperature fracture resistance of asphalt mixtures using the semi-circular bend test (with discussion). *J. Assoc. Asph. Paving Technol.* **2004**, *73*, 401–426.

25.	Li, X.; Marasteanu, M.O. Investigation of low temperature cracking in asphalt mixtures by acoustic emission. *Road Mater. Pavement Des.* **2006**, *7*, 491–512. [CrossRef]

26.	Li, X.; Marasteanu, M.O.; Iverson, N.; Labuz, J.F. Observation of crack propagation in asphalt mixtures with acoustic emission. *Transp. Res. Record J. Transp. Res. Board* **2006**, *1970*, 171–177. [CrossRef]

27.	Li, X.; Marasteanu, M.O.; Turos, M. Study of low temperature cracking in asphalt mixtures using mechanical testing and acoustic emission methods (With Discussion). *J. Assoc. Asph. Paving Technol.* **2007**, *76*, 427–453.

28.	Li, X.; Marasteanu, M. The fracture process zone in asphalt mixture at low temperature. *Eng. Fract. Mech.* **2010**, *77*, 1175–1190. [CrossRef]

29. Nesvijski, E.; Marasteanu, M. Wavelet transform and its applications to acoustic emission analysis of asphalt cold cracking. *The e-Journal of Nondestructive Testing & Ultrasonics* **2001**, *12*, 6. Available online: http://citeseerx.ist.psu.edu/viewdoc/download?doi=10.1.1.128.1542&rep=rep1&type=pdf (accessed on 20 February 2018).

30. Nesvijski, E.; Marasteanu, M. Spectral analysis of acoustic emission of cold cracking asphalt. *The e-Journal of Nondestructive Testing & Ultrasonics* **2006**, *11*, 10. Available online: http://citeseerx.ist.psu.edu/viewdoc/download?doi=10.1.1.216.6582&rep=rep1&type=pdf (accessed on 20 February 2018).

31. Behnia, B.; Buttlar, W.G.; Reis, H.; Apeagye, A.K. Determining the embrittlement temperature of asphalt binders using an acoustic emission approach. Proceedings of Structural Materials Technology Conference, New York, NY, USA, 16 August 2010; pp. 318–325.

32. Behnia, B.; Dave, E.V.; Buttlar, W.G.; Reis, H. Characterization of embrittlement temperature of asphalt materials through implementation of acoustic emission technique. *Constr. Build. Mater.* **2016**, *111*, 147–152. [CrossRef]

33. Behnia, B.; Buttlar, W.G.; Reis, H. Nondestructive low-temperature cracking characterization of asphalt materials. *J. Mater. Civ. Eng.* **2016**, *29*, 04016294. [CrossRef]

34. Hill, B.; Oldham, D.; Behnia, B.; Fini, E.; Buttlar, W.; Reis, H. Low-temperature performance characterization of biomodified asphalt mixtures that contain reclaimed asphalt pavement. *Transp. Res. Record J. Transp. Res. Board* **2013**, *2371*, 49–57. [CrossRef]

35. Hill, B.; Oldham, D.; Behnia, B.; Fini, E.H.; Buttlar, W.G.; Reis, H. Evaluation of low temperature viscoelastic properties and fracture behavior of bio-asphalt mixtures. *Int. J. Pavement Eng.* **2018**, *19*, 362–369. [CrossRef]

36. Buttlar, W.G.; Behnia, B.; Reis, H. *An Acoustic Emission-Based Test to Determine Asphalt Binder and Mixture Embrittlement Temperature*; Final Report, NCHRP IDEA Project No. 144; National Cooperative Highway Research Program: Washington, DC, USA, 2011.

37. Buttlar, W.G.; Reis, H.; Behnia, B. *Development and Implementation of the Asphalt Embrittlement Analyzer*; Final Report, NCHRP IDEA Project No. 170; National Cooperative Highway Research Program: Washington, DC, USA, 2015.

38. Behnia, B.; Buttlar, W.G.; Reis, H. Spiral cracking pattern in asphalt materials. *Mater. Des.* **2017**, *116*, 609–615. [CrossRef]

39. Behnia, B.; Dave, E.; Ahmed, S.; Buttlar, W.; Reis, H. Effects of recycled asphalt pavement amounts on low-temperature cracking performance of asphalt mixtures using acoustic emissions. *Transp. Res. Record J. Transp. Res. Board* **2011**, *2208*, 64–71. [CrossRef]

40. Dave, E.V.; Behnia, B.; Ahmed, S.; Buttlar, W.G.; Reis, H. Low temperature fracture evaluation of asphalt mixtures using mechanical testing and acoustic emissions techniques. *J. Assoc. Asph. Paving Technol.* **2011**, *80*, 193–220.

41. Hill, B.; Behnia, B.; Buttlar, W.G.; Reis, H. Evaluation of warm mix asphalt mixtures containing reclaimed asphalt pavement through mechanical performance tests and an acoustic emission approach. *J. Mater. Civ. Eng.* **2012**, *25*, 1887–1897. [CrossRef]

42. Hill, B.; Behnia, B.; Hakimzadeh, S.; Buttlar, W.; Reis, H. Evaluation of low-temperature cracking performance of warm-mix asphalt mixtures. *Transp. Res. Record J. Transp. Res. Board* **2012**, *2294*, 81–88. [CrossRef]

43. Arnold, J.W.; Behnia, B.; McGovern, M.E.; Hill, B.; Buttlar, W.G.; Reis, H. Quantitative evaluation of low-temperature performance of sustainable asphalt pavements containing recycled asphalt shingles (RAS). *Constr. Build. Mater.* **2014**, *58*, 1–8. [CrossRef]

44. Behnia, B.; Buttlar, W.G.; Reis, H. Cooling cycle effects on low temperature cracking characteristics of asphalt concrete mixture. *Mater. Struct.* **2014**, *47*, 1359–1371. [CrossRef]

45. Hellier, C.J. *Handbook of Nondestructive Evaluation*; McGraw-Hill: New York, NY, USA, 2001; ISBN 978-0-070-2812-19.

46. McGovern, M.E.; Behnia, B.; Buttlar, W.G.; Reis, H. Characterization of oxidative ageing in asphalt concrete–Part 1: Ultrasonic velocity and attenuation measurements and acoustic emission response under thermal cooling. *Insight-Non-Destr. Test. Cond. Monit.* **2013**, *55*, 596–604. [CrossRef]

47. McGovern, M.E.; Behnia, B.; Buttlar, W.G.; Reis, H. Characterization of oxidative ageing in asphalt concrete–Part 2: Estimation of complex moduli. *Insight-Non-Destr. Test. Cond. Monit.* **2013**, *55*, 605–609. [CrossRef]

Appl. Sci. **2018**, *8*, 306

48. McGovern, M.E.; Behnia, B.; Buttlar, W.G.; Reis, H. Use of nonlinear acoustic measurements for estimation of fracture performance of aged asphalt mixtures. *Transp. Res. Rec. J. Transp. Res. Board* **2017**, *2631*, 11–19. [CrossRef]

49. Brown, E.R. *Preventive Maintenance of Asphalt Concrete Pavements*; Report No. 88-1; National Center for Asphalt Technology: Auburn, AL, USA, 1988.

50. Sun, Z.; Farace, N.; Behnia, B. Buttlar, W.G.; Reis, H. Quantitative evaluation of rejuvenators to restore embrittlement temperatures in oxidized asphalt mixtures using acoustic emission. *J. Acoust. Emiss.* **2015**, *33*, 1–15.

51. Sun, Z.; Behnia, B.; Buttlar, W.G.; Reis, H. Acoustic emission quantitative evaluation of rejuvenators to restore embrittlement temperatures to oxidized asphalt mixtures. *Constr. Build. Mater.* **2016**, *126*, 913–923. [CrossRef]

52. Geiger, L. Probability method for the determination of earthquake epicenters from the arrival time only. *Bull. St. Louis Univ.* **1912**, *8*, 60–71.

53. McGovern, M.E.; Buttlar, W.G.; Reis, H. Effectiveness of rejuvenators on aged asphalt concrete using ultrasonic non-collinear subsurface wave-mixing. *Mater. Eval.* **2015**, *73*, 1367–1378.

54. McGovern, M.E.; Buttlar, M.E.; Reis, H. Damage Evaluation and Life Extension of Asphalt Pavements using Rejuvenators and Non-Collinear Ultrasonic Wave Mixing—A Review. *ASME J. Nondestruct. Eval. Diagn. Progn. Eng. Syst.* **2018**, *1*, 011002:1–011002:13. [CrossRef]

 *applied sciences*

Article

Acoustic Emission Monitoring of Industrial Facilities under Static and Cyclic Loading

Sergey V. Elizarov *, Vera A. Barat, Denis A. Terentyev, Peter P. Kostenko, Vladimir V. Bardakov, Alexander L. Alyakritsky, Vasiliy G. Koltsov and Pavel N. Trofimov

INTERINIS-IT LLC, Moscow 101000, Russia; baratva@interunis-it.ru (V.A.B.); terentyevda@interunis-it.ru (D.A.T.); kostenkopp@interunis-it.ru (P.P.K.); bardakovvv@interunis-it.ru (V.V.B.); sasa@interunis-it.ru (A.L.A.); koltsovvg@interunis-it.ru (V.G.K.); pav.trof@yandex.ru (P.N.T.)
* Correspondence: serg@interunis-it.ru; Tel.: +7-495-361-1990

Received: 12 July 2018; Accepted: 19 July 2018; Published: 26 July 2018

Featured Application: Structure health monitoring for refinery equipment, bridges, rotating machines, cranes and draglines.

Abstract: Acoustic emission (AE) testing is traditionally carried out on non-operating objects. Such requirement is associated both with the Kaiser effect, leading to the necessity for exceeding the test load above the working one and with a high level of noise during object operation. However, AE testing could be performed under operating conditions, if the AE data acquisition period is increased and a specialized method is developed, which should take account of the effect of various noise, features of the object loading under operating conditions, the effect of damaging factors and possible destruction mechanisms. This paper investigates the possibility to carry out structural health monitoring (SHM) of hydrotreaters, highway bridges, high-temperature pipelines, gas adsorbers, roller bearings of rotary kilns and draglines on the basis of AE method. Architecture of SHM-system and specific data analysis procedures are proposed.

Keywords: structural health monitoring; acoustic emission; non-destructive testing; hydrotreater; bridge; high temperature; gas adsorber; rotary kiln; dragline

---

1. Introduction

Nowadays the status of complex (especially dangerous) industrial structures is widely assessed by structural health monitoring (SHM), that is the continuous evaluation of their structural heath [1]. The main reasons to apply SHM are:

- access to the structure during operation is difficult or impossible;
- periodic nondestructive testing (NDT) is not possible or is time-consuming;
- temporary stoppage of equipment for its testing induces large financial losses;
- rapid development of operational defects and, as a result, low damage tolerance of the structure;
- the consequences of structural failure, which can lead to large material losses and a danger for maintenance personnel;
- the presence of known defects when it is necessary to extend the lifetime of the equipment.

SHM of equipment allows the early detection of most of the failures and alerts maintenance personnel about developing malfunction that already exists but is not yet dangerous and does not disrupt the operability of the equipment. Among various NDT methods, the acoustic emission (AE) method is the most effective one in the monitoring mode. This method due to the possibility of remote testing and high sensitivity in the crack detection allows real-time monitoring of the structures with a length of up to several hundred meters. It is important to emphasize that the detectability of a defect is not affected by its orientation. AE method makes it possible to detect the following types of defects: cracks, microcracks, various types of corrosion damage and leaks [2]. Despite all the advantages of AE method, other types of testing should be used to provide a more accurate and objective assessment of the structural health [3–7]. This is caused by several reasons: firstly, in the case of usual AE testing, a temporarily decommissioned structure must be additionally loaded to stimulate the growth of the defects, while in the monitoring mode this need disappears, since testing is performed continually and AE sources emit under the influence of variations in operational loads, the value of which is not precisely known. Therefore, it must be ensured that the almost unlimited time of the AE testing can compensate the reduced magnitude of the load changes. Secondly, AE testing in the monitoring mode is complicated by increased process acoustic noise of the equipment. Experience in the use of monitoring systems shows that under such conditions defects are reliably detected only at the final stages of destruction, when the number of AE signals and their amplitudes increase by more than an order of magnitude. All this leads to the need of complex monitoring, that is the use of AE method with other NDT methods, as well as with the tracking of operational parameters.

Also, the design and implementation of structural health monitoring system (SHM-system) should always take into account the specificity of the monitored structures. It is necessary first to study the design features of the structure and the results of previous investigations, analyze the types of loads and factors that generate defects [8]. For the determination of the zones in the structure, in which the occurrence of cracks is most likely, the stress-strain state of the structure can be numerically simulated. Then a complete investigation of the structure is carried out, usually with trial AE testing. As a result, the acoustic parameters of the structure are determined, with the types of possible defects, their nature of developing, reasons of the formation, their most probable positions and their influence on the lifetime of the structure are revealed. Sufficiency of load changes for the stimulation of AE sources is estimated. After that, the selection of methods and instruments of NDT and the positions of the sensors are determined. In addition, in the course of preliminary work, noise filtering and data analysis procedures are developed for each specific structure (due to the large volume of primary data).

2. Methods

«INTERUNIS-IT» Limited Liability Company has developed SHM-systems for various structures and plants on the basis of A-Line AE system. SHM-system hardware is implemented as a distributed system for data collection and processing with digital data transfer. All components of SHM-system are built from the unified measuring devices and have a common control core. The proposed concept of SHM-system design implies the unification of different components into a single system:

- Monitoring by multiple NDT methods.
- Monitoring of the structure's stress-strain state.
- Tracking of the working parameters of the industrial process [9]

SHM-system is a system designed to assess the diagnostics of hazardous matters in real time without stopping, dismantling and shut down. SHM-system provides information about the status of the monitoring facility in the required quantity and quality for testing. The structure of SHM-system includes the following functional elements: equipment for primary data collection, multifunctional data acquisition and transmission modules, concentrators, galvanic isolation switching cabinet and central computer station (Figure 1).

The equipment for primary data collection may include one or several types of acoustic emission sensors, strain gauges, corrosion rate meters; pH meters, vibration sensors, displacement meters and inclinometers, crack opening sensors; level gauges, thermocouples, pressure sensors, gas analyzers, video recorders and weather station. The specific composition varies depending on the structure and the goal of the monitoring.

Multifunctional data acquisition and transmission module is the main element of this architecture and is a separate device in an explosion-proof enclosure with installed cable entries. Each module contains several measuring units that collect data from external sensors of various types, and, if necessary, supply power to them. In particular, analogue filtering, digitization and digital filtering of signals, as well as calculation of AE parameters (such as arrival time, amplitude, duration, etc.) and the registration of waveforms are performed in a measuring unit intended for operation with an AE sensor. The total number of sensors connected to the module can be up to 6. In addition, module includes a digital processing unit, communication unit and power unit. Multifunctional data acquisition and transmission modules are connected in series with each other by a cable that provides power and relay the data through the cascaded line. The modules are placed, as a rule, near the sensors, which together with digital data transmission, achieve high interference immunity.

The concentrator is a separate device in an explosion-proof enclosure that receives information from several (up to 4) cascaded lines of modules and sends it further via the Ethernet interface.

The switching cabinet of the galvanic isolation carries out a galvanic isolation between the primary supply voltage of 220 V/50 Hz and the secondary direct current (DC) voltage of 48 V, as well as the galvanic isolation between the measuring units.

The central computer station, which is designed for receiving, transmitting, storing and analyzing data and for measuring channels control. The central computer station includes computing devices for data processing and storing, as well as a display and input devices, or a device for communicating with a remote-control terminal.

The SHM-system can also include an automated workstation—a remote terminal used for remote access to a central computer station, monitoring data processing and performing backups of information.

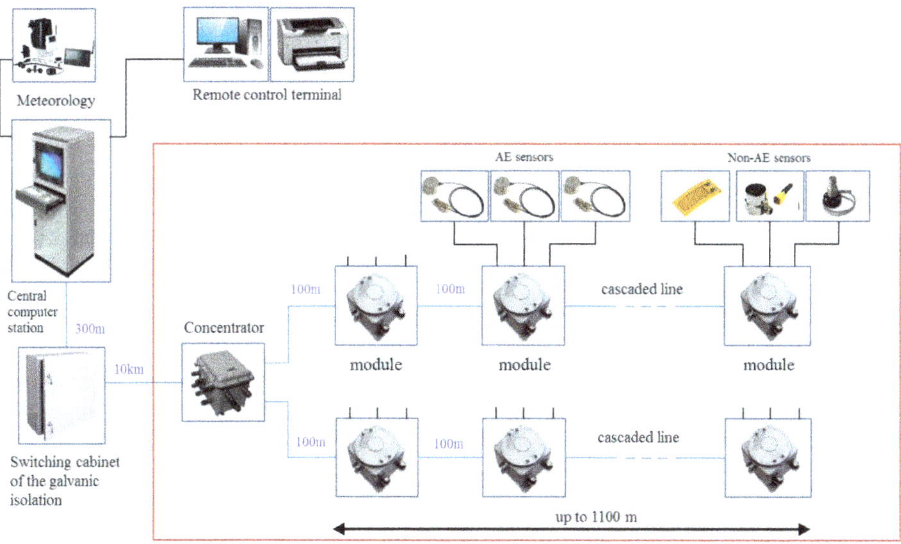

Figure 1. Example of structural health monitoring (SHM)-system structural scheme.

3. Results and Discussion

3.1. Hydrotreater of Oil Refinery Plant

3.1.1. SHM-System Location

Overheating caused by a run-away of the process induced the formation of defects at the hydrotreater at the oil refinery plant. During periodic AE testing, significant AE sources were detected and it was assumed that it is the destruction of the cladding layer. In December 2006, SHM-system was put into operation, the task of which was to ensure the safe operation of the hydrotreater with an extended lifespan until it is possible to replace it.

Structurally, the hydrotreater is a thick-walled vertical steel vessel. The inner surface of the hydrotreater is protected from the influence of an aggressive medium by a cladding layer. The high operating temperatures of the hydrotreater (from 320 °C to 425 °C), the aggressive environment and high operating pressure are factors that can lead to the generation and catastrophically rapid flaw growth, especially when the cladding layer is destroyed, which can induce hydrogen embrittlement and a decrease in the strength of the material.

The main diagnostic method of SHM-system was the method of AE testing. The AE handling part was represented by 18 AE sensors GT200UB (130–200 kHz), grouped into 6 belts, 4 on the hydrotreater shell and 1 belt on each lid. Sensors were mounted on waveguides to ensure the temperature allowed for the sensor is not exceeded. Temperature, vibration and pressure sensors were also installed. A threshold data acquisition algorithm was used. The threshold value was about 40 dB. The analog-to-digital converter (ADC) sampling rate was equal to 2 MHz.

By May 2007, the state of the hydrotreater had deteriorated. It was found to have 9 active zones, corresponding to AE sources with varying degrees of danger (Figure 2). The activity of some zones increased significantly, which indicated a progressive flaw growth and the possibility of the hydrotreater destruction. The source found in zone #5 and characterized by 370 events, was considered to be critically active.

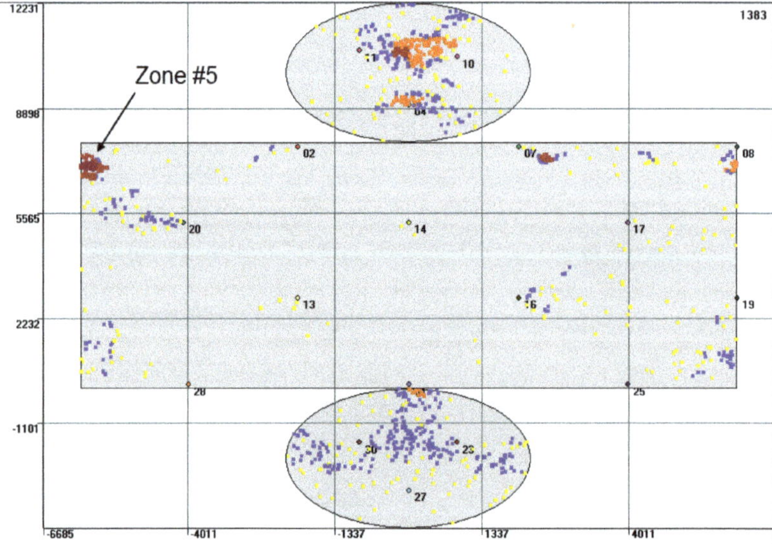

Figure 2. Location map. Color of points mean quantity of location events. Blue points: 1–2 events, yellow: 3–10 events, orange points: 11–25 events, red points: 26–100 events, brown points: more than 100 events.

In order to resolve the issue of the time of decommissioning, it was decided to conduct additional analysis of monitored AE data.

3.1.2. Cluster Analysis of AE Data

After preliminary filtering, a cluster analysis of the AE data was carried out. AE signals were clustered for each measuring channel separately. As a measure of the similarity for each pair of signals, the correlation coefficient of their waveforms was used (threshold value was set to 0.7), or a measure based on the similarity between their AE parameters. As a result, "clusters of signals" were formed [10].

In total, about 500,000 AE signals were analyzed, including 65,000 waveforms. For AE signals registered by the channels testing zone #5, 23 clusters were formed. While conducting cluster analysis, 14 clusters was attributed to process-induced AE events, 4 clusters corresponded to the correlated noise and 5 clusters were characterized as the sources of AE with high probability.

The most representative clusters were formed for signals registered by channels № 8 and № 17 (Figure 3, Table 1). These signals had an amplitude of more than 70 dB$_{AE}$ with a high probability they were generated by AE source located in zone #5.

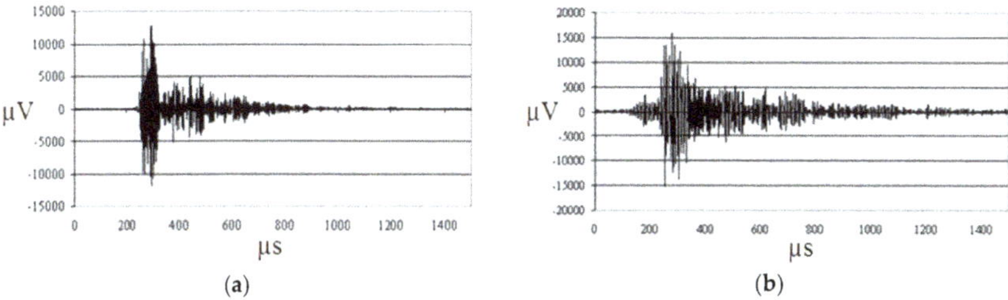

(a)   (b)

Figure 3. Waveforms corresponding to clusters centers of mass, for acoustic emission (AE) signals from channels: (**a**) channel #8; (**b**) channel #17.

Table 1. Parameters of clusters for channels 8 and 17.

Parameters	Channel № 8	Channel № 17
Signals in the cluster	284	138
Mean amplitude, dB$_{AE}$	72	69
Mean rise time, µs	240	379
Mean duration, µs	893	1057

3.1.3. Testing of the Dismantled Object

Thus, the results of the cluster analysis confirmed a previously determined degree of danger of AE source located in zone #5. A final decision to stop the hydrotreater was made. The dismantled hydrotreater was investigated by NDT in order to establish the reasons of its transition to the critical state. Internal examination revealed numerous disintegrations of the cladding layer (up to its complete absence in some areas) (Figure 4). Dimensions of damages were up to 5 mm. Ultrasonic testing revealed numerous discontinuities in welds and the base metal shell. The lengths of discontinuities were 100 mm along and 15 mm in depth. In one of the most active zones, discontinuities were detected both in welds and in the heat-affected zone. Certain discontinuities occurred at different depths, from 26 to 80 mm. These damages demonstrated significant degradation with loss of mechanical properties of the metal, which leads to brittle fracture of the structure [11].

Thus, the use of SHM-system allowed, on the one hand, to extend the lifetime of the hydrotreater by 6 months and on the other hand, to prevent the destruction of the hydrotreater during operation with all the ensuing consequences.

Figure 4. Breach of the cladding layer.

3.2. AE Method Application for Bridge Monitoring

3.2.1. Structure Description

Experimental studies were conducted on a highway bridge across a river. It is a 7-span metal bridge with reinforced concrete supports. The foundations for pylons are piled. The length of the bridge is about 700 m. The bridge has 2 traffic lanes. As a result of the abnormal operating impact, the supports #1 and #5 sank in by 112 mm and 44 mm, respectively. After that it was decided to install an SHM-system on this bridge.

Preliminary studies consisted of 3 main stages, the objectives of which were: determination of the AE parameters of noise signals, determination of propagation parameters of the AE signals and determination of the optimal arrangement of the sensors on the bridge construction elements [12,13]. Investigation of the possibility of AE signals detection used Hsu-Nielsen source for location and filtration.

3.2.2. Analysis of Noise Parameters

The first experiment consisted in the recording and analysis of noise at various sections of the bridge. As a data collection system, a 4-channel external USB I/O module "E20-10", manufactured by "L-Card" company (Moscow, Russia), was used. This system provides the consistently recording data with the sample rate 2.5 MHz for every channel. Two sensors of GT200 type (130–200 kHz) were connected to the module, which were installed on the bridge. The signal from the sensor went to the preamplifier with a gain of 26 dB and a frequency filter of 30–500 kHz.

The AE data were taken in 3 different zones of the bridge structure (Figure 5).

Zone I: In the area of the expansion joint near the shore support (Figure 5a), sensors were installed inside the beam structure near the bolted joints. The distance between the sensor and the expansion joint is 5 m. Zone II: In the area of attachment of the beam and the support (Figure 5b), AE sensors were installed on the lower girder and on the support. Zone III: In area of the exit hatch section of the box beam (Figure 5c), AE sensors were installed inside the beam structure at a distance of 3 m from each other symmetrically with respect to the hatch.

For all signals from passing vehicles, the energy is concentrated in the frequency region up to 60 kHz.

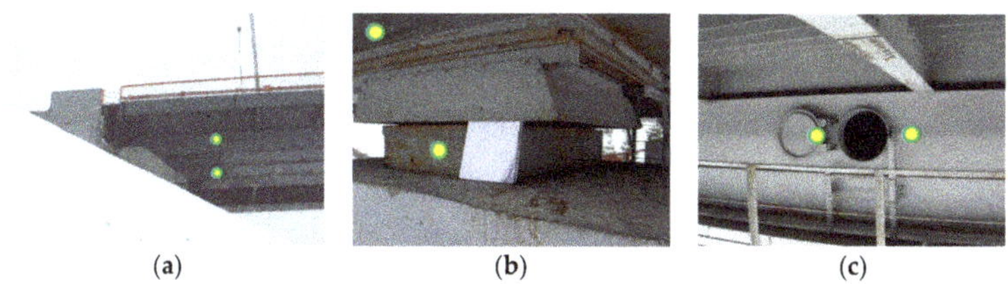

Figure 5. Experiment scheme: (**a**) Zone I; (**b**) Zone II; (**c**) Zone III. AE sensors are marked as yellow-green circles.

Vibration of the bridge from passing vehicles was the most common type of noise (Figure 6). Some characteristic features of these signals are a long duration (1–3 s) and relatively high amplitudes of about 56–90 dB_{AE}. The primary reason for the appearance of these signals is the fact that the passage of the cars leads to vibration and friction of poorly fixed structural elements.

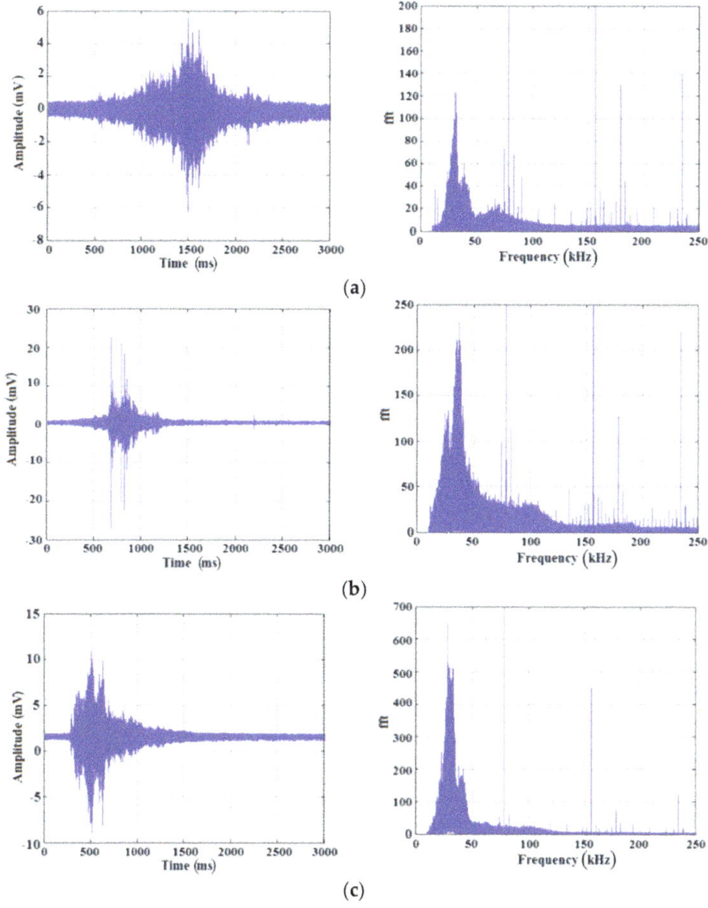

Figure 6. Acoustic emission (AE) signals from passing vehicles: (**a**) Zone I; (**b**) Zone II; (**c**) Zone III.

Signals caused by impacts of structural elements have amplitudes similar to signals from passing vehicles, of the order of 56–90 dB$_{AE}$ but they have 1–2 times shorter duration, not exceeding 50,000 μs (Figure 7). The energy of the signals, as a rule, lies in the frequency range up to 200 kHz. The primary cause of these signals is the vibration of weakly fixed structural components of the bridge, as a result of which their collision occurs. An example may be the impact of a ladder or hatch on the bridge girder or the impact of wheel sets on the expansion joint [14].

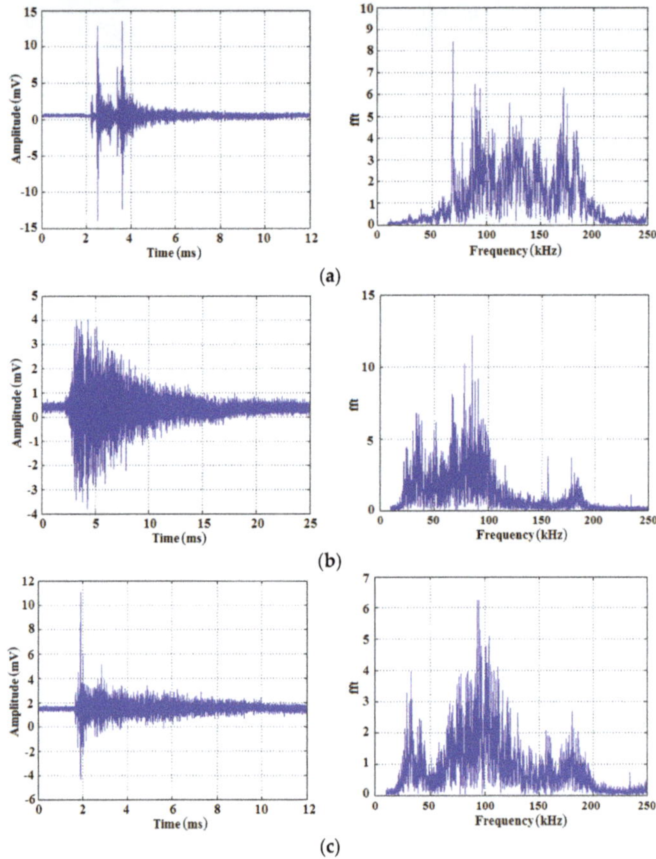

Figure 7. AE signals caused by impacts of structural elements: (**a**) Zone I; (**b**) Zone II; (**c**) Zone III.

3.2.3. Analysis of AE Signals Propagation

Hsu-Nielsen source was used to estimate signal propagation properties. The measurements in this experiment were carried out using the A-Line 32D system (Interunis-IT) with threshold data acquisition. The ADC sampling frequency was 2 MHz. The following values were obtained: AE wave propagation speed is 2838 m/s, attenuation coefficient in the far field zone is 1.65 dB/m. Based on the obtained values of the attenuation coefficient, noise level and amplitudes of signals from the AE signal simulator, the maximum distance between the AE sensors for linear or planar location was determined. This value lies in the range from 11 to 16 m.

An acoustic contact quality was also estimated in zone II (Figure 5b). In the course of the experiment it was revealed that the signals simulated in the girder region are not detected by the AE sensor mounted on the support, which indicates that there is no acoustic contact. Thus, for a full monitoring of the bridge, separate AE testing of both the supports and the main upper part of the bridge is necessary.

3.2.4. Identification of AE Signals from the Hsu-Nielsen Source

Simulation of AE signals by the Hsu-Nielsen source was carried out at various points of the lateral beam of the bridge structure inside one of the bridge sections. The distance from the sensor to the Hsu-Nielsen source was from 0.15 to 15 m. Totally about 100 pulses were excited (Figure 8).

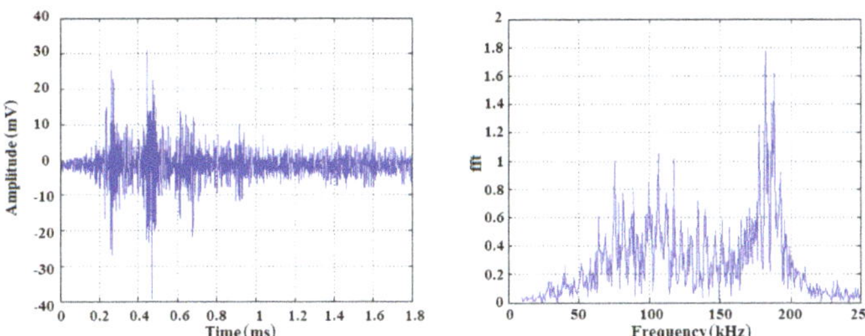

Figure 8. Signals from Hsu-Nielsen source at 1 m distance from AE sensor.

For detection of signals from the Hsu-Nielsen source, a planar location was constructed (Figure 9). It is revealed that the signals from Hsu-Nielsen source are with certainty located in their actual places and form compact clusters, in contrast to noise. This confirms the possibility of identifying signals from defects based on location data.

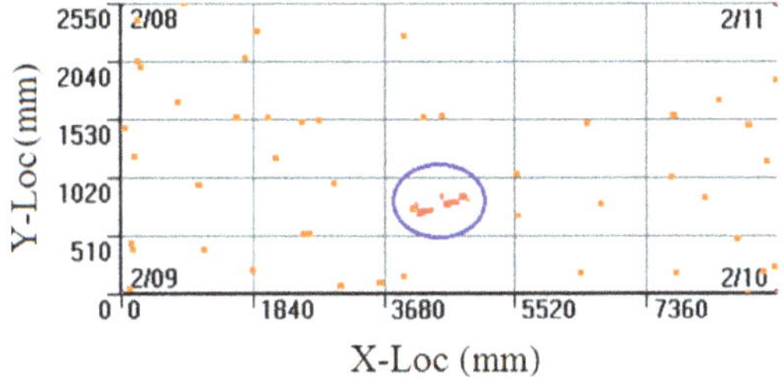

Figure 9. Planar location. Hsu-Nielsen source signals are marked by red, noise is marked by orange. Color of points indicates total number of events in this coordinates. Yellow points: 3–10 events, orange points: 11–25 events, red points: 26–100 events. Blue circle – location cluster of signals from Hsu-Nielsen source.

Since the bridge is a structure of testing characterized by a high level of acoustic noise, a locational cluster corresponding to a defect can be masked by a large number of false locations of noise events. The complexity of the detection algorithm lies in the fact that the values of the AE parameters of the noise and useful signals vary in fairly wide and overlapping ranges. This is due to the variety of noise sources and different distances between the sensor and points where the defect or simulator is placed - from 0 to 15 m.

The difference of parameters is best shown in the two-dimensional plots of characteristics, for example, on the scatter plot diagram "Counts/Duration versus Duration" (Figure 10a) and the scatter plot diagram "Energy versus Counts/Duration" (Figure 10b). As can be seen from the chart, the AE impulses emitted at the distance of 7 m from the sensor have the parameter distribution other than the noise distribution. When the distance increases over 7 m, the differences disappear, the AE impulse parameters become close to those of noise [13].

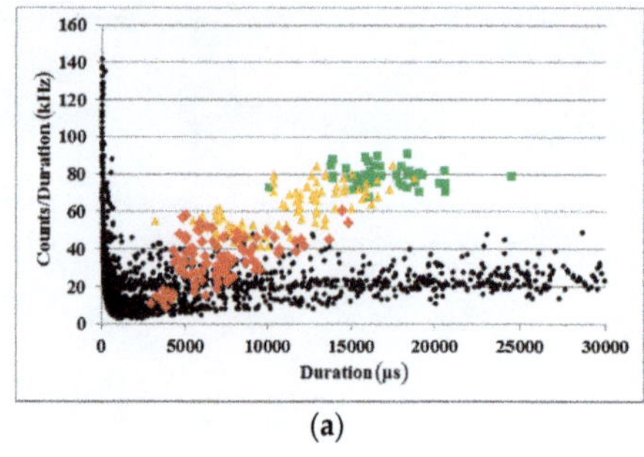 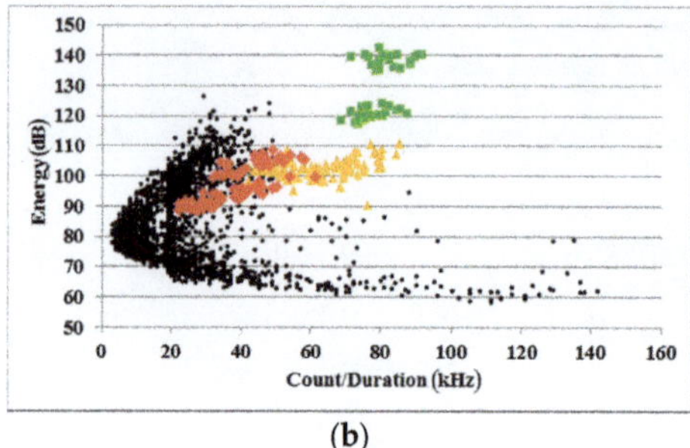

(a)   (b)

Figure 10. (a) Scatter plot diagram of AE signal energy and duration; (b) scatter plot diagram of amplitude and parameter "counts/duration." Impulses corresponding to noise are marked black, while impulses from the Hsu-Nielsen source emitted at distances of 0–2 m, 2–7 m and 7–15 m from AE sensors are marked green, yellow and red, respectively.

3.2.5. SHM-System

SHM-system for this application was designed and installed on the bridge. AE sensors, vibration transducers, strain gauges, temperature sensors, modules for collecting, measuring and processing signals were installed on the span of the bridge structure (box beams). Low-frequency AE sensors, inclinometers, crack opening sensor were installed on the supports. In addition, SHM-system is equipped with video cameras to monitor the traffic situation, a weather station to monitor wind loads and to determine the presence of precipitation for AE signals filtering and a communication system.

3.3. High-Temperature Process Pipeline

On the active high-temperature pipeline without decommissioning, studies were produced to assess the feasibility of conducting AE monitoring. Application of AE monitoring of high-temperature or cyclic thermal loaded structures is an effective method for defects detection. Several interesting applications are described in the papers [15–18]. The tested object consisted of pipe sections with a diameter of 50 to 80 mm with a total length of 8 m. Individual pipe sections were connected with both flanged and welded joints. The working medium of the pipeline was molten salt, the temperature of which could reach 600 °C. To synchronize the AE data with the operating mode of the pipeline, one thermocouple was installed. 8 AE sensors were installed through waveguides. Simulations of AE signals were carried out once per day by emitting from a sensor to monitor the attenuation factor.

When the pipeline was cold, the propagation velocity and attenuation coefficient were measured, which were 4900 m/s and 1.7–2.7 dB/m, respectively. Then the pipeline was put into operation. AE data were taken in various operating modes of the pipeline for 63 days, including in the total amount for 15.5 days in a stationary high-temperature regime. The process of heating the pipeline was long and took up to several days. During the heating process, the melt began to flow into the pipeline. The most suitable was the use of the frequency range 100–500 kHz and the threshold in the region of 40–55 dB_{AE}. It was found that the activity and amplitudes of noise increased significantly with the pipeline temperature rising at a rate exceeding 1 °C/min, especially when clamping fixtures for waveguides were used (Figure 11a). Amplitudes of noise reached values of up to 100 dB_{AE}. It is assumed that the occurrence of noise is associated with the thermal expansion of the pipeline, which induced friction on the thermocouple and other items on the surface of the pipe. During the work in the stationary mode, the noise level was reduced to an acceptable one and amounted to about 50 dB_{AE}.

It is revealed that the attenuation coefficient increases with the accumulation of sediments on the inner surface of the pipeline and can reach 58 dB/m. With this level of attenuation, the maximum distance between the AE sensors could not be more than 0.7 m. The value of the propagation velocity for acoustic waves was unchanged.

In the course of the work, the melt leakage was observed as a result of the destruction of the flange joint gasket. In this case, the AE system recorded an increase in the duration of the AE signals a few hours before the leak invisible under the layer of thermal insulation began to be observed visually (Figure 11b).

As a result of the conducted studies, the possibility of AE monitoring of the most critical sections of the pipeline was confirmed: the detection and location of crack-like defects and detection of leaks on this pipeline by the AE method can be accomplished.

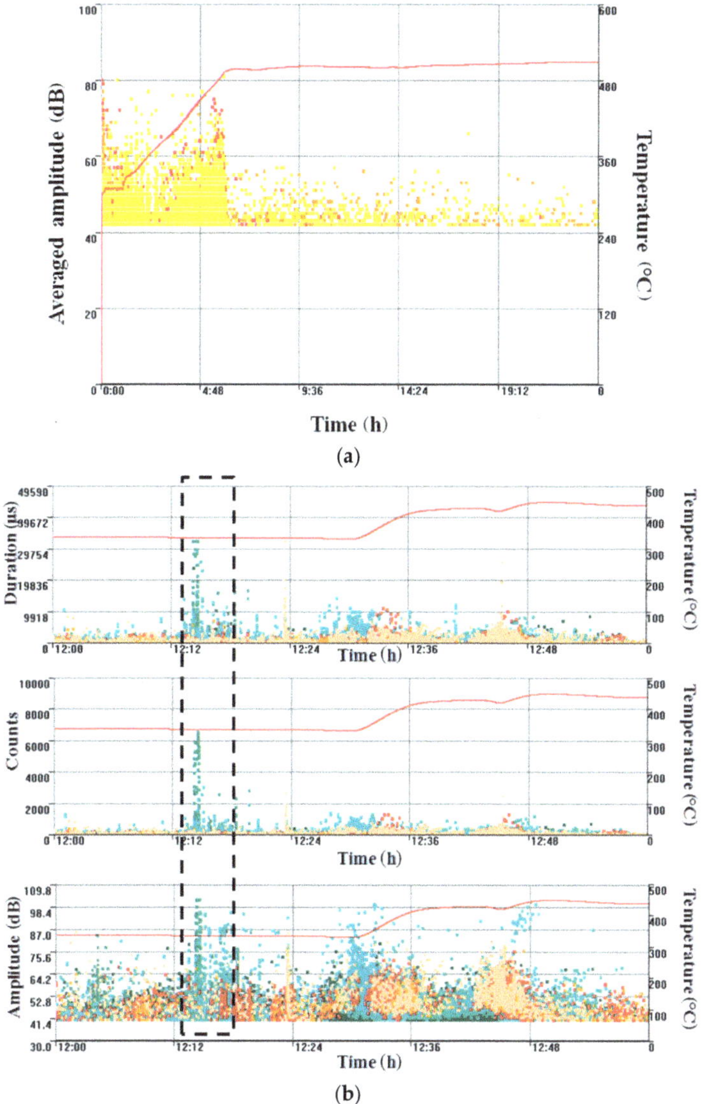

Figure 11. Amplitudes of noise depend on the pipeline temperature (**a**); the melt leakage (**b**). Colors of points indicates different channels.

3.4. SHM-System for Oil Refinery Equipment with Defects Presence

At an oil refinery, non-permissible defects were found in the upper part of the gas adsorber of the circulating gas (commissioned in 1970). From 1999 to 2009, the lamination was known to exist in the main upper end plate (the area of the lamination ~0.4 m^2, the depth of the lamination: 6.1–12.4 mm). On the same unit, the defect of the planar type (crack) in the weld of the shell to the end plate (length: 140 mm, depth: 12–14 mm) was also present (Figure 12). This gas adsorber has the following parameters: working pressure: 4.2 MPa, operating temperature +45 °C, wall thickness: 22.0–24.0 mm, the inner diameter: 1.156 m and the height: 16.45 m.

A preliminary decision was made to install a monitoring system for the upper part of the adsorber. The main task of the monitoring system was to ensure safe operation of the adsorber before its replacement within a year.

Prior to installation of the monitoring system during the overhaul in the spring of 2012, its AE testing was carried out (Figures 12 and 13). AE sources of class I and II were registered, the locations of which were correlated with the place of previously detected defects on the vessel head of the adsorber (Figure 12). This allowed us to assume the possibility of further expansion of the lamination zone during the life extension period of the adsorber.

Based on the results of the AE testing, the previously approved decision on the installation of the monitoring system for the vessel head of the gas adsorber was confirmed. The expanded SHM-system included 4 AE sensors and a weather station (Figure 14). The threshold data acquisition approach was used, and the threshold value was about 45 dB$_{AE}$.

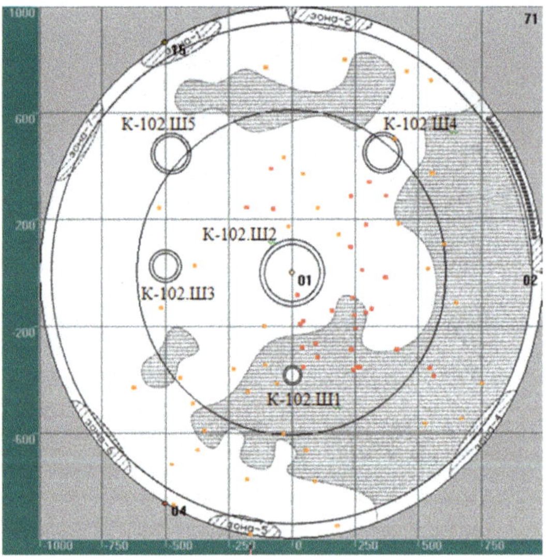

Figure 12. Previously found defects of gas absorber head plate: lamination in the vessel head (gray areas) and locations of the registered AE source (orange). Color of points indicates total number of events in this coordinates. Yellow points: 3–10 events, orange points: 11–25 events, red points: 26–100 events.

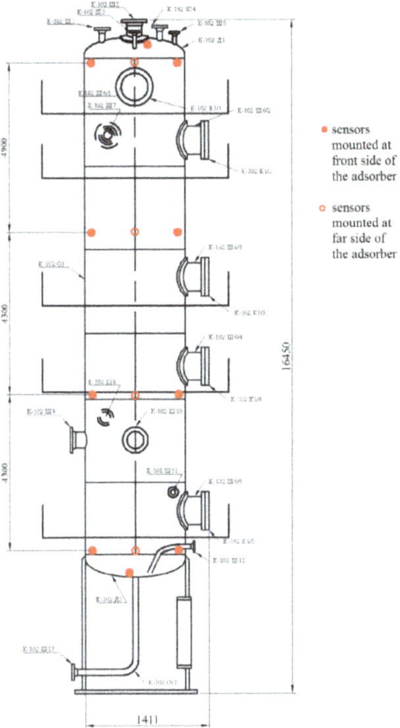

Figure 13. Gas adsorber schematic drawing with places of AE sensors mounting.

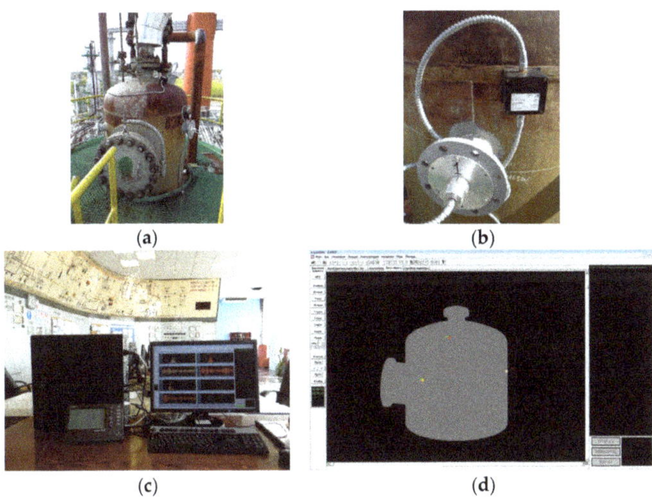

(a)

(b)

(c)

(d)

Figure 14. SHM-system for gas adsorber: (**a**) overall view of the head part; (**b**) AE sensor mounting place; (**c**) workplace of operator; (**d**) dialogue box of the software A-Line Mon. Color of points indicates total number of events in this coordinates. Green points: 1–2 events, yellow points: 3–10 events, orange points: 11–25 events, red points: 26–100

3.5. Roller Bearings of Rotary kilns

3.5.1. Tested Object

The possibility of conducting AE testing of roller bearings of rotary kilns without their decommissioning was investigated (Figure 15). Using AE monitoring for rotating equipment is an alternative solution to vibrodiagnostics [19,20]. In the case of slowly rotating elements, the use of the AE method is more effective for the diagnosis of rolling bearings, rotating shafts, rollers and others [21,22]. Experimental work was carried out at an alumina plant in the sintering and calcination shops and also in the repair shop.

Figure 15. Rotary kiln on paired roller bearings. 1: kiln; 2: roller bearings; 3: support foundation.

The kiln is a cylinder with a diameter of up to 5 m and a length of up to 185 m, installed with a slope. The kiln is supported by 7–8 pairs of roller bearings. Each pair of roller bearings consists of 2 support blocks, mounted on a common foundation. Each block consists of a shaft pressed onto it by a supporting roller and 3 bearing assemblies mounted in 2 bearing housings (Figure 16). The rotation period of the kiln is 50 s; the rotation period of the support roller is 12.5 s. The lifetime of the monitored supports was from 1 to 13 years.

Roller support is a complex load carrying element of the kiln. In case of its breakdown, an emergency stop of the equipment occurs, followed by the replacement of defective components. Of the 200 supporting blocks at the plant, on average 5 blocks per year are replaced because of broken bearings and 1 more block due to the fracture of the shaft. Therefore, it is necessary to ensure the detection of defects of all its elements such as shaft, roller and bearings.

The roller is a part of the support block and is in direct contact with the kiln. Thus, the main defects on it are scoring and metal chipping from rolling contact under load.

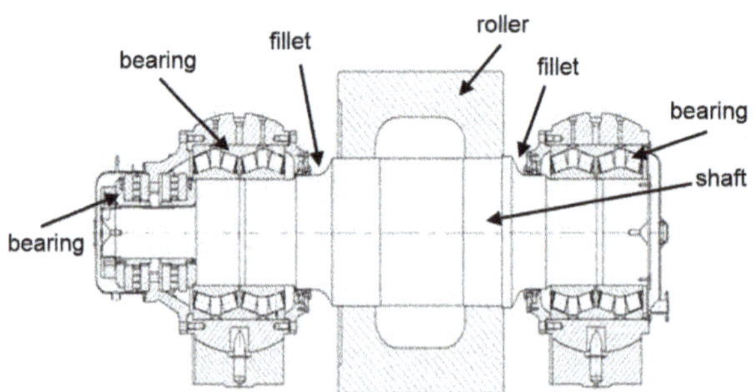

Figure 16. Roller support.

The shaft is a cast steel component. Since roller bearings are operated under unfavorable temperature conditions, with a high level of vibration and under cyclic loading, the most likely defect of the shaft is the formation of fatigue cracks. Since the shaft is a monolithic all steel construction with numerous protrusions and thickness differences, it is difficult to carry out the NDT of shaft by traditional scanning methods, even for roller bearings in decommissioned state. The main bearing elements include an outer ring, an inner ring, cylindrical rolling bodies and a separator. Destruction of the bearing occurs for 3 main reasons: incorrect mounting of the bearing, insufficient quantity or low quality of lubricants, as well as fatigue failure of the bearing. Due to the low angular velocity (1 rotation in 12.5 s), a well-known method of vibrometry is not quite suitable for testing the roller bearings and AE is expected to provide a good solution.

3.5.2. Preliminary Research

AE sensors GT200 was used with a working frequency range of 130–200 kHz. The signal from the sensor was fed to the preamplifier with a gain of 26 dB and a frequency filter of 30–500 kHz. As a data collection system, a 4-channel external USB I/O module "E20-10," manufactured by "L-Card" company (Moscow, Russia), was used. The output data of this module was long time waveforms of the AE signal with a sampling frequency of 2.5 MHz.

3.5.3. Noise Investigation

In the case of roller bearings of a rotary kiln in the operating mode, the noise level in the audible range is quite high. However, in the range above 30 kHz, due to the point dry contact between the shroud of the kiln and the roller, the acoustic noise, as a rule, was no more than 32 dB$_{AE}$ at the most unfavorable point of the structure: at the feed point of the batch. The only significant external source of acoustic noise is scoring and roughness on the surface of the roller or kiln bandage. When it touches the place of scuffing, an AE signal with amplitude of up to 100 dB$_{AE}$ arises. However, such noise signals can easily be excluded from further analysis, since their registration occurs with a certain period equal to the period of rotation of the roller support, or the kiln (Figure 17).

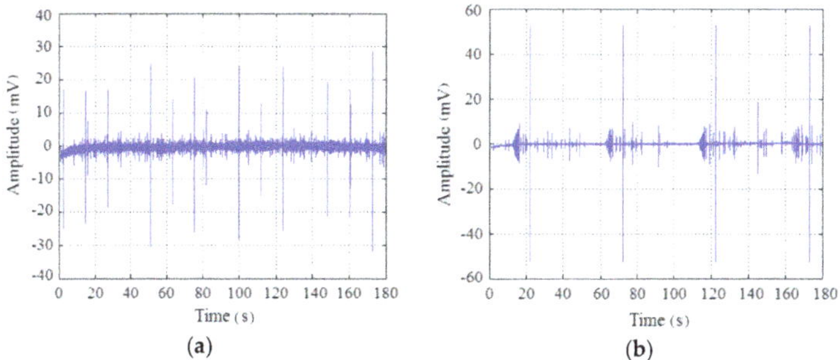

Figure 17. AE signals with periods (**a**) 12.5 s and (**b**) 50 s associated with scoring and roughness on the roller surface or a kiln bandage.

3.5.4. Calibration Measurements

In the repair shop, measurements were made to determine attenuation and propagation velocity of the acoustic signal on the decommissioned support block with the end caps removed. To simulate the sources of AE, an electronic simulator was used. The emission was produced on the shaft in the region of the fillet and at the ends of the shaft, as well as on the inner and outer rings of the bearings (Figure 18).

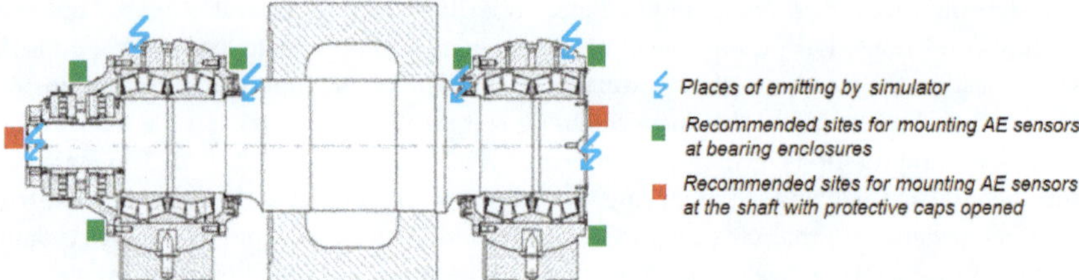

Figure 18. Scheme of the roller bearing with the protective caps removed. The points of emitting by simulator and the recommended positions for the sensors mounting.

Within the shaft, the attenuation coefficient was 5 dB/m. The presence of a high attenuation value (30–35 dB) is revealed at the boundary between bearing housings and the shaft.

The obtained results revealed that the AE testing should be performed as follows (Figure 18). For full AE testing of all 3 bearings, use the zone location with the accuracy of the element of a specific bearing. For this task, it is required to place the AE sensors at all available access points, namely 3 sensors on each of the 2 bearing housings (Figure 19a). Since the length of the shaft is small (4050 mm), it is enough to provide access to the shaft by removing the covers and to install one sensor at each of 2 ends (Figure 19b) for its full AE testing including linear location. Slow rotation of the sensor at the shaft ends does not cause significant technical difficulties with the duration of the AE testing of about 1 h.

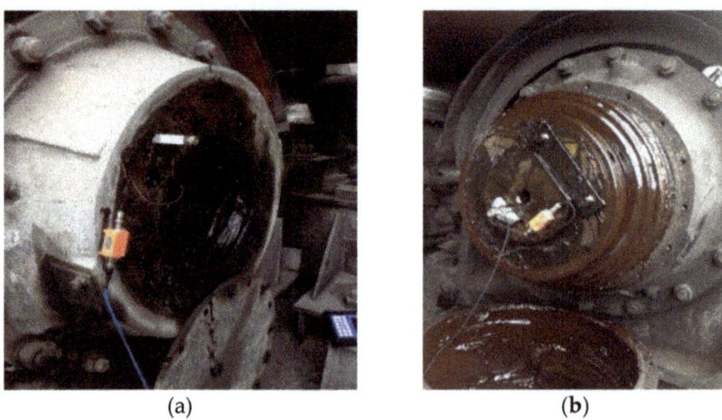

(a) (b)

Figure 19. Sensors mounted at bearing housing (**a**) and at the end of the roller bearing (**b**) with the protective caps removed.

3.5.5. AE Testing

The next stage of the work is to carry out AE testing of the support without taking the object out of operation. The AE data were taken with a regular rotation of the kiln and the support for 1–2 h. Five support blocks with open caps were examined, where both the bearings and the shaft were directly accessible. Also, 10 support blocks with closed caps were tested, where access was available only to the bearings.

3.5.6. AE Testing of Bearings

In one of the bearings, the lubricant dried, which led to intense friction and a rattling in the audible range. AE sensors also recorded signals with an amplitude of about 100 dB_{AE} and a duration of more than 100,000 μs (Figure 20).

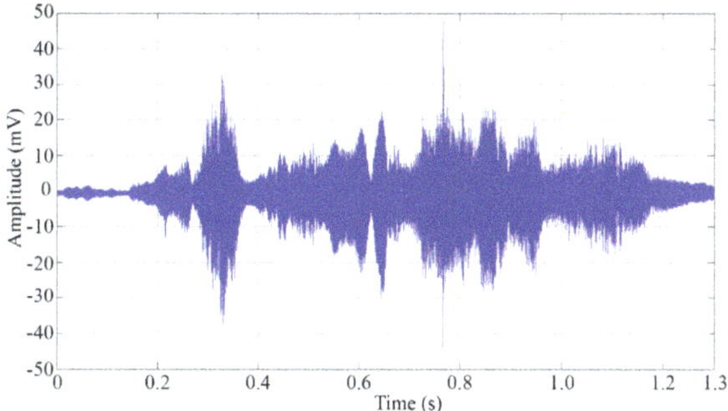

Figure 20. AE signal recorded on the bearing housing. Drying of lubricant.

When defects are formed, for example, fatigue cracking or crack, a significant change in the nature of the AE signal flow occurs such as the activity grows, amplitude trends appear, and, most importantly, there is a periodicity of appearance due to the fact that acoustic signals at this stage occur during periodic collision of the bearing roller with place of damage [23].

A moderately active source of AE was detected on the bearing of one of the supports (Figure 21a), which, due to a low characteristic amplitude level, did not undergo the confirmation procedure. A more powerful periodic source of AE (about 100 dB$_{AE}$), found on another support (Figure 21b), was a crack in the outer ring of the bearing.

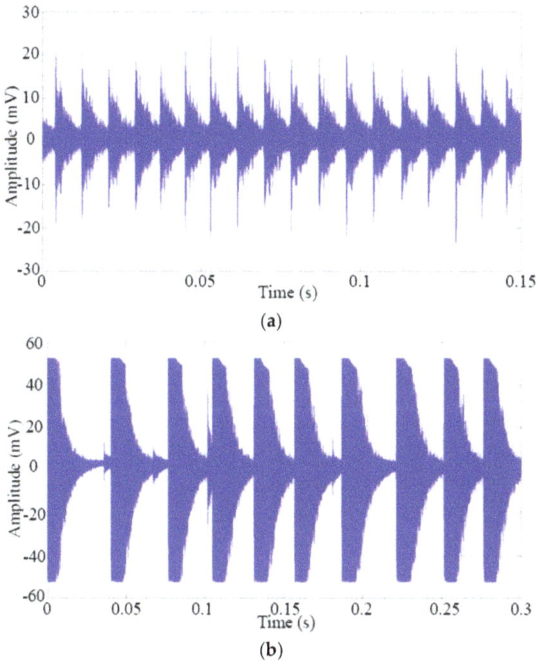

Figure 21. Periodic AE signals registered on the bearing housing. (**a**) Moderately active source of AE; (**b**) external ring crack.

3.5.7. AE Testing of the Shaft

The AE signal, characteristic for the friction process, is a continuous random process characterized by a distribution of pulse amplitudes close to normal, which is not typical for cracks. It is found that higher RMS amplitudes of the acoustic signal are observed in roller bearings with a long service life, which allows us to conclude that the RMS amplitude of the AE signal characterizes the integral degree of wear of the shaft surface [24]. Figure 22 shows the signals obtained in the diagnosis of roller shafts that were in service for 2 and 9 years. The corresponding RMS amplitudes are 0.9 mV and 2.2 mV.

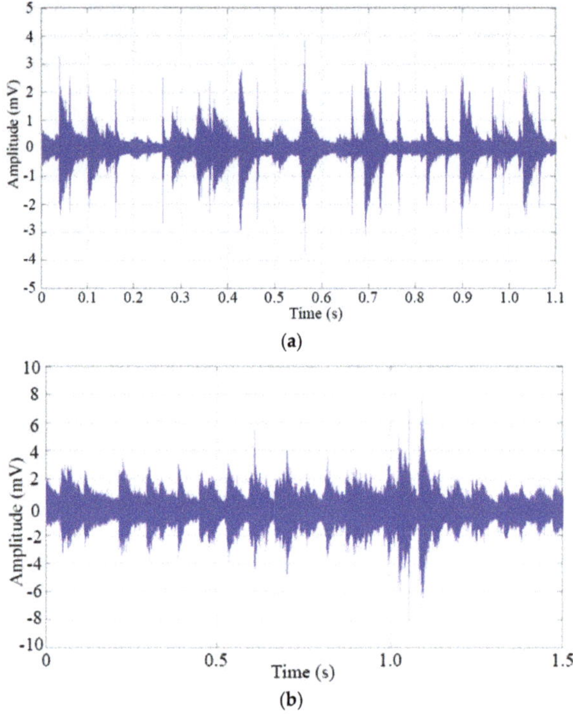

Figure 22. AE signals obtained during the diagnostics of shafts of roller bearings being in operation for (**a**) 2 and (**b**) 9 years. The corresponding RMS amplitudes are 0.9 mV and 2.2 mV.

The appearance of individual pulses of large amplitude (up to 81 dB$_{AE}$) against the background of a continuous signal may indicate the beginning of the destruction of the surface of the shaft in the support block during frictional contact (Figure 23).

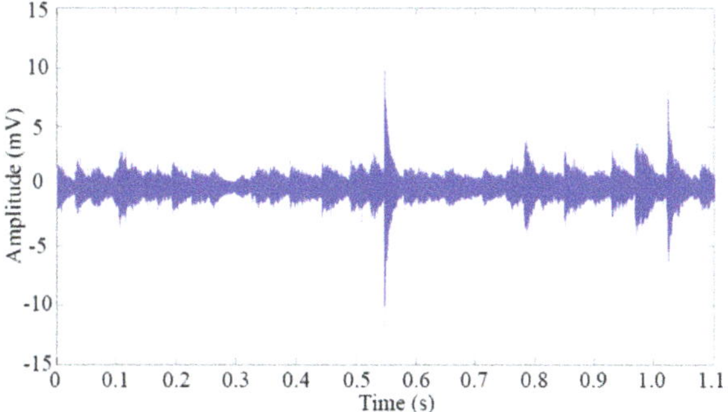

Figure 23. AE signals received during the diagnosis of shafts. The beginning of the destruction of the shaft surface during frictional contact.

Acoustic emission that is caused by the formation and growth of cracks has a fundamentally discrete and aperiodic nature. When the crack grows, as a rule, high-amplitude signals are generated, with a certain characteristic distribution of the parameters of the AE data stream. AE testing of the support blocks on 2 shafts being in operation for 11 and 13 years revealed AE sources with amplitudes up to 78 dB_{AE} and activity of about 10–20 s^{-1} (Figure 24), which, judging from the waveforms and the distribution of the AE parameters, correspond to active fatigue cracks. Analysis of the linear location showed that the sources of AE are located in the region of the fillets, that is in places with the greatest concentration of stresses. The identified sources of AE turned out to belong to the I and II hazard classes and for this reason did not undergo the confirmation procedure.

At present, the design of SHM-system for the roller bearings testing is under consideration.

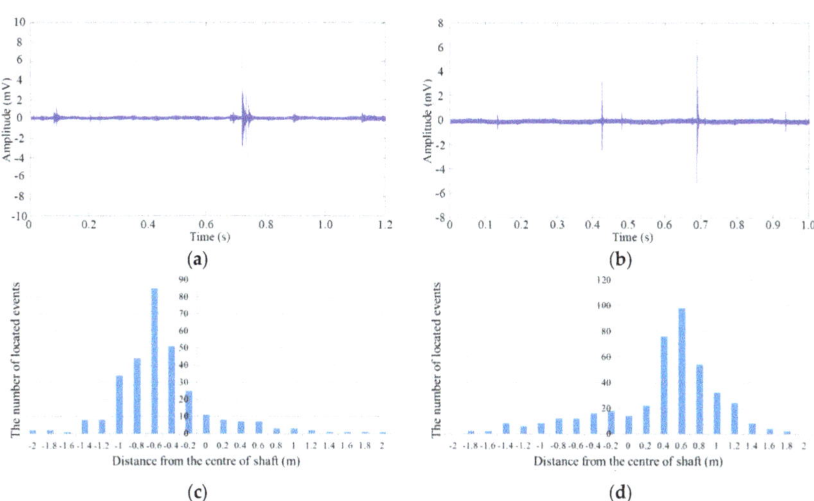

Figure 24. Waveforms of AE signals recorded on shafts of 2 roller bearings (**a**,**b**) and linear location results (**c**,**d**).

3.6. Dragline Excavator

On the territory of Russia, handling equipment is often used in difficult climatic conditions. Fluctuations of the temperature, cyclic operating loads lead to the nucleation and propagation of the defects in a load-carrying structure component, which reduces its residual life and, as a result, can lead to an emergency destruction of the structure.

Using AE monitoring can prevent a sudden occurrence of an accident, reduce economic damage and prevent human casualties. Despite the fact that the operation of the lifting equipment is accompanied by a high level of acoustic noise, AE monitoring is possible by optimizing the location of the sensors and applying special methods of data processing [25,26].

Studies were conducted to assess the feasibility of developing an AE monitoring of metal structural elements of a walking dragline excavator. The work of such dragline excavator is carried out all year round, continuously in 3 shifts. Most of the dragline excavators on the mine where the studies were conducted worked out their normal service life; however, it continues to be actively used.

The most stressed and, as a consequence, the most susceptible to the formation of defects are the struts. Emergency situations with dragline excavator lead to downtime, whose duration, for example, in 2016 at one mine was 812 h. Currently, in accordance with the recommendations of manufacturers two times a year periodic diagnostic tests of dragline excavator by ultrasonic method are performed.

Research work was carried out on booms and struts of dragline with bucket capacity—20 m^3 and boom length—90 m, which previously had damage at the left strut near the weld (Figure 25). The work was carried out without dragline excavator decommissioning. A 30-channel A-Line DDM AE system and GT200 AE sensors with a working frequency range of 130–200 kHz were used. A-Line DDM system provides the principle of digital data transmission. The system consists of several sequentially connected measuring lines. The amplification of AE signals, filtering, AD conversion and threshold impulse detection and calculation of AE parameters are carried out in the AE module, which is located next to the AE sensor directly in the tested structure. AE sensors were installed on the main structural elements of the dragline: 3 on the left and right struts, 8 on the upper boom belt and 4 on the lower boom belt, 4 on the lower struts and 3 on the column. The distances between the AE sensors were from 5 to 15 m. In addition, 5 AE sensors were mounted on the anchor links where the zone location was carried out, since they consist of structural elements between which there is no acoustic contact. AE monitoring of a rope was not carried out, as the damage to the rope does not lead to prolonged downtime. In addition, the rope is a movable part, that greatly complicates its AE testing. Data was collected directly during the work of the dragline.

At the first stage, the frequency range and threshold were selected and the acoustic properties of the monitored object were determined. Two frequency ranges were tested: 30–500 and 150–500 kHz. When using a wide frequency range (30–500 kHz), an extremely high AE activity was observed, which did not have a visible dependence on the operating mode of the dragline excavator, that is, on the value of the load. Such noise is associated with the continuous operation of the equipment (working winch, loose structural elements, engine operation, vibration, etc.).

Therefore, a filter of 150–500 kHz was used. Complete noise detuning by the threshold method was still impossible, since the noise reached 100 dB$_{AE}$. As a compromise, a threshold value of 50 dB$_{AE}$ was chosen. With such settings, the nature of AE activity became synchronized with the cycle of operation of the dragline excavator (Figure 26). The greatest activity was observed with excavation of the ground and with unloading the bucket, that is at the loading of the structure, when cracks could grow and at structure unloading, when the crack edges can rub each other.

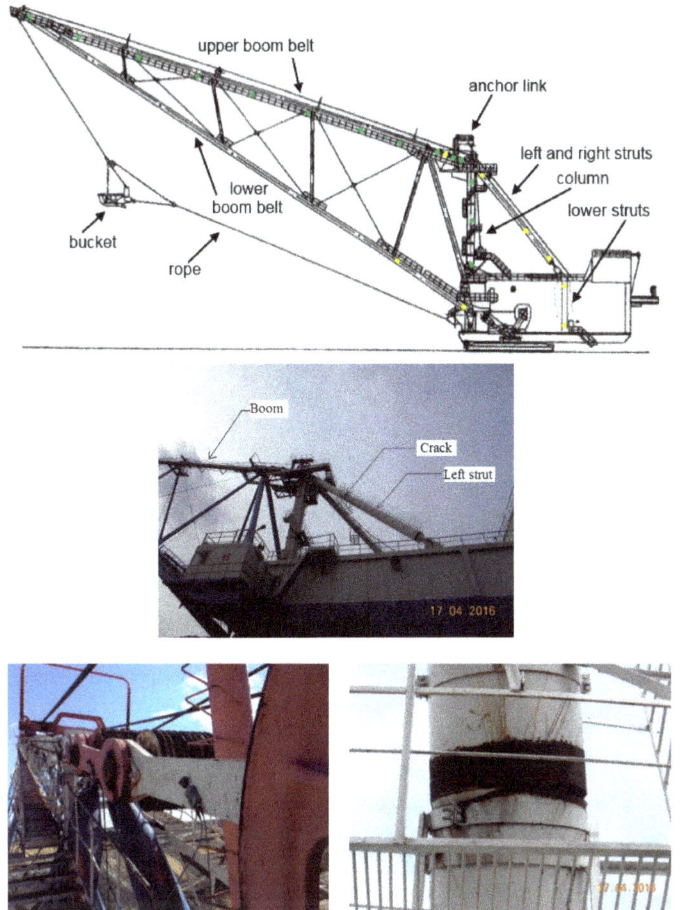

Figure 25. Tested dragline excavator.

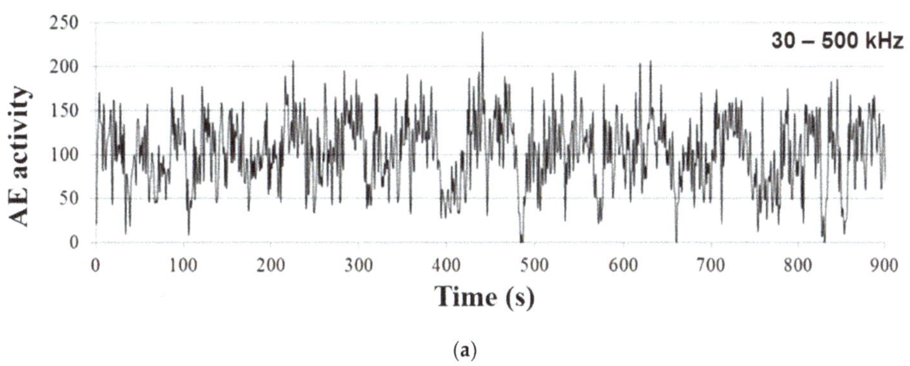

(a)

Figure 26. *Cont.*

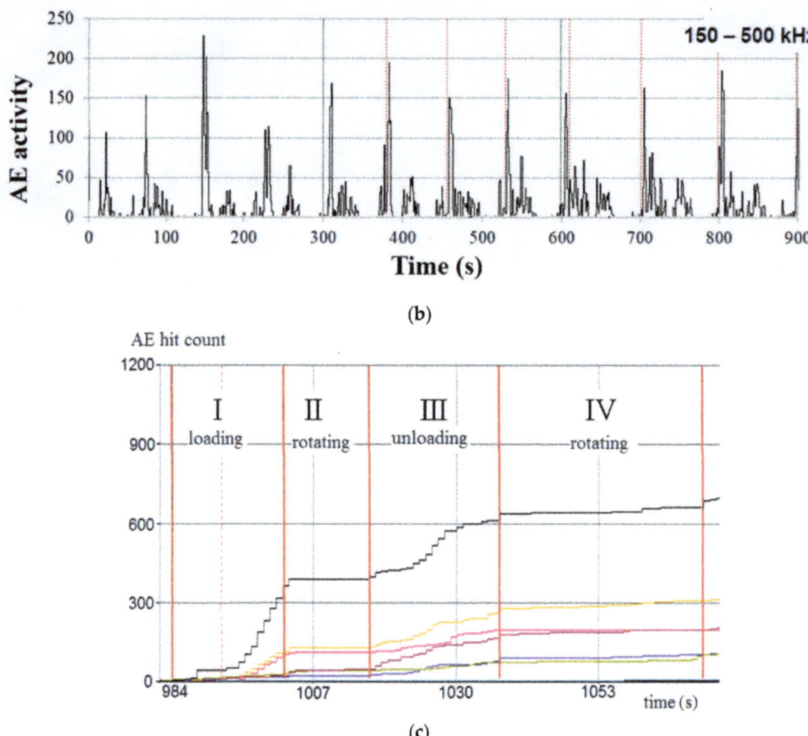

(b)

(c)

Figure 26. AE activity when using 30–500 kHz (**a**) and 150–500 kHz (**b**) frequency filters. Correlation of AE hit count with the cycle of operation of the dragline excavator when using 150–500 kHz frequency range (**c**).

Measurement of the acoustic properties of the monitored excavator was carried out using the Hsu-Nielsen source. The following values were obtained: velocity was 3200–3700 m/s and attenuation was 1.5–2.8 dB/s. The recommended maximum distances between AE sensors that ensure a linear location are from 4.5 to 9 m.

At the second stage, within 2 days, the AE data was collected directly for subsequent detailed analysis. The distribution of the AE signal amplitudes was analyzed. For all structural elements except for the struts, the differential distribution of the amplitudes was exponential, that is corresponding to a random noise process. On the left and right struts in the range of amplitude values of 80–85 dB_{AE}, the distribution was significantly different from the exponential, which indirectly indicated the presence of a defect.

The results of linear location analysis were as follows (Figure 27). On the lower boom belt, there were no sources of AE. Several sources were found on the column and the upper boom belt, which turned out to be noise, since they were localized near the winch attachment points, or were inactive. On the left and right struts, a linear location revealed 3 sources of AE on each, near which there were no potential sources of noise. Signals appeared at the stages of loading and unloading. The amplitude distributions presented above also make it possible to classify these sources as defects.

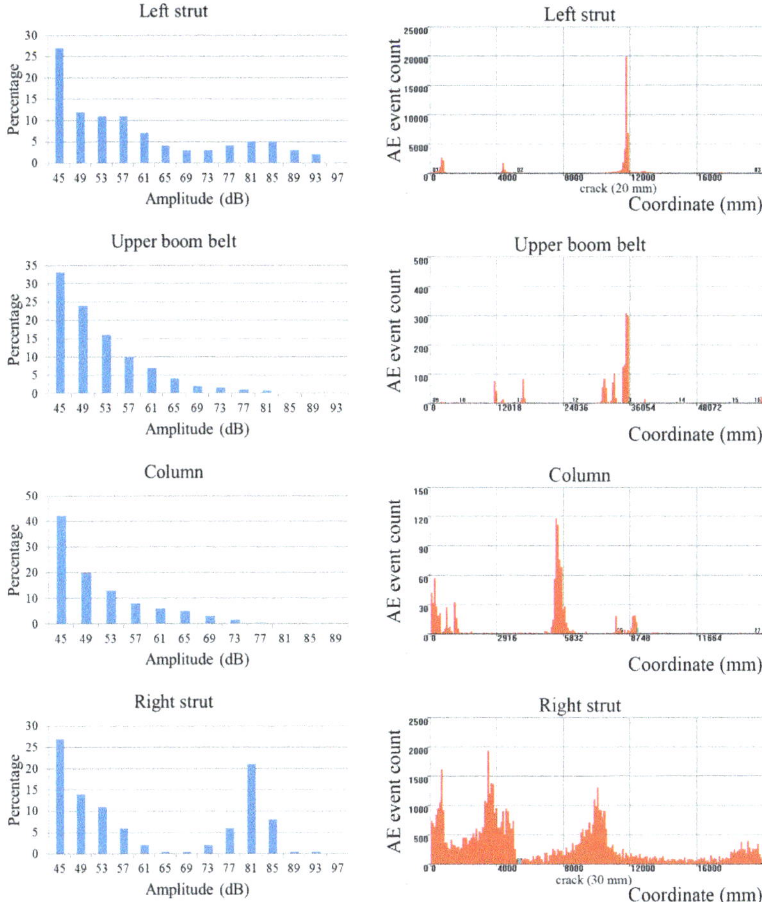

Figure 27. Results of acoustic emission location.

In addition, the spectrograms of the signals on the struts were analyzed, which made it possible to determine the arrival times of different frequency components (Figure 28). It was confirmed that the signals were emitted by impulsive sources and propagated along the walls of the object as Lamb waves. With the help of the spectrogram analysis, the distances between the AE sources and AE sensors were obtained, which coincided with the results of the usual linear location based on the analysis of the arrival times difference [27,28].

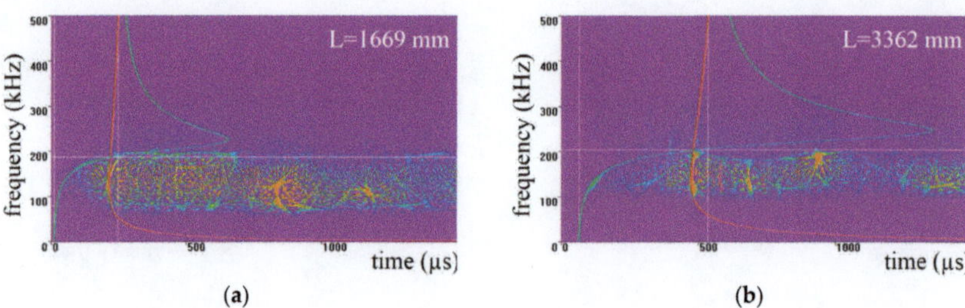

Figure 28. Dispersion curves for channel 29 (**a**); and for channel 30 (**b**).

At the points on the struts revealed as defective by the AE testing, an additional ultrasonic testing was carried out. On the left strut, 1 crack was found with the size of 20 mm, on the right—1 crack with the size 30 mm.

At present, the question of the SHM-system design for the testing of draglines is under consideration.

4. Conclusions

1. Based on our improved AE system, structural AE monitoring scheme is developed as SHM-system. Examples of its successful application are given.

2. Oil refinery: Overheating induced the formation of defects at the hydrotreater at the oil refinery plant. The use of SHM-system allowed to extend the lifetime of the hydrotreater by 6 months and to prevent the destruction of the hydrotreater during operation. Cluster analysis method of the AE data was developed and carried out.

3. Bridges: As a result of a series of experimental studies, the basic principles have been formulated, which make it possible to improve the efficiency of AE method as applied to the SHM of bridge structures. The solutions for their implementation have been suggested. The acoustic noises arising in monitoring of bridge structures have been analyzed and the causes of noise such as vibrations and friction due to the passing vehicles and wind gusts have been revealed. The multi-stage algorithm of noise filtering has been offered that includes selection by AE parameters.

4. Pipelines: Experimental work was carried out on the AE monitoring of the high-temperature process pipeline. The most suitable was the use of the frequency range 150–500 kHz and the threshold in the region of 40–55 dB_{AE}. It was found that the activity and amplitudes of noise increase significantly when the rate of pipeline temperature change exceeds $1°C/min$. It is revealed that the attenuation coefficient essentially increases with the accumulation of sediments in the internal volume of the pipeline. The possibility of detecting melt leakage through flange joints is shown, even if it is invisible under a layer of thermal insulation and is not noticeable.

5. Gas adsorbers: In the upper part of the gas adsorber at the refinery, non-permissible defects were found. A decision was made to install a system for SHM for the upper part of the gas adsorber. The main task of the installed monitoring system was to ensure safe operation of the gas adsorber until it can be replaced in a year.

6. Support rollers of rotary kiln: The possibility to conduct AE testing of support rollers of rotary kilns in operating mode has been investigated. It was found that in the range above 30 kHz the acoustic noises generally are no more than 32 dB_{AE}. For full testing of all 3 bearings it is necessary to mount 3 AE sensors on each of 2 bearing enclosures. To perform full AE testing of shaft it is sufficient to provide access to the shaft by removing protective caps and to mount 2 AE sensors: one on the side of thrust bearing and one on the opposite side. The signals with amplitude of about 100 dB_{AE}, duration of more than 100,000 µs allows us to estimate the needed lubrication intervals. The higher values of AE signal RMS are observed on the shafts of support rollers with

long lifetime. While testing the support rollers, the AE sources activity was registered on 2 shafts in the area of fillet. They presumably correspond to active fatigue cracks.

7. Dragline excavator: Work was carried out on AE monitoring of boom, column and struts of a walking dragline excavator. To remove noise, the use of a frequency range of 150–500 kHz and a threshold of 50 dB$_{AE}$ was most appropriate. With these settings, the AE activity acquires a cyclic character synchronized with the excavator operation cycle. By the results of attenuation measuring the recommended distance between the AE sensors is from 4.5 to 9 m. The AE data was collected for 2 days. Six sources of AE have been detected on the left and right struts. When performing ultrasound testing, 2 of them were classified as 20 and 30 mm cracks.

Author Contributions: S.A.E. managed the project and developed the software for the SHM-system. V.A.B. carried out the data acquisition and provided data filtering and clusterization. D.A.T. wrote the text of the article, participated in the data processing and verification of the obtained results. P.P.K. wrote the text of the article, and took part in the data analysis and verification of the results. V.V.B. carried out the data acquisition and the analytical research for all described applications. P.N.T. and A.L.A. created the original concept of the SHM-system and developed its hardware. V.G.K. took part in the design of the hardware of the SHM-system and developed the design of all units.

Funding: This research received no external funding.

Conflicts of Interest: The authors declare no conflict of interest.

References

1. Ono, K. Application of acoustic emission for structure diagnosis. *Struct. Health Monit.* **2011**, *2*, 3–18.
2. Kishi, T.; Ohtsu, M.; Yuyama, S. *Acoustic Emission—Beyond the Millennium*; Elsevier Science: New York, NY, USA, 2000.
3. Bryla, P.; Walaszek, H.; Herve, C.; Catty, J. Real time & long term acoustic emission monitoring: A new way to use acoustic emission—Application to hydroelectric penstocks and paper machine. In Proceedings of the 31st European Conference on Acoustic Emission Testing, Dresden, Germany, 3–5 September 2014.
4. Svoboda, V.; Zemlicka, F. Investigation of fatigue crack growth on material for reactor pressure vessel by acoustic emission. In Proceedings of the 14th International Conference of the Slovenian Society for Non-Destructive Testing "Application of Contemporary Non-Destructive Testing in Engineering", Bernardin, Slovenia, 4–6 September 2017.
5. Baran, I.; Nowak, M.; Jagenbrein, A.; Bulglacki, H. Acoustic Emission Monitoring of Structural Elements of a Ship for Detection of Fatigue and Corrosion Damages. In Proceedings of the 30th European Conference on Acoustic Emission Testing & 7th International Conference on Acoustic Emission University of Granada, Granada, Spain, 12–15 September 2012.
6. Serris, E.; Cameirao, A.; Gruy, F. Monitoring industrial crystallization using acoustic emission. In Proceedings of the 32nd European Conference on Acoustic Emission Testing Prague, Prague, Czech Republic, 7–9 September 2016.
7. Tscheliesnig, P.; Lackner, G.; Jagenbrein, A. Corrosion Detection by Means of Acoustic Emission (AE) Monitoring. In Proceedings of the 19th World Conference on Non-Destructive Testing (WCNDT 2016), Munich, Germany, 13–17 June 2016.
8. Alyakritskiy, A.; Elizarov, S.; Shaporev, V.; Trovimov, P. Overview of A-Line32D Series AE Systems, Produced by INTERUNIS, Ltd. In Proceedings of the 9th European Conference on NDT, Berlin, Germany, 25–29 September 2006.
9. Kharebov, V. Integrated Diagnostic Monitoring of Hazardous Production Facilities. In Proceedings of the 10th European Conference on Non-Destructive Testing, Moscow, Russia, 7–11 June 2010.
10. Barat, V.; Alyakritskiy, A. Automated method for statistic processing of AE testing data. *J. Acoust. Emiss.* **2008**, *26*, 311–317.
11. Barat, V.A.; Alyakritsky, A.L. Method of statistical processing of AE monitoring data on the example of the hydrotreater of the Mozyr refinery. *NDT World* **2008**, *42*, 52–55.
12. Shiotani, T.; Aggelis, D.; Makishima, O. Global Monitoring of Concrete Bridge Using Acoustic Emission. *J. Acoust. Emiss.* **2007**, *25*, 308–315.

13. Bardakov, V.; Elizarov, S.; Terentyev, D.; Chernov, D. Features of AE Method Use in Monitoring of Bridge Structures. Progress in Acoustic Emission XVIII. In Proceedings of the 23rd IAES, IIIAE and 8th ICAE, Kyoto, Japan, 5–9 December 2016; pp. 99–104.

14. Barat, V.A.; Chernov, D.V.; Elizarov, S.V. Discovering Data Flow Discords for Enhancing Noise Immunity of Acoustic-Emission Testing Russian. *Nondestruct. Test.* **2016**, *52*, 347–356. [CrossRef]

15. Por, G.; Bereczki, P.; Fekete, B.; Trampus, P. Heat Treatment and Tension Curves in Contemporary Steel Materials Monitored by Acoustic Emission. In Proceedings of the 19th World Conference on Non-Destructive Testing, Munich, Germany, 13–17 June 2016; pp. 1–8.

16. Babak, V.; Filonenko, S.; Kalita, V. Acoustic emission under temperature tests of materials. *Aviation* **2005**, *9*, 24–28.

17. Herve, C.; Dahmene, F.; Laksimbis, A.; Jaubert, L.; Hariri, S.; Cherfaoui, M.; Mouftiez, A. Acoustic Emission Monitoring of High Temperature Crack Propagation in AISI 304L and P265GH. In Proceedings of the 30th European Conference on Acoustic Emission Testing & 7th International Conference on Acoustic Emission University of Granada, Granada, Spain, 12–15 September 2012.

18. Boon, M.J.; Zarouchas, D.; Martinez, M.; Gagar, D.; Rinze, B.; Foote, P. Temperature and Load Eects on Acoustic Emission Signals for Structural Health Monitoring Applications. Le Cam, Vincent and Mevel, Laurent and Schoefs, Franck. EWSHM. In Proceedings of the 7th European Workshop on Structural Health Monitoring, Nantes, France, 8–11 July 2014.

19. Al-Ghamd Abdullan, M.; Mba, D. A comparative experimental study on the use Acoustic Emission and vibration analysis for bearing defect identification and estimation of defect size. *Mech. Syst. Signal. Process.* **2006**, *20*, 1537–1571. [CrossRef]

20. He, Y.; Zhang, X.; Friswell, M.I. Defect Diagnosis for Rolling Element Bearing Using Acoustic Emission. *Vib. Acoust.* **2009**, *131*. [CrossRef]

21. Bruzelius, K.; Mba, D. An initial investigation on the potential applicability of Acoustic Emission to rail track fault detection. *NDT E Int.* **2004**, *37*, 507–516. [CrossRef]

22. Miettinen, J.; Pataniitty, P. Acoustic Emission in Monitoring Extremely Slowly Rotating Rolling Bearing. In Proceedings of the 12th International Conference on Condition Monitoring and Diagnostic Engineering Management (COMADEM '99), University of Sunderland, Sunderland, UK, 6–9 July 1999; pp. 289–297.

23. Elizarov, S.; Barat, V.; Bardakov, V.; Chernov, D.; Terentyev, D. Features of the AE testing of equipment on operating mode. In Proceedings of the 32nd European Conference on Acoustic Emission Testing, Prague, Czech Republic, 7–9 September 2016; pp. 115–124.

24. Baccar, D.; Schiffer, S.; Söffker, D. Acoustic Emission-based identification and classification of frictional wear of metallic surfaces. In Proceedings of the 7th European Workshop on Structural Health Monitoring, Nantes, France, 8–11 July 2014; pp. 1178–1185.

25. Proust, A. In service acoustic emission monitoring of harbor cranes in order to program maintenance operations and insure safety management. In Proceedings of the 32nd European Conference on Acoustic Emission Testing, Prague, Czech Republic, 7–9 September 2016; p. 10.

26. Tamutus, T. Structural Health Monitoring Case Studies from In-Service Structures. In Proceedings of the 5th International CANDU In-Service Inspection Workshop in conjunction with the NDT in Canada 2014 Conference, Toronto, Canada, 16–18 June 2014; p. 88.

27. Terentyev, D.A.; Popkov, Y.S. Determination of the parameters of the dispersion curves of lamb waves with the use of the Hough transform of the spectrogram of an AE signal/ Russian. *Rus. J. Nondestruct. Test.* **2014**, *50*, 19–28. [CrossRef]

28. Elizarov, S.; Bukatin, A.; Rostovtsev, M.; Terentyev, D. New Developments of Software for A-Line Family AE Systems. *J. Acoust. Emiss.* **2008**, *26*, 132–141.

 *applied sciences*

Article

Acoustic Emission Method for Locating and Identifying Active Destructive Processes in Operating Facilities

Grzegorz Świt

Faculty of Civil Engineering and Architecture, Kielce University of Technology, Aleja Tysiąclecia Państwa Polskiego 7, 25-314 Kielce, Poland; gswiter@gmail.com; Tel.: +48-881-290-171

Received: 2 July 2018; Accepted: 2 August 2018; Published: 3 August 2018

Featured Application: The data collected can be the basis for determining the structural condition of the structure.

Abstract: Durability, safety, and usability are the three foundations of structural reliability, vital in the economic and social context. As the locating and tracking of potential damage and evaluating its impact on the condition of the structure are part of service life assessment, relevant methods should be developed that would detect the onset of the deterioration process and enable the monitoring of its progress within the entire volume of the structure, not only in the areas selected in a subjective way. The acoustic emission (AE) method relying on the analysis of active destructive processes can be the best choice. This article reports the results of the application of the AE method for identifying active destructive processes and tracking their development during the routine operation of various types of structures.

Keywords: acoustic emission (AE); non-destructive methods (NDT); diagnostic methods; bridges; structural health monitoring (SHM)

---

1. Introduction

Structural condition diagnostics and monitoring are two important issues in the economic and social context. Aging infrastructure, deteriorating environmental conditions, and increasing operational loads are the primary stimuli for fast-progressing research on a new interdisciplinary field of technical knowledge called "Structural Health Monitoring (SHM)", closely connected to the safe service life of structures [1,2].

Two different problems are involved in the durability assessment, one associated with the analysis of existing structures designed according to the standards previously in force, and the other associated with ensuring the anticipated useful life, taking into account the mechanical and strength-related characteristics of the newly designed structures.

In this work, the author's attention will be focused on assessing the structural integrity of existing facilities and its impact on their durability.

Durability, safety, and usability are the three foundations of structural reliability. The durability factor is, unlike in the past, equally important and has to include the evaluation of environmental impacts (moisture, frost, CO_2, de-icing agents, etc.) on the present and future health of the structure. Ability to assess the environmental effects on the degradation level is critical for the durability and resistance of structures. Research programs conducted by the international ISO standardization organization [3], the American Concrete Institute [4], or those implemented within the Basic Research in Industrial Technologies/ European Research in Advanced Materials (BRITE/EURAM) framework [5] by the European Union countries, revealed that environmental changes—on a local, regional, and global scale—were faster than initially assumed, thereby making the environmental assessment an urgent issue.

It is particularly important to determine the onset of the damaging process in the materials and structural elements, because attempts to stop once-initiated processes are often inefficient, leading to the failure or catastrophic collapse of the entire structure. The Eurocodes propose four approaches to design for safety, thus providing a certain amount of autonomy to the designers. Niriaki in [6] eliminates this autonomy by proposing his original approach independent of structural safety but based on the durability and resistance analysis.

The demand for the method capable of detecting the onset of the deterioration process and monitoring its progress within the entire volume of the structure, not only in the subjectively chosen parts, is fulfilled by the technique of acoustic emission (AE). The method involves performing comparative analysis of the acoustic emission signals recorded during the tests and those collected in the database of reference signals corresponding to particular destructive processes. The results are used to identify and locate active destructive processes in reinforced concrete structures (IADP—Identification of Active Damage Processes), in prestressed concrete structure (RPD—Recognition of Destructive Processes), and in steel structures, thereby allowing the global monitoring across the whole member and recording only active deterioration processes developing under actual loading conditions [7–17]. The AE method is being increasingly developed for use in the construction of gas pipelines.

This article aims to discuss examples of AE monitoring of different engineering structures, including the AE-based Structural Health Monitoring System (SHMS) on the My Thuan Bridge.

2. Materials and Methods

2.1. Materials

2.1.1. Steel Bridge

Selected structural members of the bridge were examined. The damage assessment was based on the results of AE signal analysis.

Visual inspection (Figure 1) revealed surface corrosion of the majority of the bridge steel elements (Figure 2), being especially intensified on top surfaces in the area of bottom chords, crossbars, stringers, and lower joints of the truss. Corrosion was also found at the interface between profiles and steel plates of posts, diagonals, and flanges, as well as at the connections between steel angles and web plates, crossbars, and stringers. Due to inadequate drainage of the steel orthotropic plates on the bridge roadway, additional corrosion sites were also detected.

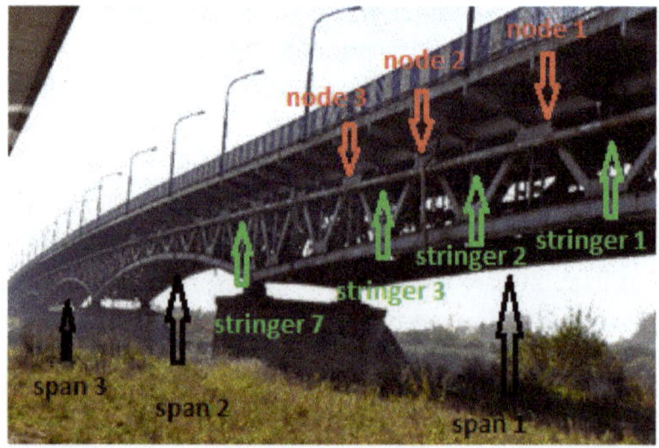

Figure 1. Side view of three bridge spans subjected to tests.

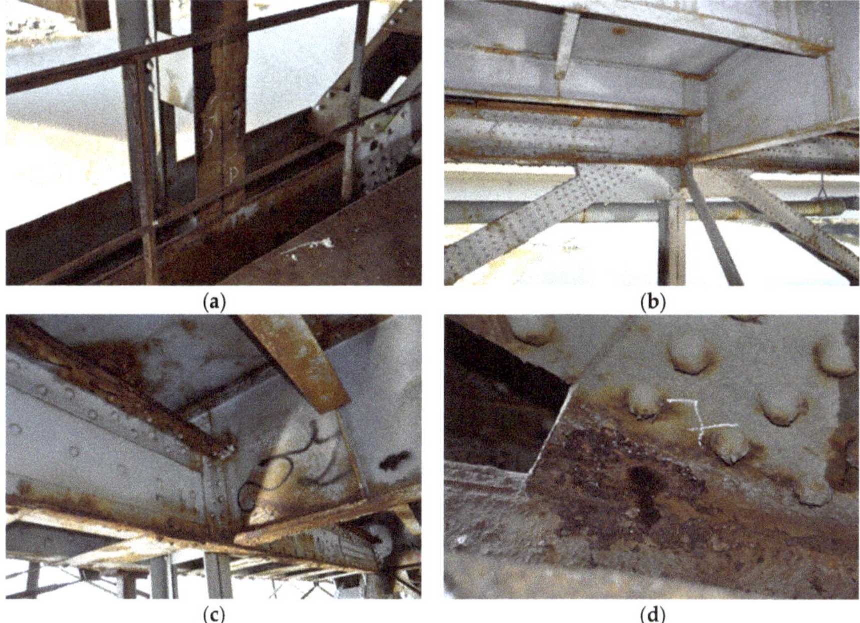

Figure 2. (**a**) View of the connection between the truss lower chord with the diagonals and the vertical member, with visible advanced surface corrosion. (**b**) View from the inside to the top node, visible surface corrosion of structural elements of the truss, as well as corrosion at the joints at the contact surface of connected profiles. (**c**) View from the inside—the upper node, visible surface and pitting corrosion on the stringers, crossbars, and orthotropic slabs. (**d**) View of the node with signs of pitting and surface corrosion on steel plate and rivets.

The damage was defined as caused by stagnant water. The method of constructing the nodes, the lack of appropriate falls, the counter-falls formed due to bridge deflection, and the lack of openings at the connection of the columns with the lower chord facilitated water accumulation, which, in the absence of a proper protective coating on all steel elements of the bridge, promoted rapid corrosion progression and degradation of the material.

The measurement set included 12 sensors spaced 2–10 m apart. Determining the distance between the sensors in the location involves the measurement of the Hsu–Nielsen signal source. This measurement was also used to determine the speed of wave propagation in the material.

2.1.2. Steel Columns

Two steel piles (Figures 3a and 4a) constituting the supporting structure for the cable car were examined. Visual inspection of the piles revealed surface corrosion of most steel members, deformations of some elements, as well as pitting corrosion and fatigue cracks on sections of those elements (Figures 3b and 4b). Corrosion intensified on the central surfaces of the flanges in the column structure, in particular in the area of gusset plates at welded and bolted joints. Corrosion was also found at the interface between the profiles and flanges of posts, crossbars, and chords, as well as at the place where the angles were connected to the webs and flanges. Corrosion processes and the loosening of bolts securing the column elements caused friction between some of the steel elements.

(a)   (b)

Figure 3. (**a**) View of column 1 with a safety platform. (**b**) View of anchor block and flanges at the connection between column 1 and foundation; signs of surface corrosion and pitting.

(a)   (b)

Figure 4. (**a**) Side view of column 2. (**b**) External view of the upper chord; signs of surface corrosion on truss members.

The damage was caused by rainwater and insufficient use of protective and anti-corrosive agents (no paint coating; sites of strongly developed corrosion). The photographs show changes in the structure of the material of the columns. Observed delamination of the steelwork and intergranular corrosion require additional maintenance and improved supervision.

Twelve sensors were installed with a spacing of 2 m.

2.1.3. Gas Pipeline

A section of steel high-pressure gas pipeline 400 mm in diameter and 200 m in length was examined (Figure 5). The age of the pipes is estimated at approximately 60 years. The pipeline operates under environmentally difficult conditions with wetlands underneath, electrical power lines, and a lumber transport route running nearby. Two 30–80 kHz sensors were installed in a linear configuration on the structure.

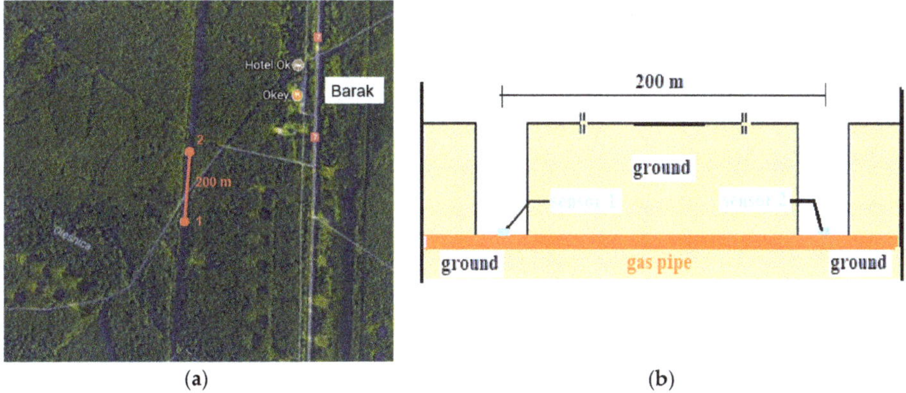

(a)   (b)

Figure 5. (a) Arrangement of measuring points on the gas pipeline. (b) Measuring scheme.

2.1.4. My Thuan Bridge

The My Thuan cable-stayed bridge connects the banks of the Tien Giang River flowing between two large provinces in the Mekong basin and and is part of Vietnam highway network. Completed in 2000, the bridge was designed to Australian and Vietnamese standards.

The total length of the bridge is 1535.2 m. The main bridge has three continuous spans with dimensions 150 + 350 + 150 m supported by two main pylons. The main spans are made of prestressed concrete class 50 (f_c = 50 MPa) and have two main girders internally attached by a set of floor beams. The bridge was constructed with the use of the cantilever method. The height of the main spans is 1760 mm with the width ranging from 1200 to 1400 mm. The deck slab is 250 mm thick. The end spans are suspended on 4 × 32 cable stays and each stay is composed of 22–29 seven-wire strands 15.2 mm in diameter—a system of dampers was installed on each side of the cable stays to control vibration. The two pylons made of reinforced concrete grade 50 have a splayed H-frame and the height of 123.5 m (from the foundation base) and 84.43 m (from the deck). Each pylon is supported on 16 bored piles 2.5 m diameter driven into the riverbed to the depth of −90 m (northern pylon) and −100 m (southern pylon).

Except for the weather stations (measurement of humidity level, wind speed and direction—installed on the main deck and on the northern pylon, plus two cameras monitoring the water level and traffic), no condition assessment systems for determining the integrity of the bridge were implemented during the design and construction stages. The current condition of these devices is not known.

2.2. Methods

Basic acoustic emission monitoring is carried out during the routine operation of the facility under load (in exceptional cases under the test load).

The tests performed during the routine operation of the facility aim to:

- look for active damage processes (progressing under service conditions) in the structure;
- identify and locate the damage;
- estimate the risk potential of the damage;
- determine the extent of the risk posed to the safety of the facility under routine operation.

The μSamos system (PAC MISTRAS Corp., 195 Clarksville Rd, Princeton Jct, NJ 085, USA) with three PCI-8 cards was used to develop the reference signal databases (24 measurement channels) used in subsequent stages to assess the condition of steel and concrete structures. Flat type sensors operating in the frequency range of 30–80 kHz was used for all structures. A 55 kHz sensor was also used in the concrete structures. Signal amplification was 40 dB. AEWin software was used for the measurements. To analyze the measurement data, Noesis 5.8—ADVANCED DATA ANALYSIS PATTERN RECOGNITION SOFTWARE (PAC MISTRAS Corp., Princeton, NJ, USA) was applied. The reference signal database was created using 13 parameters of acoustic emission: rise time (μs), counts to peak amplitude, counts, energy (EC), amplitude (dB), average frequency (kHz), root-mean square (RMS) (V), reverberation frequency (kHz), initiation frequency (kHz), absolute energy (aJ), signal strength (pVs), duration (μs), and average signal level (ASL) (dB).

The Identification of Active Damage Processes (IADP) method is based on the analysis of acoustic waves generated by active destructive processes developing under operational loads in buildings and engineering structures. The signals received by acoustic sensors mounted on the object are compared to the database of reference signals created earlier for defined destructive processes. The identified destructive processes are then localized by analyzing the difference in time of signal arrival to individual sensors. The identification and the location of active damage processes make it possible to assess the technical condition of structures, and thus can be the basis for the diagnosis and management strategies.

The big advantage of the IADP method is that the AE sensors can be arranged to cover the entire structure being measured and that the measurements can be performed under actual operational loads.

In its original version, the IADP method was developed to analyze destructive processes in prestressed concrete elements. It was then expanded and applied in the analysis of steel elements for which a new database was created with pattern signals of destructive processes resulting from cyclic and monotonic loads.

Patterns of AE signals were obtained as a result of tests carried out at temperatures ranging from $-60\,^{\circ}$C to $+60\,^{\circ}$C on monotonically loaded V-notch specimens made with St3S and 18G2A grade steels and with the steel recovered from an old bridge, on bent models of V-notch elements under monotonic and cyclic loading at $+20\,^{\circ}$C, and models of welded, riveted, and bolted nodes made of the same grade steel.

The following destructive processes being the AE signal sources can be distinguished in prestressed concrete structures:

- microcracks;
- friction between crack surfaces;
- formation and development of cracks in concrete;
- cracking at the concrete–reinforcement interface;
- concrete crumbling;
- friction at the concrete–reinforcement interface;
- corrosion;
- plastic deformation and cracking of cables and other reinforcements.

The values of the AE signal parameters allow the classification of the signals into classes, each of which is characterized by dominant destructive processes and different risk levels. The signals characteristic of each class create the reference databases capable of identifying a destructive process, e.g., "cracking at the concrete-reinforcement interface" corresponds to the database that groups selected signal parameters assigned to the specific process.

Databases for individual processes (or their groups) are determined on the samples of materials, on models used in special laboratory tests (where a given destructive process or a group of processes predominates) and on full-size structural elements during strength tests, attenuation tests, and during normal operation of the facility.

The database of AE reference signals makes it possible to identify active destructive processes over the entire element under test. Long-term measurements can be used to determine the damage under true loading conditions, with the external factors such as rain, frost, or wind taken into account.

Appropriate installation of AE sensors allows the measurement of the entire volume of the examined element and the location of the emission source (the point of damage).

To build the reference signal database with the Recognition of Destructive Processes (RPD) method, the NOESIS 5.8 program was used with hierarchical and non-hierarchical statistical grouping methods and neural networks. For the RPD database, two versions of the pattern recognition method were used: with arbitrary division into classes—unsupervised pattern recognition (USPR) and self-learning, in which the division into classes was carried out using the reference signals—supervised pattern recognition (SPR).

In the first case, analysis of arbitrary patterns is mainly used for creating a reference signal database if the number of classes is unknown. The second method is applied when model signals characterizing the destructive processes are available. The reference signals are the signals previously collected in databases generated during independent experiments.

While the USPR method classifies AE sources based on the similarity of signals without assigning appropriate mechanisms to the groups, the SPR method assigns specific processes to the groups, provided they have a base of reference signals.

Creating a reference signal database is organized in several stages. These are:

- generating signals in the laboratory while destroying specially designed specimens (bars, cubes, rolls) in a specific way,
- comparison of signals received from the specimens with the signals generated during the destruction of model beams (reinforced concrete and pre-tensioned concrete),
- verification of reference signals based on the monitoring results obtained for various types and lengths of prestressed concrete girders loaded to failure,
- final verification of selected elements of the bridge during its normal operation.

2.2.1. Acoustic Emission Evaluation of Steel Structures

Assessment involves the analysis of changes in the intensity of acoustic emission signals generated in particular zones of structural elements under routine use. Recorded AE signals are grouped into classes to which various destructive mechanisms are assigned. The number of parameters used to build the database of reference signals must be consistent with the number of previously registered parameters of AE signals [11,12].

The risk posed by generating processes within one class is determined by the so-called intensity code of destructive processes. These processes are best illustrated by graphs where each AE signal is assigned to one point. The color and shape of the point indicate the class to which the given AE signal belongs. The classes, symbols, and codes are summarized in Table 1.

Table 1. Classes, symbols, and risk codes for steel bridges.

Colour					
Class	No.1	No.2	No.3	No.4	No.5
Risk code	0	1	2	3	4
Risk	Very high	High	Medium	Low	No risk

Training patterns are used for the grouping and classification of AE signals.

Class No.5 signals are generated by the constant AE noise. Classes No. 4, No. 3, and No. 2 of signals are generated by single destructive mechanisms such as:

No.4—yielding of steel at the crack tip;

No.3—crack initiation;

No.2—crack propagation.

Class No.1 includes the signals resulting from the superposition of waves generated by more than one destructive process and by crack surface friction.

Occurrence of signals of all classes during the monitoring period is regarded as another item among the codes defining the influence of defects on the structural condition and indicates the presence of destructive processes in the structure.

The extent of damage is assessed using the zone location results and AE signal classification in the zones.

The individual signal classes mean (Table 2):

No.1—rupture;

No.2—friction;

No.3—crack propagation;

No.4—crack initiation;

No.5—perforation/deformation;

No.6—material losses;

No.7—surface corrosion;

No.8—work in the elastic range.

Table 2. Classes, symbols, and risk codes for steel engineering constructions.

Colour								
Class	No.1	No.2	No.3	No.4	No.5	No.6	No.7	No.8
Risk code	0	1	1	2	3	3	4	5
Risk	Very high	High	High	Mid-to-high	Medium	Medium	Low	No risk

2.2.2. Acoustic Emission Evaluation of Gas Pipelines

Classes, symbols and codes of AE signals together with the risk level for gas pipeline are compiled in Table 3.

Table 3. Classes, symbols, and risk levels for steel gas pipelines.

Colour								
Class	No.1	No.2	No.3	No.4	No.5	No.6	No.7	No.8
Code	0	1	1	2	2	3	4	5
Risk	Very high	High	High	Mid-to-high	Mid-to-high	Medium	Low	No risk

2.2.3. Acoustic Emission Evaluation of Reinforced Concrete Structures; Classes, Symbols, and Risk Codes for Reinforced Concrete Structures

The database of reference signals was created by conducting a series of tests on various types of reinforced concrete beams and specimens under various loads, including cyclic loading for the modeling of the car's passage [18–24]. The tests aimed at finding predominant destructive processes that might occur during testing of reinforced concrete structures under operation. The database was verified on real facilities [25–32].

The reference databases were classified on the basis of 12 AE parameters (rise time (μs), counts to peak amplitude, counts, energy (EC), amplitude (dB), average frequency (kHz), RMS (V), reverberation frequency (kHz), initiation frequency (kHz), absolute energy (aJ), signal strength (pVs), duration (μs), ASL (dB)) and marked for reinforced concrete (RC) structures as Class (Table 4):

- Class No.1 Crack formation in the paste;
- Class No.2 Crack formation at the paste—aggregate boundary;
- Class No.3 Microcrack formation;
- Class No.4 Crack growth;
- Class No.5 Loss of adhesion around the cracks;
- Class No.6 Buckling of compressed bars/crushing of compressed concrete/rupture of rebar.

The results of the cracking tests under monotonic, cyclic, and variable loading were used to develop the following criteria of structural damage:

- Classes No.6 and No.5—safe behavior of the structure;
- Class No.4—warning;
- Class No.3—threat to durability;
- Class No.2—threat to load carrying capacity;
- Class No.1—loss of safety.

Table 4. Classes, symbols, and risk levels for reinforced concrete structures.

Colour						
Class	No.1	No.2	No.3	No.4	No.5	No.6
Code	0	1	2	3	4	5
Risk	Very high	High	Medium	Medium	Low	No risk

3. Results and Discussion

3.1. Steel Bridge

The following is the AE measurement data processed for selected groups of components.

The bottom chord (span 1): Analysis of the recorded AE signals against the reference signal database reveals the presence of signals representing all classes. Signal duration is variable and reaches 350,000 µs. The energy emitted when signals are generated is low and reaches 20,000 ec (Figure 6). Individual signals attain slightly higher values. The generation of AE signals is not continuous; the signals are initiated by passing vehicles with specific characteristics (tractors with semi-trailers moving at a speed of more than 60 km/h or heavy trucks moving in the column at speeds exceeding 60 km/h). This is due to the condition of expansion joints and dynamic loads initiating AE signals at the seats of diagonals and vertical members in the gusset plates. The resulting signals suggest that stresses increase in riveted and welded joints due to significant surface corrosion and in some locations due to pitting, slight loosening of rivets, and their movement. The number of signals describing the yielding and crack development is small and concentrated in three zones of the chord. These signals are generated when trucks traveling at excessively high speeds pass through.

Nodes 1–3 (span 1): Analysis of the recorded AE signals against the reference signal database reveals the presence of signals representing all classes. Signal duration does not exceed 50,000 µs. Energy is low, 25,000 ec (Figure 7). Individual signals reach slightly higher values. The generation of AE signals is not continuous; the signals are initiated by the movement of vehicles with specific characteristics or originate from the work of the element working both in tension/compression. This should be interpreted to mean that the change in stress level will result in slight movement at riveted connections and at the connections of the gusset plate with the upper chord. The recorded signals inform about the behavior of elements in the elastic range, and individual signals inform about local yielding and corrosion at the nodes.

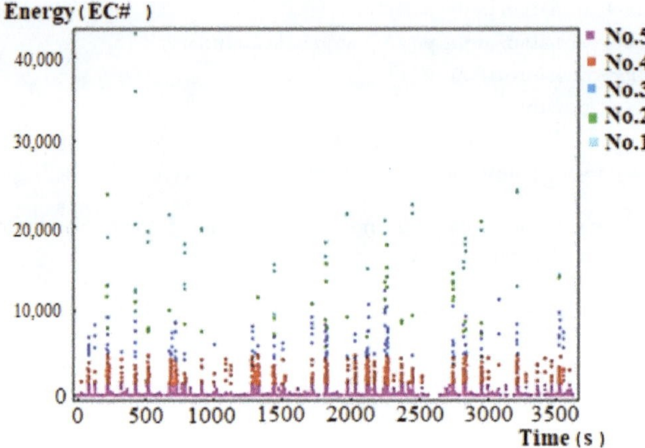

Figure 6. Energy–time relationship for the bottom chord of the truss in span 1 with the numbers of signal classes marked.

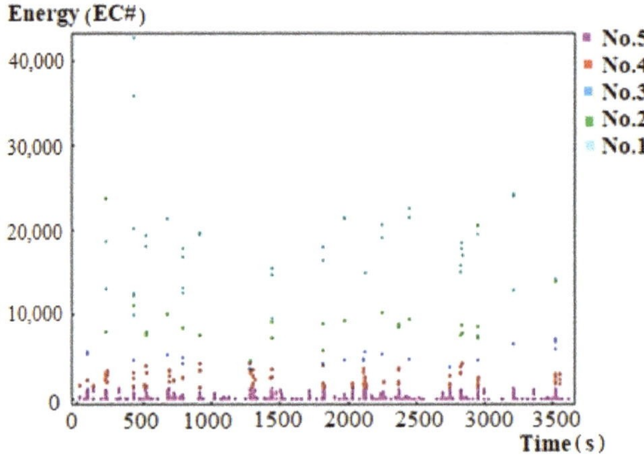

Figure 7. Energy–time relationship for node 1 in span 1 with the numbers of signal classes marked.

Stringers 1–6 (span 1): Analysis of the recorded AE signals against the reference signal database reveals the presence of signals representing all classes. Signal duration is long, up to 360,000 µs, and the energy emitted during signal generation reaches a value of up to 40,000 ec (Figure 8). Individual signals have slightly higher energy values of up to 60,000 ec. High parameter signals are not emitted continuously. The signals are initiated by passing vehicles with specific characteristics (tractors with semi-trailers moving at a speed of more than 60 km/h or heavy vehicles moving in the column at speeds above 60 km/h). Signals located and recorded originate mainly in the central zone of the stringers. The high rise time of 25,000 µs suggests that the signals originate from corrosion processes and friction occurring on the surface of the upper flange of the stringer and on the orthotropic plate. The signals indicate that corrosion processes are advanced and that dynamic loads cause delamination cracks on the corroded surfaces and friction between them. The number of signals from higher classes 1, 2, 3 is significant, which suggests that these processes do not pose a risk to the bridge although strong corrosion processes and the formation of fatigue microcracks have begun, which left without repair, can lead to fatigue cracks, thus weakening the entire structure.

Stringers 7–8 (span 2): Analysis of the recorded AE signals against the reference signal database reveals the presence of signals representing only class 5. Signal duration is short, 15,000 µs and energy emitted when signals are generated reaches 2500 ec (Figure 9). The rise time of 4500 µs confirms good performance of the members in the elastic range.

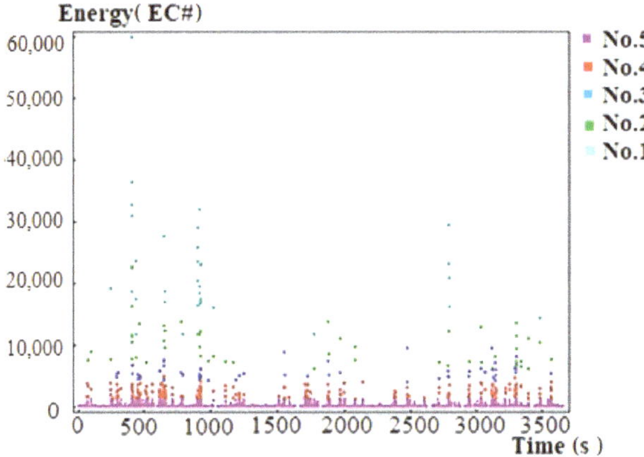

Figure 8. Energy-time relationship for stringer 2 in the span 1 with the numbers of signal classes marked.

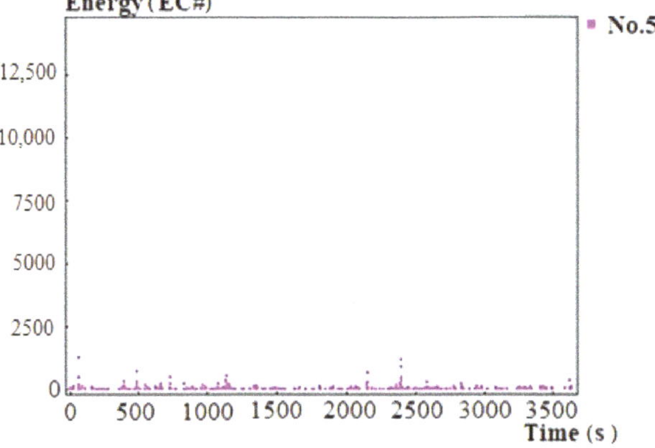

Figure 9. Energy–time relationship for stringer 8 in span 2 with the numbers of signal classes marked.

3.2. Steel Columns

Twelve sensors were installed on column 1. Spatial location demonstrated that the column of the supporting structure was working in the range that might, sometime in the future, pose a threat to the safety of the cableway. The tests were performed when the cable cars were stationary and during their regular operation. When stationary, no processes generating dangerous phenomena were recorded, but when the cable cars started moving (within the normal operating envelope), destructive processes Class 2–8 were revealed (Table 2), which indicated the growth of microcracks in the members. This demonstrates that some of the members were exposed to fatigue-induced destructive processes. Surface corrosion attack were found at the bolted connections and at half height of the vertical elements of the column.

Figures 10–13 show damage classes occurring in the tested elements. Data from 12 sensors were combined using the supervised pattern recognition method.

Analysis of the recorded AE signals against the reference signal database (Figure 10) reveals the presence of signals representing classes from 2 to 8 (Table 2). The signal strength is high, up to 2.6×10^9 pVs. Continuous emission of the signals is initiated by passing trolleys with weight of 2500 kg, which translates to the presence of locations with fatigue and overload cracks and corrosion centers. The welded and bolted connections in the nodes may be affected due to substantial surface corrosion and pitting. The condition of the members revealed during the examination was an indication for further observation of the column and for placing spacer blocks between the column and the safety platform. A portion of damage may result from additional load caused by the movement of the platform (fatigue).

Fatigue cracks were detected in the indicated places on the basis of magnetic particle tests (Figure 11). It allowed to verify the visual assessment and AE measurements with the actual condition of the element.

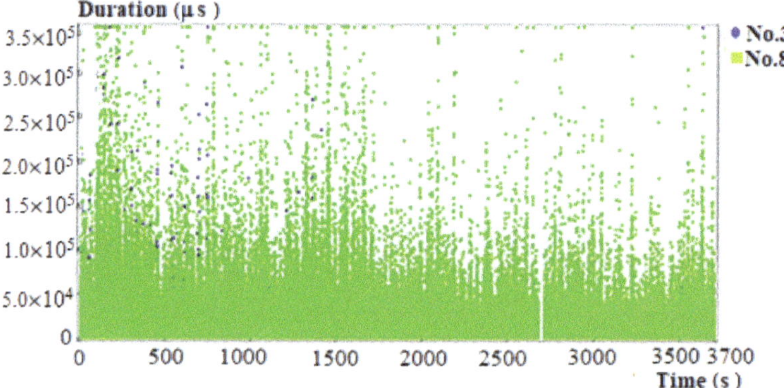

Figure 10. Signal strength versus time for sensors 1–12 on column 1 with the numbers of signal classes marked.

Figure 11. (a) Photograph of the tested node. on column 1 (b) A photograph of the crack detected on column 1.

Twelve sensors were installed on column 2. Spatial location indicated that the column of the supporting structure worked in the range that may in the future cause a cableway safety risk. The tests were performed during routine operation of the facility. Two processes that generated events within the elastic range were revealed, as were the locations where the steel yielded, described as Class 4 destructive processes. In the case of column 2, yielding was probably caused by the additional load due to vibrations of the steel platform securing the road. The lack of spacers at the connection of these two structural elements may cause steel yielding and, as a consequence, fatigue microcracks and cracks. This indicates that some elements are exposed to destructive processes caused by fatigue. Locations of surface corrosion were also detected. The most strained locations were discovered at the bolted connections and at the mid-height of the column vertical elements.

The recorded AE signals analyzed using the reference signal database (Figures 12 and 13) indicate that in the node under analysis only Class 3 and 8 signals appear. The duration of these signals is long, up to 350,000 µs, while the signal energy is low, 30,000 ec on average and up to 65,000 ec locally. The acoustic emission signals are generated dis-continuously and are not initiated by passing trolleys weighing approximately 2500 kg, which indicates the existence of sites with plastic deformations, which may degenerate into fatigue microcracks and corrosion sites. Welded and bolted connections within the nodes may be affected due to surface corrosion, with some bolts (plates) loosened. Those destructive processes were the reason for the observation of this column and for the use of spacer blocks between the column and the safety platform. Part of the damage may result from the additional load caused by the movement of the platform contributing to material fatigue.

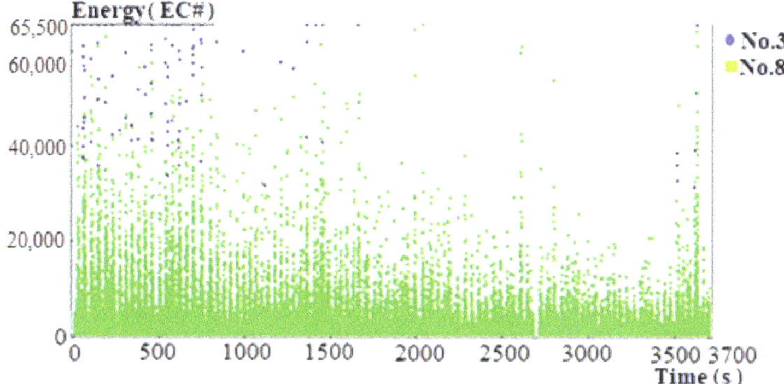

Figure 12. Energy versus time for sensors 1–12 on column 2 with the numbers of signal classes marked.

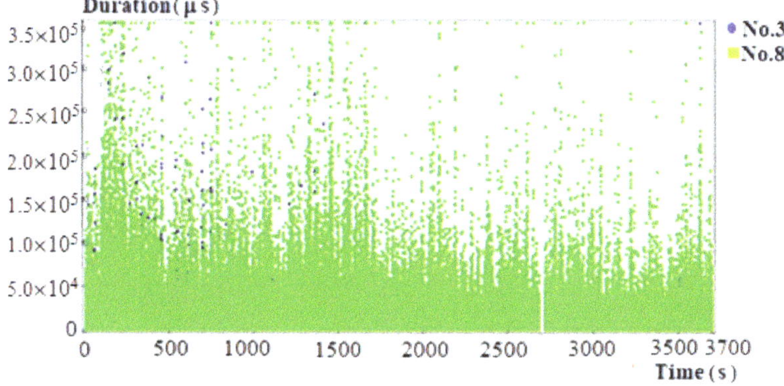

Figure 13. Acoustic emission (AE) signal duration versus time for sensors 1–2 on column 2 with the numbers of signal classes marked.

3.3. Gas Pipeline

The results of the acoustic emission evaluation of the high-pressure gas pipeline were analyzed by grouping using non-hierarchical methods (k-means) and by comparison to the reference signals (Table 3). The results showed reduced thickness of the pipe wall due to pitting and deformations and perforations of the structure due to changing working and corrosion conditions.

Analysis of absolute energy parameter as a function of time, correlated with parameters such as duration, amplitude, rise time, and AE counts, shows that destructive processes exhibit low intensity across a small area of the gas pipeline (Figure 14).

It should be emphasized that the destructive processes do not proceed continuously, which indicates the local character of these phenomena. The next step in the analysis is to locate places generating destructive processes.

Linear location of destructive process sources shows the area of increased intensity of AE signals extending between 55.79 m and 59.66 m, which corresponds to the pipes laid under the woodland path (Figure 15). Further analysis of destructive process location on the examined section of the pipeline indicated two points, at 57 m and 59.5 m.

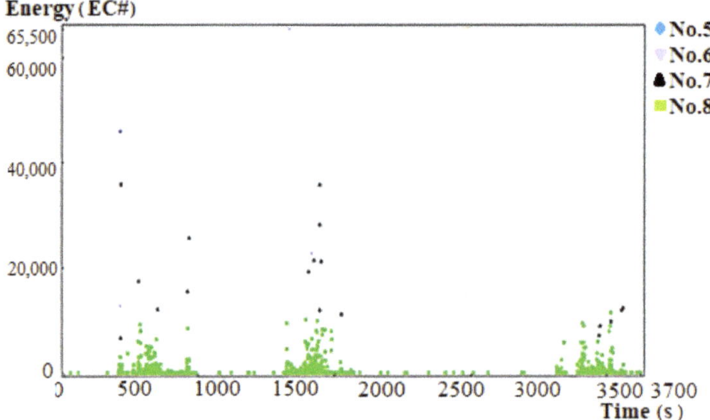

Figure 14. AE signal energy as a function of time for the gas pipeline with the numbers of signal classes marked.

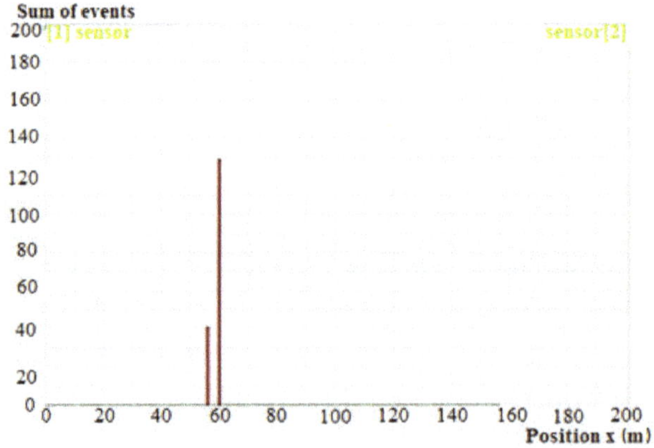

Figure 15. Exact location of destructive process sources.

Analysis of the amplitude versus time graph for AE signals recorded in the area beyond the indicated points shows that no signals indicating destructive processes were recorded in that area.

3.4. My Thuan Bridge

Examples of localized concrete cracks within compression cables are shown in Figure 16.

Figure 16. Examples of localized concrete cracks within the compression cables on My Thuan Bridge.

Described below are two complete measurement systems being the elements of the condition assessment system that helps decision makers plan maintenance and repairs [30].

The vertical displacement measurement system (Figure 17) measures the condition of and long-term changes in the span, and determines support settlements and their effect on the span vertical alignment. The static measurements are performed every hour or every 30 min, whereas the dynamic measurements are made when the accelerometer reading is higher than the value determined in the design calculations or on demand. The system comprises seven measurement points along the span and, if needed, a hydro profile meter and a hydro level (GEOKON) can be mounted on the pylons and the Global Navigation Satellite System (GNSS) can be used both along the span and on the pylons.

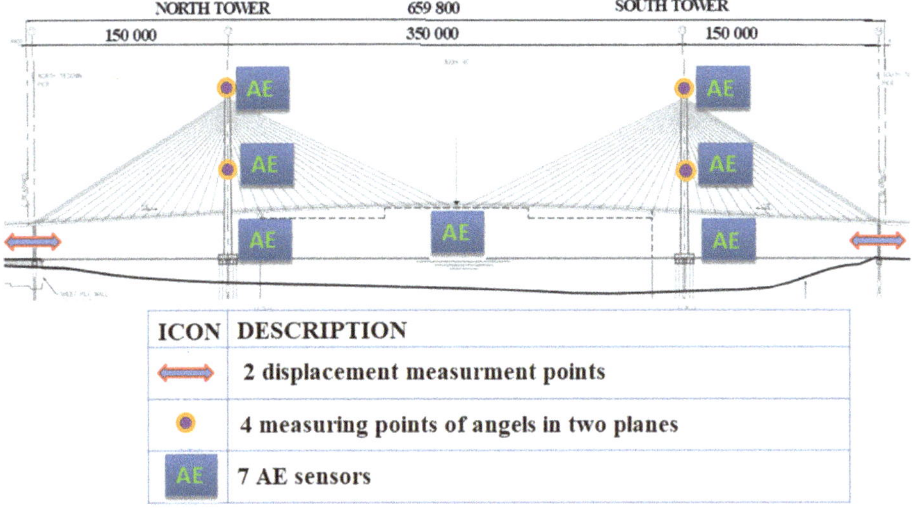

ICON	DESCRIPTION
⟷	2 displacement measurment points
●	4 measuring points of angels in two planes
AE	7 AE sensors

Figure 17. Arrangement of measurement points in the vertical displacements system.

The strain measurement system (Figure 18) measures long-term strain variations in the span and pylons. The sensors are mounted on the rebar, first uncovered and subsequently secured again. The system comprises seven measurement points with two sensors at each point to reduce the risk of error. The location of the measurement points should be established to the accuracy of ± 1.0 cm. As the bridge is large, the wind profile and its effective velocity need to be determined.

The angle measurement system measures torsion in the spans generated by all types of loads, span rheology, and workmanship quality. The system comprises seven measurement points located along the span length. At each point, torsion and bending are measured.

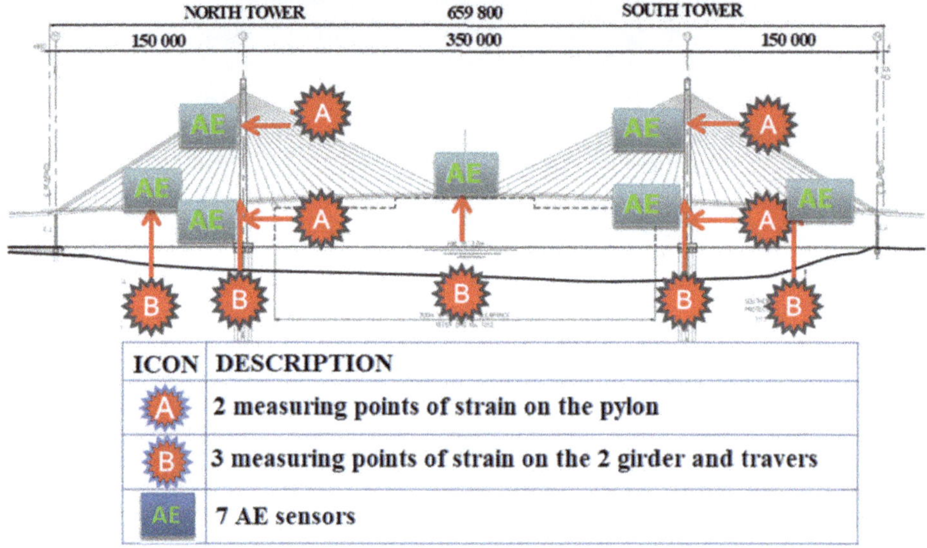

Figure 18. Arrangement of measurement points in the strain measurement system.

Prior to the design of the indicated system, for a period of 1.5 years, every 6 months AE was measured at selected points, which confirmed the development of microcracks and cracks in the concrete elements. This supported the idea of applying the AE method for monitoring the facility using the proposed prototype. Measurements were carried out in March 2016, October 2016, and March 2017. The measurement in each case lasted an hour and was conducted in the afternoon of the traffic peak. Below are sample graphs of signal energy versus time for each measurement period and one graph showing the duration of signals against time for one measurement period, because it was similar in each research period.

Analysis of the graphs (Figures 19–22) indicates that in each case the signal energy is not very high and reaches the values of 5000 ec and 6000 ec in the first and the following two measurement periods, respectively. Also, the values of the signal duration are low and reach a level of approximately 1500 µs in each analyzed period. While analyzing the graphs, it can be also observed that the data indicates generation of Class 3–6 signals (Table 4). The duration of the signals is relatively short—it is assumed that the signals are induced by atmospheric conditions and by vehicle traffic. Further test results are dependent on the implementation of the prototype installation. Currently, the research team is working on creating the reference base.

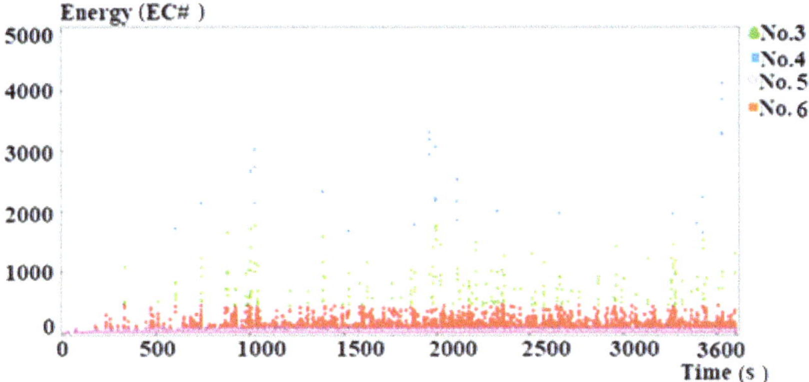

Figure 19. Energy versus time for the first measurement period—March 2016.

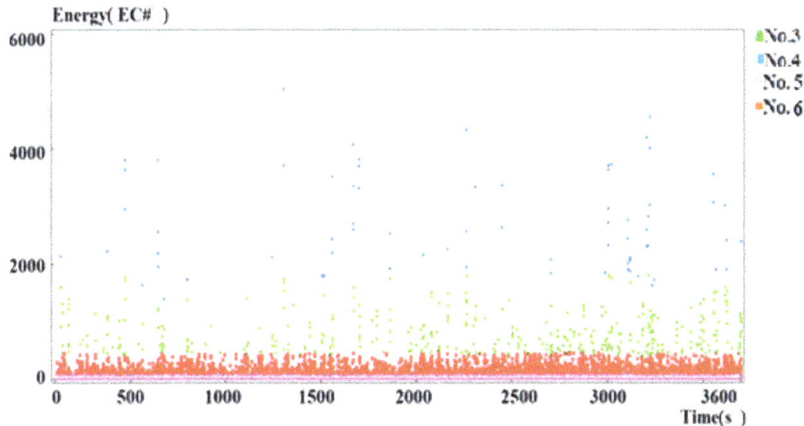

Figure 20. Energy versus time for the second measurement period—October 2016.

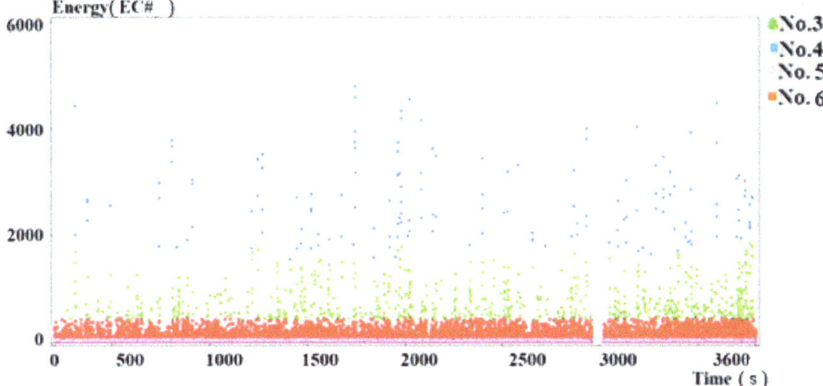

Figure 21. Energy versus time for the third measurement period—March 2017.

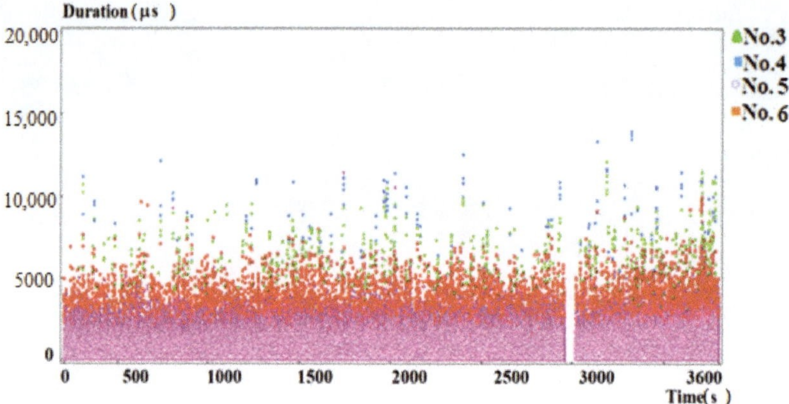

Figure 22. Signal duration versus time for three measurements.

4. Conclusions

The examples of acoustic emission evaluation of various types of engineering structures and the proposed global monitoring system based on the measurement of acoustic emission signals accompanying destructive processes, covers the entire volume of the element under test or its selected part, and allows locating and identifying active destructive processes and their dynamics in real time. The data collected can be the basis for determining the structural condition of the structure [33–39].

The system is a useful tool for:

- the assessment of the health of a structure and identification of potential risks;
- the monitoring of the dynamics of destructive processes;
- the support in decision-making process concerning the risk management;
- the assessment of the repair work quality and outcome;
- the assessment of non-standard vehicle traffic.

The findings presented in this article indicate that the use of the acoustic emission method with the systems of the global structural health monitoring ensure insight into the condition of structures.

5. Patents

"The system for detecting and locating active destructive processes in communication road infrastructure structural" Wiesław Trąmpczyński, Grzegorz Świt, Barbara Goszczyńska.

"Method for testing and/or monitoring of destructive processes in steel structures subjected to loads" Leszek Gołaski, Barbara Goszczyńska, Wiesław Trąmpczyński, Grzegorz Świt.

"Method for diagnosis and/or monitoring of technical condition of reinforced concrete and prestressed concrete structures and a system for diagnosing the condition of reinforced concrete and prestressed concrete structures" Wiesław Trąmpczyński, Grzegorz Świt, Leszek Gołaski, Barbara Goszczyńska, Kanji Ono.

Funding: This research received no external funding.

Conflicts of Interest: The author declares no conflict of interest.

References

1. Ono, K. Application of Acoustic Emission for Structure Diagnosis. *Diagnostics* **2011**, *2*, 3–18.
2. Marsh, B.K.; Nixon, P.J. Assuring performance of concrete structures through a durability audit. Int. Conf. Concrete in the Service of Mankind. In *Appropriate Concrete Technology, Proceedings of the Concrete in the Service of Mankind, Dundee, UK, 24–26 June 1996*; CRC Press: Boca Raton, FL, USA, 1996; pp. 49–59.
3. *Guide for Service Life Design of Buildings: Part 1—General Principles*; ISO Draft No. 2; International Standard Organization: Geneva, Switzerland, 1995.
4. *Service Life Design ACI 365 (1R-00)*; State-of-the-Art-Report; ACI: Farmington Hills, MI, USA, 2000.
5. BRITE/EURAM (Basic Research in Industrial Technologies/European Research in Advanced Materials) Program; DuraCrete 1996–99, DuraNet 1998–01, Darts 1997–2004.
6. Niraki, T. Service life design. *Constr. Build. Mater.* **1996**, *10*, 403–406. [CrossRef]
7. Houston, J.T.; Atimtay, E.; Ferguson, P.M. *Corrosion of Reinforcing Steel Embedded in Structural Concrete*; Research Report 112-1F; Center for Highway Research: Austin, TX, USA, 1972.
8. Yuyama, S.; Okamoto, T.; Shigeiski, M.; Ohtsu, M.; Kishi, T. *A Proposed Standard for Evaluating Integrity of Reinforced Concrete Beams by Acoustic Emission: Standard and Technology Update*; American Society for Testing and Matrials: West Conshohocken, PA, USA, 1999.
9. Lawson, R.M. *Sustainability of Steel in Housing and Residential Buildings (P370)*; The Steel Construction Institute: Ascot, UK, 2007.
10. Lawson, R.M.; Francis, K. *Energy Efficient Housing Using Light Steel Framing (P367)*; The Steel Construction Institute: Ascot, UK, 2007.
11. Lu, Y.; Michael, J. A methodology for structural health monitoring with diffuse ultrasonic waves in the presence of temperature variations. *Ultrasonics* **2005**, *43*, 717–731. [CrossRef] [PubMed]
12. Su, Z.; Ye, L.; Lu, Y. Guided Lamb waves for identification of damage in composite structures: A review. *J. Sound Vib.* **2006**, *295*, 753–780. [CrossRef]
13. Ebrahimkhanlou, A.; Salamone, S. Single-Sensor Acoustic Emission Source Localization in Plate-Like Structures Using Deep Learning. *Aerospace* **2018**, *5*, 50. [CrossRef]
14. Ebrahimkhanlou, A.; Salamone, S. A probabilistic framework for single-sensor acoustic emission source localization in thin metallic plates. *Smart Mater. Struct.* **2017**, *26*, 095026. [CrossRef]
15. Ebrahimkhanlou, A.; Salamone, S. Acoustic emission source localization in thin metallic plates: A single-sensor approach based on multimodal edge reflections. *Ultrasonics* **2017**, *78*, 134–145. [CrossRef] [PubMed]
16. Carpinteri, A.; Lacidogna, G.; Puzzi, S. From criticality to final collapse: Evolution of the "b-value" from 1.5 to 1.0. *Chaos Solitons Fractals* **2009**, *4*, 843–853. [CrossRef]
17. Carpinteri, A.; Corrado, M.; Lacidogna, G. Three different approaches for damage domain characterization in disordered materials: Fractal energy density, b-value statistics, renormalization group theory. *Mech. Mater.* **2012**, *53*, 15–28. [CrossRef]
18. Balageas, J.; Fritzen, C.; Guemes, A. *Structural Health Monitoring Systems*; ISTE: Washington, DC, USA, 2006.
19. Adams, D. *Health Monitoring of Structural Materials and Components*; Wiley: New York, NY, USA, 2007.
20. Swamy, R.N. Durability of Rebars in Concrete. *Spec. Publ.* **1992**, *131*, 67–98.
21. Vennesland, O.; Gjorv, O.E. Effect of Cracks in Submerged Concrete Sea Structures on Steel Corrosion. *Mater. Perform.* **1981**, *20*, 49–51.
22. Alampalli, S.; Ettouney, M. Results of workshop on structural health monitoring in bridge security. In Proceedings of the 3rd International Conference on Structural Health Monitoring of Intelligent Infrastructure, Vancouver, BC, Canada, 14–16 November 2007; pp. 13–16.
23. Inaudi, D. Structural Health Monitoring of bridges: General Issues and Applications. *Struct. Health Monitor. Civ. Infrastruct. Syst.* **2009**, 339–370.
24. Zhou, H.F.; Ni, Y.Q.; Ko, J.M. Structural Damage Alarming Using Auto-Associative Neural Network Technique: Exploration of Environment-Tolerant Capacity and Setup of Alarming Threshold. *Mech. Syst. Signal Process.* **2011**, *25*, 1508–1526. [CrossRef]
25. Olaszek, P.; Świt, G.; Casas, J.R. On-site assessment of bridges supported by acoustic emission. *Proc. Inst. Civ. Eng. Bridge Eng.* **2016**, *169*, 81–92. [CrossRef]

26. Świt, G.; Adamczak, A.; Krampikowska, A. Wavelet Analysis of Acoustic Emissions during Tensile Test of Carbon Fibre Reinforced Polymer Composites. *IOP Conf. Ser. Mater. Sci. Eng.* **2017**, *245*. [CrossRef]

27. Świt, G.; Adamczak, A.; Krampikowska, A. Time-frequency analysis of acoustic emission signals generated by the Glass Fibre Reinforced Polymer Composites during the tensile test. *IOP Conf. Ser. Mater. Sci. Eng.* **2017**, *251*, 1–8. [CrossRef]

28. Goszczyńska, B.; Świt, G.; Trąmpczyński, W.; Krampikowska, A. Application of the Acoustic Emission Method of identification and location of destructive processes to the monitoring concrete bridges. In Proceedings of the 7th International Conference on Bridge Maintenance, Safety and Management (IABMAS), Shanghai, China, 7–11 July 2014; pp. 688–694.

29. Świt, G.; Krampikowska, A. Influence of the Number of Acoustic Emission Descriptors on the Accuracy of Destructive Process Identification in Concrete Structures. In Proceedings of the 2016 Prognostics and System Health Management Conference, Chengdu, China, 19–21 October 2016; pp. 6–11.

30. Świt, G.; Krampikowska, A.; Chinh, L.M. A Prototype System for Acoustic Emission-Based Structural Health Monitoring of My Thuan Bridge. In Proceedings of the 2016 Prognostics and System Health Management Conference, Chengdu, China, 19–21 October 2016; pp. 624–630.

31. Chinh, L.M.; Adamczak, A.; Krampikowska, A.; Świt, G. Dragon bridge—The world largest dragon-shaped (ARCH) steel bridge as element of smart city. In Proceedings of the International Conference on the Sustainable Energy and Environment Development (SEED 2016), Krakow, Poland, 17–19 May 2016; Volume 10, pp. 1–5.

32. Świt, G.; Krampikowska, A.; Chinh, L.M.; Adamczak, A. Nhat Tan Bridge—THE Biggest Cable-Stayed Bridge in Vietnam. *Procedia Eng.* **2016**, *161*, 666–673. [CrossRef]

33. Olaszek, P.; Świt, G.; Casas, J.R. Proof load testing supported by acoustic emission. An example of application. In *Bridge Maintenance, Safety, Management and Life-Cycle Optimization, Proceedings of the Fifth International IABMAS Conference, Philadelphia, PA, USA, 11–15 July 2010*; CRC Press: Boca Raton, FL, USA, 2010; pp. 472–479.

34. Deraemaeker, A.; Reynders, E.; De Roeck, G.; Kullaa, J. Vibration—Based structural health monitoring using output-only measurements under changing environment. *Mech. Syst. Signal Process.* **2008**, *22*, 34–56. [CrossRef]

35. Wong, K.Y. Design of a Structural Health Monitoring System for Long-Span Bridges. *Struct. Infrastruct. Eng.* **2007**, *3*, 169–185. [CrossRef]

36. Catbas, F.N. Structural Health Monitoring: Applications and Data Analysis. In *Structural Health Monitoring of Civil Infrastructure Systems*; Woodhead Publishing: Cambridge, UK, 2009.

37. Goszczyńska, B.; Świt, G.; Trąmpczyński, W. Application of the IADP acoustic emission method to automatic control of traffic on reinforced concrete bridges to ensure their safe operation. *Arch. Civ. Mech. Eng.* **2016**, *16*, 867–875. [CrossRef]

38. Świt, G. Evaluation of compliance changes in concrete beams reinforced by glass fiber reinforced plastics using acoustic emission. *J. Mater. Civ. Eng.* **2004**, *16*, 414–418. [CrossRef]

39. Goszczyńska, B.; Świt, G.; Trąmpczyński, W. Analysis of the microcracking process with the Acoustic Emission method with respect to the service life of reinforced concrete structures with the example of the RC beams. *Bull. Pol. Acad. Sci. Tech. Sci.* **2015**, *63*, 55–65. [CrossRef]

 *applied sciences*

Article

Time Series Analysis of Acoustic Emissions in the Asinelli Tower during Local Seismic Activity

Alberto Carpinteri, Gianni Niccolini and Giuseppe Lacidogna *

Department of Structural, Geotechnical and Building Engineering, Politecnico di Torino, Corso Duca degli Abruzzi, 24-10129 Torino, Italy; alberto.carpinteri@polito.it (A.C.); giuseppe.lacidogna@polito.it (G.N.)
* Correspondence: giuseppe.lacidogna@polito.it; Tel.: +39-011-090-4910

Received: 5 May 2018; Accepted: 14 June 2018; Published: 21 June 2018

Abstract: The existence of ongoing damage processes in a masonry wall of the Asinelli Tower in Bologna have been investigated by the acoustic emission (AE) technique. A time correlation between the AE activity in the monitored structural element and the nearby earthquakes has been observed. In particular, the largest cluster of AE signals has been recorded within a few hours after the main shock (4.1 magnitude) occurrence. The presented findings suggest that aging and deterioration of the monitored structural element significantly depend on the action of light earthquakes, even at considerable distance. Trends of two evolutionary parameters, the b-value and the natural time variance κ_1, have been derived from the AE time series in order to identify the approach of the monitored structural element to a "critical state" in relation to the earthquake occurrence.

Keywords: structural health monitoring; acoustic emission; time series analysis; b-value; natural time; critical phenomena

1. Introduction

Fracture precursors in metallic [1,2] and quasi-brittle materials like rocks [3,4], concrete [5–9], and masonry [10,11] can be experimentally investigated focusing on the statistical properties of an acoustic emission (AE) time series from growing micro-fractures, where the discovery of underlying scaling laws suggests a description of fracture as a critical phenomenon [12–21]. Within this context, finding fracture precursors means identifying critical scaling exponents and early indicators of the approach to a critical state [22].

The relevance of the AE applications in civil engineering for structural health monitoring is widely recognized. In this regard, it is increasingly necessary to give the highest priority to seismic risk mitigation in large portions of the Italian territory. Minor and light earthquakes can drive invisible damage processes in buildings and monuments, which eventually result in catastrophic collapses during stronger earthquakes. It is worth noting that AE can be exploited as a diagnostic tool in geophysics as well, since recent experimental evidences and theoretical studies support the hypothesis that increased AE and electromagnetic activities may be signature of crustal stresses redistribution in a large zone during the preparation of a seismic event [23–34]. Laboratory studies have been motivated by the need to provide tools for the earthquake prediction [35].

Therefore, the AE structural monitoring might potentially provide twofold information: one concerning the structural damage and the other concerning widespread micro-seismic activity, propagating across the ground-building foundation interface, for which the building foundation represents a sort of extended underground probe [10,11,21,23,24,30,33].

The presented research study was initially motivated by the debate about the alleged incompatibility between heavy vehicle traffic and Bologna's historical center, which concerns also the structural stability of two medieval towers, the Garisenda and the Asinelli (the taller) [36]. As reported in a previous publication [30], the influence of environmental phenomena on the AE activity in the Asinelli tower, such as the flux of surrounding vehicle traffic, unusual anthropogenic activities, and wind action, has been excluded. Actually, the nearby seismicity apparently has some influence on the AE activity detected in a masonry wall of the tower [30]. Here, the focus is on the time correlation observed between the largest AE cluster and the strongest (magnitude 4.1) local earthquake, with an epicenter 100 km far from the monitoring site.

2. Materials and Methods

An array of six broadband AE transducers (with a flat frequency response nominally in the range of 50–500 kHz, as declared by the manufacturer, i.e., well beyond the frequency range of signals propagating in the masonry) was fixed to the north-east angle of the Asinelli Tower at an average height of 9.00 m above ground level. The frequencies of interest in the old masonry, around 30 kHz, were still detected by the adopted transducers as already reported in previous papers for similar surveys [37]. Figure 1a–c show the tower and the transducers applied to a masonry wall portion.

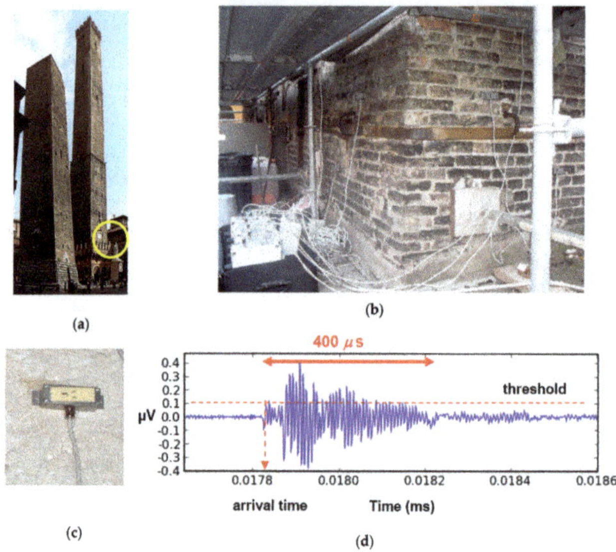

Figure 1. (**a**) The Asinelli Tower (with Garisenda Tower on the left) in the city centre of Bologna; (**b**) Monitored portion of the masonry wall with the applied AE transducers, printed from [30]; (**c**) AE transducer adopted for the monitoring; (**d**) Typical AE signal formed by the sequence of P-, S- and surface waves.

The transducers were connected to a six-channel acquisition system able to store AE signal parameters such as arrival time t (determined with accuracy of 0.2 µs), duration, peak amplitude, and ring-down count (number of times the AE signal exceeds a preset threshold). A time accuracy of 0.2 µs, equivalent to a sampling rate of five mega-samples per second, is adequate to measure frequency components up to 500 kHz (only frequencies up to 10 times smaller than the sampling rate are usually considered), covering satisfactorily the AE frequency range.

Prior to starting the monitoring, were preliminarily performed for a representative period of time, i.e., 8 h, in order to determine the level of spurious signals. Thus, the signal acquisition threshold was set to 100 µV in order to filter out the electrical noise (see a typical signal waveform in Figure 1d). Keeping fixed the threshold made it possible to capture possible variations in the noise level due to changeable environmental conditions like traffic, weekdays, weekends, etc.

AE monitoring began on 23 September 2010 at 5:40 p.m. and ended on 28 January 2011 at 1:00 p.m. for a 2915-h period [30]. All the AE signals were found to fall in the amplitude range of 100 to 12,800 µV or, equivalently, of 40 to 82 dB, if the signal peak amplitudes $A_{\max}$ are expressed in decibels (dB), $A_{dB} = 20 \log_{10}(A_{\max}/1 \text{ µV})$ [38]. Because of heterogeneity of masonry, the point location method of AE sources exploiting signals recorded by multiple transducers could hardly be applied.

3. Results

3.1. Correlation between AE and Earthquakes

The plot of the AE signals count rate over the monitoring time (Figure 2) suggests that the masonry wall is undergoing a damage process. The clusters of AE hits, especially around the time 500 h, can be regarded as signature of high crack growth rate. During the monitoring period, frequent seismic events occurred in the region [30], but only those that might have affected the stability of the tower have been considered. Basing on their magnitude and epicentral distance, we have selected the two strongest regional earthquakes (the magnitude 4.1 event that hit the Rimini area on 13 October 2010 at 11:43 p.m. with epicenter about 100 km far from Bologna, as shown in Figure 3, and the magnitude 3.4 event recorded in the Modena Apennines on 21 November 2010 at 4:10 p.m.) and the nearby earthquakes with magnitude ≥ 0.5 and epicentral distance ≤ 20 km from the monitoring site.

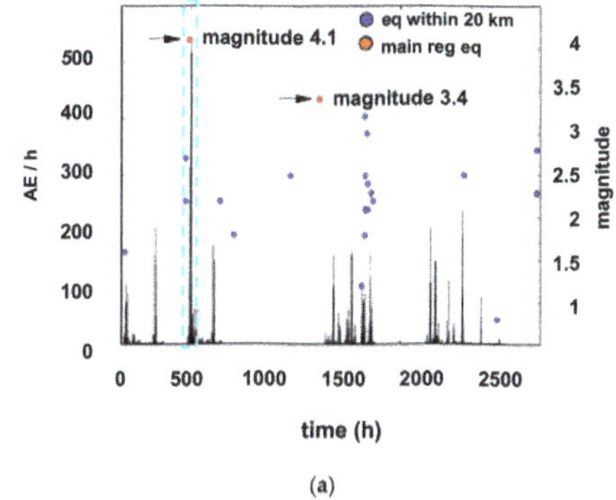

(a)

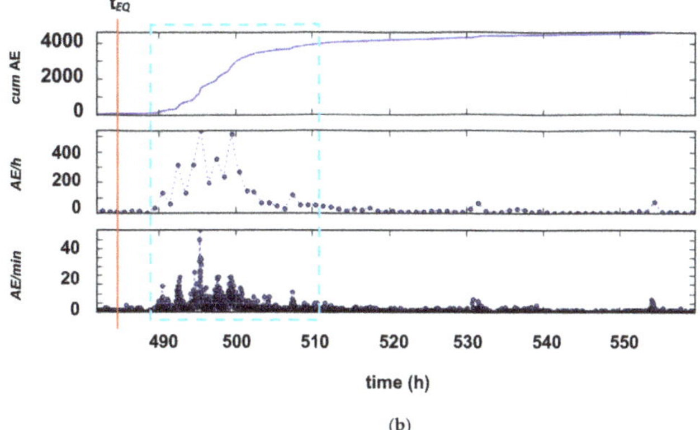

(b)

Figure 2. (a) Time series of the AE signals count rate and nearby earthquakes (extracted from http://www.ingv.it/it/, see Supplementary Materials) printed from [28]; (b) Expansion of the dashed frame: the red line marks the occurrence time t_{EQ} = 486 h of 13 October; earthquake (EQ).

Figure 2a shows the AE signals count rate and the earthquake time series, referring to the entire monitoring period, where a correlation in time between AE clusters and seismic events can be observed. A significant example is given by the seismic and the AE sequences that occurred between 1500 and 1800 h. Even more remarkably, the densest AE cluster (formed by 4000 hits and highlighted by the dashed frame in Figure 2) and the magnitude 4.1 earthquake that occurred on 13 October 2010 at 11:43 p.m. appear to be closely correlated.

Unless uncontrolled factors affecting the measurements, the main seismic shock apparently triggered temporary intensification of the AE activity, revealing local instability until the recovery of equilibrium. The comparatively longer duration of this AE cluster with respect the earthquake duration could be explained by a viscoelastic behavior, assumed for masonry structures [39], which would produce stress and strain over time in response to an impulsive load [40].

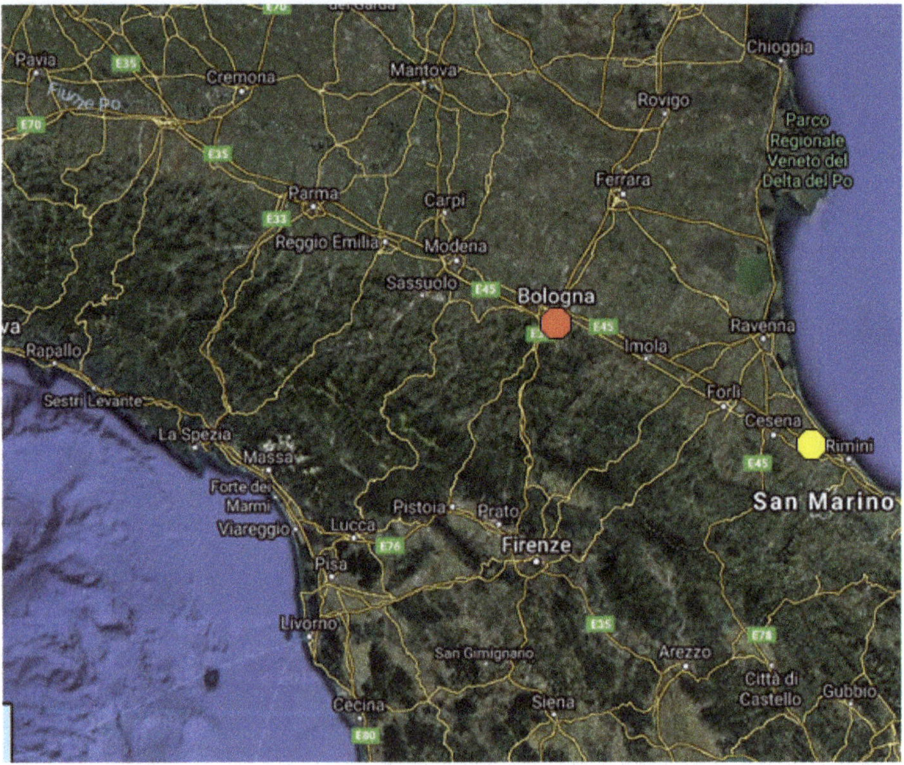

Figure 3. Map showing the epicenter of the 4.1-magnitude earthquake (yellow point) occurred on 13 October 2010 with epicenter 100 km far from the monitoring site in Bologna (red point).

3.2. b-Value versus Natural Time Analyses of AE Time Series

In order to investigate the entrance of the structural element to a critical state, we focus on the cluster of AE hits occurred in the interval 490–510 h. The temporal evolution of two AE parameters has been considered: the b-value of the Gutenberg-Richter (GR) law, and the variance κ_1 of the natural-time transformed time series [25–27].

The GR law, initially introduced in seismology [41] and then extended to the statistics of AE signals [38,42–46], is expressed by the relation

$$\log_{10} N = a - b \times M \tag{1}$$

where $M \equiv \log_{10}(A_{\max}/1\ \mu V)$ is the magnitude [38,45], i.e., the logarithm of the AE peak amplitude, N is the incremental frequency, i.e., the number of AE hits with magnitude greater than the threshold M, and a and b (termed as b-value) are empirical constants to be fitted. The standard error of b-value is $b/\sqrt{N}$ for a population of N samples and 95% confidence limits are twice this value.

The b-value is the negative slope of the linear descending branch of the GR law, and it grossly represents the relative number of micro-fractures to macro-fractures. Generally, systematic decreasing b-values are indeed observed during laboratory loading tests, with a single minimum just prior to specimen failure [38,42–46].

In the present study, the b-value analysis has been carried out twice, by partitioning the data set in groups formed by 300 and 800 hits. The b-value of each subset has been obtained through Equation (1) and regression analysis in the linear range of GR graphs, from 2.0 to 3.8 or 4.1 magnitude (i.e., from 40 to 76 or 82 dB). The related time series of b-values exhibit similar trends, especially sharing a global minimum $b = 0.9$–1 at time $t_{b\text{-}min} = 501$ h (marked in bottom diagram of Figure 4). It is worth remembering that b-values close to unity correspond to the growth of macro-fractures in the monitored element [38,39].

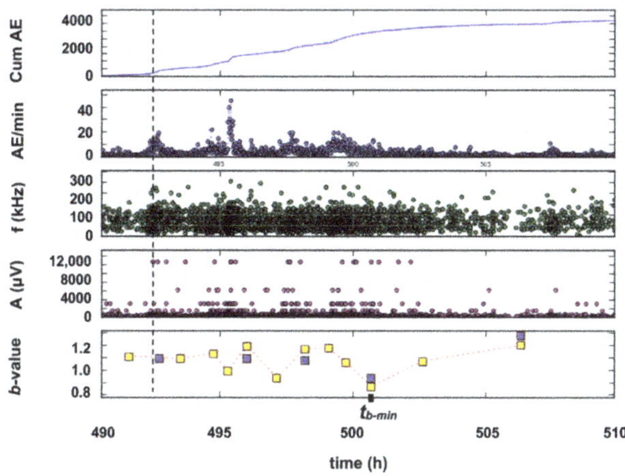

Figure 4. From top to bottom: accumulated number of AE signals; AE signals count rate; time series of AE signals frequencies (derived from signal duration and ring-down count) and amplitudes; b-value over time calculated using groups of, respectively, 300 (yellow squares), and 800 (blue squares) numbers of signals. The dashed line indicates a critical point revealed by the natural time analysis.

The results of the b-value analysis appear to be reliable, as they are substantially independent from the chosen partition. The final trend toward higher b-values suggests that the monitored masonry wall is substantially stable, with a damaging episode driven by the specific seismic event, and interpretable as a sign of enhanced structural sensitivity to light earthquakes.

In order to take into account variations in the statistical properties of the AE amplitude distributions, the "improved b-value" (Ib) was introduced and defined as follows [42,44,45]:

$$Ib = \left[\log_{10} N(\mu - \alpha_1 \sigma) - \log_{10} N(\mu + \alpha_2 \sigma)\right] / \left[(\alpha_1 + \alpha_2)\sigma\right] \tag{2}$$

where μ and σ are mean and standard deviation of each AE amplitude subset, found to be, respectively, ~45 and 8 dB, whereas α_1 and α_2 are user-defined coefficients representing lower and upper limits of the amplitude range in which the cumulative frequency–magnitude distribution properly fits a straight line. The *Ib-* and *b*-values formed by a 300-hit partitioning are plotted in Figure 5, showing a good accordance between the two analyses.

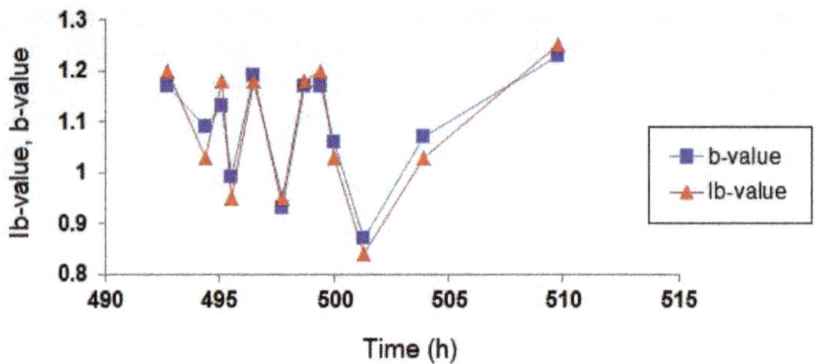

Figure 5. The *b*-value and *Ib*-value calculated using groups of 300 numbers of signals.

Besides the *b*-value, other synthetic parameters acting as potential fracture precursors can be extracted from a time series of N AE hits by the natural time analysis. This concept is introduced by ascribing the natural time $\chi_k = k/N$ to the *k*-th event of energy Q_k [25–27].

Regarding the normalized energies $p_k \equiv Q_k/\Sigma_{k=1}^N Q_i$ as the probability distribution of the discrete variable χ_k, the variance $\kappa_1 \equiv \overline{\chi^2} - \chi^2$, is considered a key parameter for identifying the approach to a critical state, and defined as follows:

$$\kappa_1 \equiv \sum_{k=1}^N p_k \chi_k^2 - \left(\sum_{k=1}^N p_k \chi_k \right)^2 \equiv <\chi^2> - <\chi^2> \tag{3}$$

As χ_k and p_k are rescaled upon the occurrence of any additional hit, κ_1 results to be an evolutionary parameter.

It has been successfully shown that a variety of dynamical systems (2D Ising model [47], Bak-Teng-Wiesenfeld sandpile model [12,47], and pre-seismic electric signals [25–27]) become critical when κ_1, evolving hit by hit, approaches the value 0.07.

Two criteria were defined to identify the entrance of a system to true critical state [47–50]:

(1)   the parameter κ_1 must approach the value 0.07 "by descending from above";
(2)   the entropies S and S_{rev} (entropy upon time reversal) must be lower than the entropy of uniform noise, $S_u = 0.0966$, when κ_1 coincides to 0.07. The entropy S is defined as

$$S \equiv <\chi \ln \chi> - <\chi> \ln <\chi> \tag{4}$$

where $<\chi \ln \chi> = \sum_{k=1}^N p_k \chi_k \ln \chi_k$.

Here, the evolution of variance κ_1 and entropies S and S_{rev} of the natural-time transformed AE time series $\{\chi_k\}$ has been studied, where the event energy Q_k is derived from the amplitude A_k through the relation $Q_k = cA_k^{1.5}$, where c is a constant of proportionality [51,52]. Plotting all natural-time quantities as functions of the conventional time t provides a visual way to reveal the possible entrance point to "critical stage," corresponding to the fulfillment of criticality Conditions (1) and (2). An entrance point to critical stage has been identified at time $t_{crit} = 492$ h (criticality initiation time, marked by a vertical dashed line in Figure 4 and by a vertical dotted line in Figure 6), i.e., 9 h before the *b*-value reaches its minimum. Remarkably, this critical point corresponds to the middle-height peak in the AE rate (highlighted by the dashed line before the highest peak in Figure 4) and to the appearance of high-amplitude AE signals.

This analysis suggests that the variance κ_1 of the natural-time transformed AE series can be identified as a pre-failure indicator, before the onset of non-reversible damage—supposedly revealed by the minimum b-value—within the bulk of the structural element.

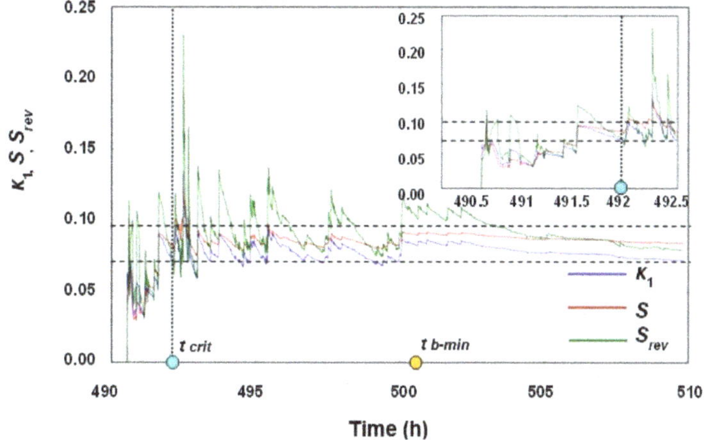

Figure 6. Time evolution of natural-time quantities κ_1, S and S_{rev}. Horizontal dashed lines represent the characteristics value $\kappa_1 = 0.07$ and $S_u = 0.0966$ defining the criticality initiation time t_{crit}. Note the relative positions of t_{crit} and $t_{b\text{-}min}$ along the time-axis (the inset shows the approach of natural-time quantities to the critical point).

4. Conclusions

The stability of a masonry wall of the Asinelli Tower has been assessed by the Acoustic Emission (AE) technique for a four-month period. The trend of the whole AE time series suggests that the monitored structural element is substantially stable, albeit with an ongoing damage process revealed by the AE activity observed over the entire monitoring period.

The observed correlation between the AE time series and the sequence of local earthquakes suggests that the local structural response is driven by the nearby seismicity. In particular, the densest AE cluster has been recorded within a few hours after the occurrence of the main local earthquake (4.1 magnitude). The b-value analysis of the AE cluster has revealed a significant, though momentary, acceleration of aging and deterioration processes in the monitored element, whose mechanical stability appears to be affected by light earthquakes.

Another emerging pattern is the transition of the monitored element to a state of criticality according to the paradigm of the natural time analysis [25–27]. The entrance to the critical state before the b-value reaches its minimum leads to consider the natural time variance κ_1 [47–50], as a possible early failure precursor. Future investigations in similar surveys would hopefully include the use of a seismometer, installed together with the AE system to record possible small seismic events affecting the observations.

Supplementary Materials: Earthquake data were taken from the website of the Istituto Nazionale di Geofisica e Vulcanologia-INGV, http://www.ingv.it/it/.

Author Contributions: A.C. conceived the research study; G.L. designed and performed the experiments; G.N. analyzed the data and wrote the paper.

Funding: This research received no external funding.

Acknowledgments: The authors wish to thank the Municipality of Bologna and Eng. R. Pisani for allowing this study on the Asinelli Tower. This research received no specific grant from any funding agency in the public, commercial, or not-for-profit sectors.

Conflicts of Interest: The authors declare no conflict of interest.

References

1. Bhuiyan, M.Y.; Giurgiutiu, V. The signature of acoustic emission waveforms from fatigue crack advancing in thin metallic plates. *Smart Mater. Struct.* **2018**, *27*, 015019. [CrossRef]
2. Bhuiyan, M.Y.; Lin, B.; Giurgiutiu, V. Acoustic emission sensor effect and waveform evolution during fatigue crack growth in thin metallic plate. *J. Intell. Mater. Syst. Struct.* **2017**, *29*, 1275–1284. [CrossRef]
3. Mogi, K. Study of elastic shocks caused by the fracture of heterogeneous materials and its relation to earthquake phenomena. *Bull. Earthq. Res. Inst.* **1962**, *40*, 125–173.
4. Aki, K. A probabilistic synthesis of precursory phenomena. In *Earthquake Prediction*; Simpson, D.W., Richards, P.G., Eds.; American Geophysical Union: Washington, DC, USA, 1981; pp. 566–574.
5. Shah, S.P.; Li, Z. Localization of microcracking in concrete under uniaxial tension. *ACI Mater. J.* **1994**, *91*, 372–381.
6. Grosse, C.U.; Reinhardt, H.W.; Dahm, T. Localization and classification of fracture types in concrete with quantitative acoustic emission measurement techniques. *NDT Int.* **1997**, *30*, 223–230. [CrossRef]
7. Ohtsu, M.; Okamoto, T.; Yuyama, S. Moment tensor analysis of acoustic emission for cracking mechanisms in concrete. *ACI Struct. J.* **1998**, *95*, 87–95.
8. Aggelis, D.G.; Mpalaskas, A.C.; Matikas, T.E. Investigation of different fracture modes in cement-based materials by acoustic emission. *Cem. Concr. Res.* **2013**, *48*. [CrossRef]
9. Carpinteri, A.; Lacidogna, G.; Niccolini, G. Damage analysis of reinforced concrete buildings by the acoustic emission technique. *Struct. Control Health Monit.* **2011**, *18*, 660–673. [CrossRef]
10. Carpinteri, A.; Lacidogna, G.; Niccolini, G. Acoustic emission monitoring of medieval towers considered as sensitive earthquake receptors. *Nat. Hazards Earth Syst. Sci.* **2007**, *7*, 251–261. [CrossRef]
11. Carpinteri, A.; Lacidogna, G. Damage evaluation of three masonry towers by acoustic emission. *Eng. Struct.* **2007**, *29*, 1569–1579. [CrossRef]
12. Bak, P.; Tang, C. Earthquakes as a self-organized critical phenomenon. *J. Geophys. Res.* **1989**, *94*, 15635–15637. [CrossRef]
13. Zapperi, S.; Vespignani, A.; Stanley, H.E. Plasticity and avalanche behaviour in microfracturing phenomena. *Nature* **1997**, *388*, 658–660. [CrossRef]
14. Guarino, A.; Garcimartin, A.; Ciliberto, S. An experimental test of the critical behaviour of fracture precursors. *Eur. Phys. J. B* **1998**, *6*, 13–24. [CrossRef]
15. Bonnet, E.; Bour, O.; Odling, N.E.; Davy, P.; Main, I.P.; Cowie, P.; Berkowitz, B. Scaling of fracture systems in geological media. *Rev. Geophys.* **2001**, *39*, 347–383. [CrossRef]
16. Bak, P.; Christensen, K.; Danon, L.; Scanlon, T. Unified scaling law for earthquakes. *Phys. Rev. Lett.* **2002**, *88*, 178501. [CrossRef] [PubMed]
17. Turcotte, D.L.; Newman, W.I.R.; Shcherbakov, R. Micro and macroscopic models of rock fracture. *Geophys. J. Int.* **2003**, *152*, 718–728. [CrossRef]
18. Corral, A. Statistical Features of Earthquake Temporal Occurrence. In *Modeling Critical and Catastrophic Phenomena in Geoscience*; Bhattacharya, P., Chakrabarti, B.K., Eds.; Springer: Berlin, Germany, 2006; Volume 705, pp. 191–221, ISBN 978-3-540-35375-1.
19. Davidsen, J.; Stanchits, S.; Dresen, G. Scaling and universality in rock fracture. *Phys. Rev. Lett.* **2007**, *98*, 125502. [CrossRef] [PubMed]
20. Kun, F.; Carmona, H.A.; Andrade, J.S., Jr.; Herrmann, H.J. Universality behind Basquin's Law of Fatigue. *Phys. Rev. Lett.* **2008**, *100*, 094301. [CrossRef] [PubMed]
21. Niccolini, G.; Carpinteri, A.; Lacidogna, G.; Manuello, A. Acoustic emission monitoring of the Syracuse Athena Temple: Scale invariance in the timing of ruptures. *Phys. Rev. Lett.* **2011**, *106*, 108503. [CrossRef] [PubMed]
22. Stanley, H.E. Scaling, renormalization and universality: Three pillars of modern critical phenomena. *Rev. Mod. Phys.* **1999**, *71*, 358–366. [CrossRef]

23. Gregori, G.P.; Paparo, G. Acoustic Emission (AE): A diagnostic tool for environmental sciences and for non destructive tests (with a potential application to gravitational a antennas). In *Meteorological and Geophysical Fluid Dynamics*; Schroeder, W., Ed.; Science Edition: Bremen, Germany, 2004; pp. 166–204.

24. Gregori, G.P.; Paparo, G.; Poscolieri, M.; Zanini, A. Acoustic emission and released seismic energy. *Nat. Hazards Earth Syst. Sci.* **2005**, *5*, 777–782. [CrossRef]

25. Varotsos, P.A.; Sarlis, N.V.; Skordas, E.S. Spatio-temporal complexity aspects on the interrelation between seismic electric signals and seismicity. *Pract. Athens Acad. Greece* **2001**, *76*, 294–321.

26. Varotsos, P.A.; Sarlis, N.V.; Skordas, E.S. *Natural Time Analysis: The New View of Time*; Springer: Berlin/Heidelberg, Germany, 2011.

27. Varotsos, P.A.; Sarlis, N.V.; Skordas, E.S.; Lazaridou, M.S. Seismic Electric Signals: An additional fact showing their physical interconnection with seismicity. *Tectonophysics* **2013**, *589*, 116–125. [CrossRef]

28. Lacidogna, G.; Carpinteri, A.; Manuello, A.; Durin, G.; Schiavi, A.; Niccolini, G.; Agosto, A. Acoustic and electromagnetic emissions as precursor phenomena in failure processes. *Strain* **2011**, *47*, 144–152. [CrossRef]

29. Niccolini, G.; Borla, O.; Lacidogna, G.; Carpinteri, A. Correlated Fracture Precursors in Rocks and Cement-Based Materials under Stress. In *Acoustic, Electromagnetic, Neutron Emissions from Fracture and Earthquakes*; Carpinteri, A., Lacidogna, G., Manuello, A., Eds.; Springer International Publishing: Cham, Switzerland, 2015; pp. 238–248.

30. Carpinteri, A.; Lacidogna, G.; Manuello, A.; Niccolini, G. A study on the structural stability of the Asinelli Tower in Bologna. *Struct. Control Health Monit.* **2016**, *23*, 659–667. [CrossRef]

31. Carpinteri, A.; Borla, O. Fracto-emissions as seismic precursors. *Eng. Fract. Mech.* **2017**, *177*, 239–250. [CrossRef]

32. Borla, O.; Lacidogna, G.; Di Battista, E.; Niccolini, G.; Carpinteri, A. Electromagnetic emission as failure precursor phenomenon for seismic activity monitoring. In *Proceedings of the Conference & Exposition on Experimental and Applied Mechanics (SEM), Greenville, SC, USA, 2–5 June 2014*; Springer International Publishing: Cham, Switzerland, 2014; Volume 5, pp. 221–229.

33. Niccolini, G.; Manuello, A.; Marchis, E.; Carpinteri, A. Signal frequency distribution and natural time analyses from acoustic emission monitoring of an arched structure in the Racconigi Castle. *Nat. Hazards Earth Syst. Sci.* **2017**, *17*, 1025–1032. [CrossRef]

34. Dobrovolsky, I.P.; Zubkov, S.I.; Miachkin, V.I. Estimation of the size of earthquake preparation zones. *Pure Appl. Geophys.* **1979**, *117*, 1025–1044. [CrossRef]

35. Lei, X.; Ma, S. Laboratory acoustic emission study for earthquake generation process. *Earthq. Sci.* **2014**, *27*, 627–646. [CrossRef]

36. Pesci, A.; Casula, G.; Boschi, E. Laser scanning the Garisenda and Asinelli towers in Bologna (Italy): Detailed deformation patterns of two ancient leaning buildings. *J. Cult. Herit.* **2011**, *12*, 117–127. [CrossRef]

37. Carpinteri, A.; Lacidogna, G.; Invernizzi, S.; Accornero, F. The Sacred Mountain of Varallo in Italy: Seismic Risk Assessment by Acoustic Emission and Structural Numerical Models. *Sci. World J.* **2013**, *2013*, 170291. [CrossRef] [PubMed]

38. Colombo, S.; Main, I.G.; Forde, M.C. Assessing damage of reinforced concrete beam using "*b*-value" analysis of acoustic emission signals. *J. Mater. Civ. Eng. ASCE* **2003**, *15*, 280–286. [CrossRef]

39. Cecchi, A.; Tralli, A. A homogenized viscoelastic model for masonry structures. *Int. J. Solids Struct.* **2012**, *49*, 1485–1496. [CrossRef]

40. Pritchard, R.H.; Terentjev, E.M. Oscillations and damping in the fractional Maxwell materials. *J. Rheol.* **2017**, *61*, 187–203. [CrossRef]

41. Richter, C.F. *Elementary Seismology*; W.H. Freeman and Company: San Francisco, CA, USA; London, UK, 1958.

42. Shiotani, T.; Fujii, K.; Aoki, T.; Amou, K. Evaluation of progressive failure using AE sources and improved *b*-value on slope model tests. *Prog. Acoust. Emiss.* **1994**, *VII*, 529–534.

43. Sammonds, P.R.; Meredith, P.G.; Murrel, S.A.F.; Main, I.G. Modeling the evolution of damage in rock containing porefluid by acoustic emission. In Proceedings of the Eurock'94, Delft, The Netherlands, 29–31 August 1994.

44. Shiotani, T.; Yuyama, S.; Li, Z.W.; Ohtsu, M. Application of the AE improved *b*-value to qualitative evaluation of fracture process in concrete materials. *J. Acoust. Emiss.* **2001**, *19*, 118–132.

45. Rao, M.V.M.S.; Prasanna Lakhsmi, K.J. Analysis of *b*-value and improved *b*-value of acoustic emissions accompanying rock fracture. *Curr. Sci.* **2005**, *89*, 1577–1582.

46. Carpinteri, A.; Lacidogna, G.; Manuello, A.; Niccolini, G.; Puzzi, S. Critical defect size distributions in concrete structures detected by the acoustic emission technique. *Meccanica* **2008**, *43*, 349–363. [CrossRef]
47. Varotsos, P.A.; Sarlis, N.V.; Skordas, E.S.; Uyeda, S.; Kamogawa, M. Natural time analysis of critical phenomena. *Proc. Natl. Acad. Sci. USA* **2011**, *108*, 11361–11364. [CrossRef] [PubMed]
48. Varotsos, P.A.; Sarlis, N.V.; Skordas, E.S.; Lazaridou, M.S. Fluctuations, under time reversal, of the natural time and the entropy distinguish similar looking electric signals of different dynamics. *J. Appl. Phys.* **2008**, *103*, 014906. [CrossRef]
49. Hloupis, G.; Stavrakas, I.; Pasiou, E.D.; Triantis, D.; Kourkoulis, S.K. Natural time analysis of acoustic emissions in Double Edge Notched Tension (DENT) marble specimens. *Procedia Eng.* **2015**, *109*, 248–256. [CrossRef]
50. Hloupis, G.; Stavrakas, I.; Vallianatos, F.; Triantis, D. A preliminary study for prefailure indicators in acoustic emissions using wavelets and natural time analysis. *Proc. Inst. Mech. Eng. Part L* **2016**, *230*, 780–788. [CrossRef]
51. Kanamori, H.; Anderson, D.L. Theoretical basis of some empirical relationships in seismology. *Bull. Seismol. Soc. Am.* **1975**, *65*, 1073–1095.
52. Turcotte, D.L. *Fractal and Chaos in Geology and Geophysics*; Cambridge University Press: Cambridge, UK, 1997.

Article

Part Qualification Methodology for Composite Aircraft Components Using Acoustic Emission Monitoring

Shane Esola [1], Brian J. Wisner [1], Prashanth Abraham Vanniamparambil [1], John Geriguis [2] and Antonios Kontsos [1,*

[1] Theoretical & Applied Mechanics Group, Mechanical Engineering & Mechanics Department, Drexel University, 3141 Chestnut Street, Philadelphia, PA 19104, USA; sae47@drexel.edu (S.E.); bjw63@drexel.edu (B.J.W.); pv35@drexel.edu (P.A.V.)

[2] General Atomics-Aeronautical Systems Inc., 9779 Yucca Rd., Adelanto, CA 92301, USA; John.Geriguis@ga-asi.com

* Correspondence: antonios.kontsos@drexel.edu; Tel.: +1-215-895-2297

Received: 31 July 2018; Accepted: 24 August 2018; Published: 29 August 2018

Abstract: The research presented in this article aims to demonstrate how acoustic emission (AE) monitoring can be implemented in an industrial setting to assist with part qualification, as mandated by related industry standards. The combined structural and nondestructive evaluation method presented departs from the traditional pass/fail criteria used for part qualification, and contributes toward a multi-dimensional assessment by taking advantage of AE data recorded during structural testing. To demonstrate the application of this method, 16 composite fixed-wing-aircraft spars were tested using a structural loading sequence designed around a manufacturer-specified design limit load (DLL). Increasing mechanical loads, expressed as a function of DLL were applied in a load-unload-reload pattern so that AE activity trends could be evaluated. In particular, the widely used Felicity ratio (FR) was calculated in conjunction with specific AE data post-processing, which allowed for spar test classification in terms of apparent damage behavior. To support such analysis and to identify damage critical regions in the spars, AE activity location analysis was also employed. Furthermore, recorded AE data were used to perform statistical analysis to demonstrate how AE datasets collected during part qualification could augment testing conclusions by providing additional information as compared to traditional strength testing frequently employed e.g., in the aerospace industry. In this context, AE data post-processing is presented in conjunction with ultimate strength information, and it is generally shown that the incorporation of AE monitoring is justified in such critical part qualification testing procedures.

Keywords: part qualification; structural design; composites; acoustic emission; nondestructive evaluation (NDE)

---

1. Introduction

Current aircraft structural design and qualification methodologies require large amounts of testing in bottom-up type approaches that typically start at the coupon level and extend to full aircraft evaluation [1–7]. The prescribed assessment process is, therefore costly and time consuming, which additionally makes the adoption of new materials or design modifications difficult. This is especially challenging in the case of composite materials for which slight changes in manufacturing parameters can invalidate prior test data and require re-qualification of the material performance [8,9]. Typically, once extensive coupon testing is completed, design limit loads are computed for specific critical components. In this process, safety factors are added to account for reliability and uncertainty effects. Furthermore, pass/fail qualification criteria are implemented during component testing. This approach, however, limits the information that engineers can gather during full-scale part qualification testing that can be related to material performance.

In an effort to gather supplemental data, prior research has explored nondestructive evaluation (NDE) methods, including acoustic emission (AE), thermography, shearography, X-ray computed tomography (XCT), and ultrasonic testing (UT), for use during composite end-item qualification testing [10,11]. NDE data can, in general, augment existing structural test protocols by providing additional datasets, while also offering insight into damage initiation and progression. Modern NDE tools, however, have not yet been fully integrated into legacy aircraft qualification testing protocols, partially due to a lack of expertise, equipment, and knowledge on the usefulness of such methods, or even confidence that NDE could assist in this process. In this context, the U.S. Federal Aviation Administration has only recently proposed that the use of NDE tools for on-board structural health monitoring (SHM) could provide more confidence in complex aircraft designs while decreasing risk and maintenance costs [12].

AE is a NDE tool with the immediate potential to augment existing airframe strength test protocols, which is the focus of this investigation. Specifically, AE is a versatile, passive NDE method that senses pressure waves emitted from a variety of sources, predominantly linked to damage, across length scales and materials [13–16]. A physical analogue for AE is seismic activity resulting from tectonic plate motion of the Earth's crust. Provided that the appropriate sensing equipment and data-processing is applied, AE can provide real time volumetric information about dynamic changes related to damage. Based on this method, material damage processes at the coupon [17–20], component [21–24], and structural [25–27] scales have been reported. AE datasets can be further leveraged for advanced analytics such as data-driven models and machine learning [28–32], which may lead to a validated SHM methodology with potential even to be applied on-board, e.g., on aircraft [33].

Structural monitoring using AE is most commonly found in civil infrastructure [34–37], where it can be employed to passively collect data while a structure remains in use. The AE method has also been applied in wind turbine applications to prevent catastrophic failure of the large composite blades [38–41]. In addition, there have been some attempts to use AE in aircraft structural monitoring and component testing. Both military and civilian aircraft examples have been reported [42]. Other aircraft-focused investigations have used AE to examine damage localization and ultimate failure in landing gear [43], assess skin-spar bond-line integrity [44], evaluate impact damage locations [45,46], assess damage modes in composite fuselage under complex loading [47], determine the failure modes in composite airframe parts under quasi-static and cyclic loading [48], and assess rotorcraft composite fuselage durability and damage tolerance [49]. However, more progress is needed to gain the confidence required to integrate AE into structural qualification test standards [50,51].

In general, AE analysis and post-processing can leverage the recorded waveforms and a number of useful parameters extracted from them. Common AE waveform parameters are described in Shull [13] and ASNT's NDT Handbook [52]. Certain AE patterns can provide information about the test specimen's structural integrity, beyond pass/fail load criteria. For example, AE activity may not be observed until load levels are reached that previously induced damage. Specifically, if a structure is progressively loaded, new damage may not be initiated until the prior maximum load has been exceeded. This observation is known as the Kaiser effect [13,53], which, however, may not hold for composite materials similar to those evaluated in this manuscript [54,55]. In cases where AE activity is detected at loads that are lower than the previous maximum, then a specific number is computed called the Felicity ratio (FR)—defined as the load at which significant AE reinitiates, divided by the previous maximum load. Materials that consistently emit AE at loads below their previous maximum, thus, have a FR value of less than one, and investigations have shown that they exhibit progressive damage behavior. In fact, changes in observed Kaiser and FR effects as well as other AE activity patterns have been correlated with the presence of damage [56–59].

Based on this introduction, this research seeks to evaluate the progressive failure of full-scale, composite aircraft spars using AE during structural testing, and associate such information with ultimate failure prognosis. The insight gained from AE is intended to augment current test methodologies in order to capitalize on readily-available data and decrease the time taken to qualify novel airframe materials. The overall approach described includes the following parts: (1) characterization of damage progression by examining AE activity trends; (2) identification of probable damage regions; (3) ultimate failure load prognosis; and (4) statistical evaluation of the spar static strength requirements using AE data.

2. Technical Approach

2.1. Experimental Setup

The testing methodology applied generally follows the MIL-A-8867C(AS) protocol [1], and it is similar to the wind turbine blade qualification testing approach [60]. Sixteen composite spars were tested to bending failure while instrumented with AE sensors. Load was applied to a cantilevered spar in a stepwise repeated (pseudo-cyclic) loading manner while test samples were monitored by an AE sensing network. All tests were performed with flight hardware bushings installed. Slack was taken out of the loading apparatus before the load was applied to a spar. A given load was applied with a load rate of approximately 200 lbf/s. Once the target load was reached, it was held for approximately 150 s and then released before the spar was loaded to the following load level. Care was taken to apply the load slowly and steadily, to avoid imparting sudden impact loading. The applied load was distributed in the spar using the "whiffle-tree" device shown in Figure 1; a concentrated crane-load applied at the top of the loading apparatus was distributed through a series of beams and connectors to the eight ribs on the spar test bed. The load was distributed along the spar at the locations indicated by yellow boxes in Figure 1. The spar was pinned in two locations at the root, and laid along the test bed. By loading in this manner, the highest shear stress and moments occurred at the spar root, while gradually reducing toward the spar tip, to accurately represent the load distribution observed by the component during flight. Four target crane-load levels were tested for each spar (identified in the text as LL1, LL2, LL3, and LL4). After the final target crane load was achieved, loading continued at a rate of 200 lbf/s until part failure.

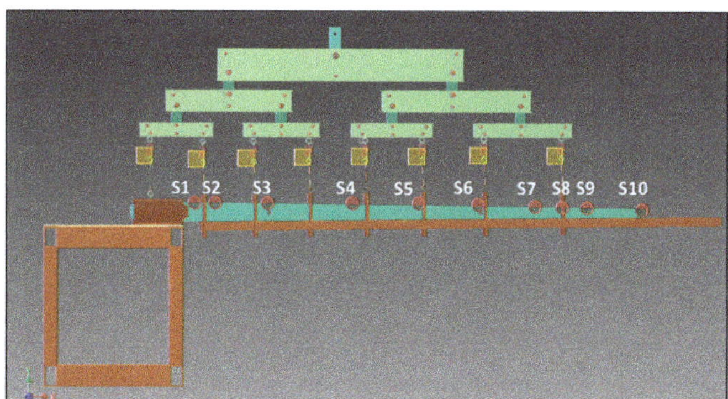

Figure 1. Schematic of the "whiffle-tree" loading apparatus to simulate distributed loading. Load cell locations are indicated by yellow boxes, while acoustic emission (AE) sensor locations are indicated as red circles.

Acoustic emission activity was recorded using eight to 10 commercially available R15I resonant piezoelectric sensors (operating frequency range of 50–400 kHz, manufactured by Physical Acoustics,

Princeton Junction, NJ, USA), distributed along the top of the spar from root to tail. Sensors were placed on the top surface of the spar since the top surface showed less attenuation in pre-testing and the mounting surface was relatively flat. The approximate sensor locations along the length of the spar measured as a distance from the spar root are given in Table 1 for each of the 16 tests. It should be noted that sensor locations varied by several inches from test-to-test, while no data was recorded during the first test, which was used to validate the loading method.

Table 1. Sensor locations along the length of each spar.

Test ID	Acoustic Emission Sensor ID and Location (in)									
	S1	S2	S3	S4	S5	S6	S7	S8	S9	S10
1	-	-	-	-	-	-	-	-	-	-
2	3	9	23.5	36	48	66.5	83.5	99.5	114	130
3	1	6	25	41.5	56.5	73	83.5	-	112.5	-
4	2	6	25	42	57	74	93.5	-	114	-
5	1	7	23.5	38	55	70	86	-	101	-
6	1	7	24	39.5	55	69	84	-	100	-
7	1.5	6	25.5	40.5	56.5	74	93	-	112.5	-
8	2	6.5	26	41.5	56	72.5	93	-	113.5	-
9	1	6	27	43	61	76	93	-	113.5	-
10	1.5	6.5	24.5	40.5	56.5	73.5	93.5	-	113.5	-
11	1	6	26	41.5	57	72	88	-	105.5	-
12	1	6.5	25.5	40.5	57	73.65	90.5	-	112.5	-
13	1	6.75	23.5	38	55	70	86	-	101	-
14	2	6.5	25.5	41	56.5	79	91	-	111.5	-
15	3	13.5	26	36	48	66.5	83.5	99.5	114	130
16	1.5	7.25	25	40.5	56	70	86	-	101	-

Sensors were bonded firmly in accordance with ASTM E650 and secured with tape in order to minimize the sensor loss during ultimate structural failure. Sensor cables were also taped, and sufficient cable slack was left free to allow movement during loading, unloading, and failure. Sensors were placed with sufficient spacing such that signal attenuation would not negatively impact results and they were field-calibrated in accordance with ASTM E976/1106 using a pencil lead source.

2.2. Data Acquisition and Processesing

AE signals were monitored and recorded using a Physical Acoustics PCI8-Express data acquisition board. Prior to testing for record, pre-tests were conducted to calibrate the AE sensors, determine sensor placement, rehearse the test procedure, and collect noise data. Ambient noise levels and the effect of the loading fixture (e.g., friction) were tested and adequately controlled via test procedures (e.g., slow and steady load application, lubricated connections, etc.). In addition to the procedural controls, the data acquisition amplitude threshold was set above the noise floor and AE hit (i.e., wave packet) record timing parameters were set to minimize typical noise waveforms recorded during pre-testing. AE signals were uniformly pre-amplified across all sensor frequencies using 40 dB sensor-internal pre-amplification. Crane load-cell data was synchronously collected and correlated with AE signal data to aid in AE activity trend observation.

As an example of the collected AE activity, Figure 2 showed representative waveforms collected from Test 8. Specifically, Figure 2a shows a low amplitude continuous AE signal, which is representative of noise recorded under zero load and demonstrably different compared to the burst-type signals typically associated with damage observed in Figure 2b,c. Figure 2b is an example of the type of AE signal recorded during the loading step, while Figure 2c shows an example of the type of waveforms recorded during unloading. In addition, Figure 2d shows the highest amplitude signal that was recorded at the time of final failure. Figure 2b,c both show similar wave characteristics including a burst of energy that occurs at 150 kHz while the signal obtained under no load showed that its energy was distributed across many frequency values, similar to the final fracture signal, which however, had significantly higher amplitude and broader frequency content.

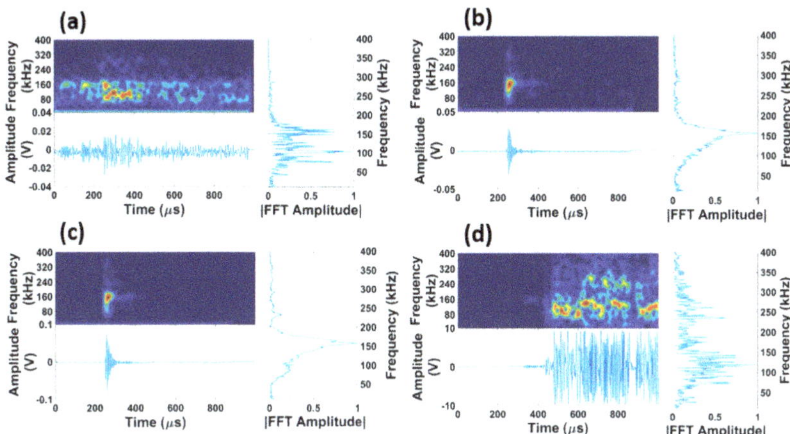

Figure 2. Sample AE activity observed during Test 8 at four different load points. (**a**) Zero load, (**b**) during the first load step, (**c**) during the second load step, and (**d**) at final failure.

As mentioned in the introduction, in general, there were two main AE data analysis approaches. The first leveraged the full recorded waveforms and the second relied on post-processing extracted features from such waveforms. For the analysis presented herein, raw AE signals similar to the ones shown in Figure 2 were parameterized via MISTRAS Group Inc. Noesis software (version 5.3, MISTRAS Group Inc., Princeton Junction, NJ, USA). The authors examined extracted AE features correlated with load data, including amplitude and absolute energy, to analyze AE activity trends and draw conclusions. AE signal energy was conceptually represented by the area under the rectified waveform, and was practically computed for discrete signal data by summing the square of the voltage amplitude over the signal length. To arrive at absolute energy with appropriate units (aJ), the squared voltage amplitude was further divided by the reference impedance over the signal duration. The cumulative absolute energy was the sum of the absolute energy for each AE hit, across all recorded hits during a test. The authors used this feature to characterize the test item performance, as it related to damage accumulation and ultimate failure. Additionally, the authors calculated the FR by identifying the load at which AE activity re-initiates during a loading cycle, divided by the previous maximum load reached. As discussed in the introduction, FR values less than one have been correlated with damage, and hence the authors used this parameter to further assess spar behavior.

3. Results and Discussion

3.1. Progressive Damage Characterization

Acoustic emission activity trends were used to infer progressive damage during pseudo-cyclic loading of the composite spar test specimens. Representative AE data is provided in Figure 3 for four different spar tests; the grey line provides the applied loading, while the red diamonds represent the recorded AE events correlated with the corresponding load values at which they occur. In addition, the blue markers indicate the AE activity amplitude distribution. To complement the graphical AE data trends, FR values were calculated and are reported in Table 2. All reported cases showed some AE activity prior to achieving the previous peak load, in accordance with the Felicity effect. However, Figure 3a,b shows AE events at comparatively lower loads, marked by black ellipses, compared to the test cases in Figure 3c,d. Additionally, in Figure 3b–d, it is noticeable that AE events occurred during the unloading portion of the second loading sequence. AE events during unloading can indicate the presence of damage [13,33,47,54,56], and are highlighted by green ellipses in Figure 3. Furthermore, the amplitude distributions in Figure 3a,b show relatively large spikes during the second cycle with values greater than 95 dB, while all other activity is below 70 dB for both Test 14 (Figure 3a) and Test 8 (Figure 3b). Sudden spikes of high-amplitude AE activity can correspond to material damage initiation or progression. In contrast, Tests 10 and 4 (Figure 3c,d) show a gradual increase in amplitude with each loading sequence. Based on the information presented in Figure 3 and Table 2, it can be concluded that the presented AE data visualization as a function of applied loading, in combination with the FR analysis, can provide some preliminary assessment of test item damage progression. Specifically, damage appears to initiate after the second target load (LL2) for most tests, as indicated by both the FR values, as well as the appearance of significant AE activity in the unloading portion of the applied cyclic loading. Furthermore, two distinct AE dataset classes existed among the 16 tests, including those that showed some abrupt and high-amplitude activity starting early in the loading sequence (e.g., Figure 3a,b) and those that demonstrated a more gradual AE activity, which was expected, as the loading increased with time.

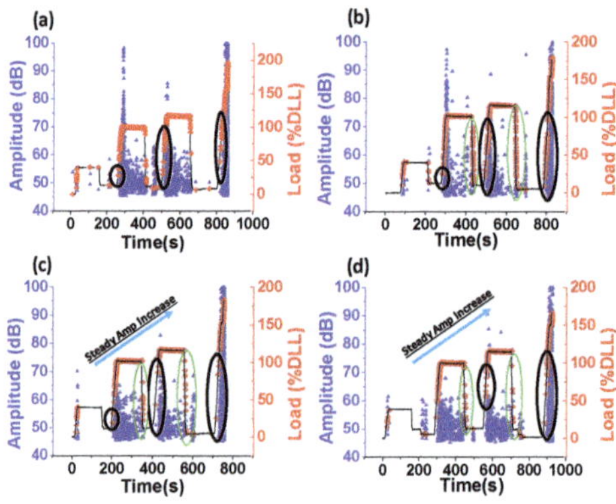

Figure 3. Sample AE activity observed during four separate spar tests. (**a**) Test 14, (**b**) Test 8, (**c**) Test 10, and (**d**) Test 4.

Table 2. Felicity ratio values calculated based on the four applied loading cycles.

File Name	Felicity Ratio			
	Load 2	Load 3	Load 4	Average
1	N/A	N/A	N/A	N/A
2	1.08	0.93	0.87	0.96
3	1.09	0.73	0.86	0.89
4	1.25	0.91	0.83	1.00
5	1.46	0.96	0.75	1.06
6	1.15	0.91	0.89	0.98
7	1.28	1.03	0.99	1.10
8	0.59	0.51	0.42	0.51
9	1.13	0.61	0.56	0.77
10	1.15	1.00	0.72	0.96
11	1.06	0.82	0.80	0.94
12	1.32	0.99	0.94	1.08
13	1.08	0.72	0.93	0.91
14	1.27	0.72	0.82	0.94
15	0.97	1.02	0.99	0.99
16	0.73	1.08	1.00	0.94

Supplemental to the AE activity trends shown in Figure 3, cumulative absolute AE energy is plotted against crane load in Figure 4. In all loading cases observed in Figure 4, the cumulative AE energy increases during loading, while remaining relatively constant during load holds and unloading. Similar to the AE amplitude spikes shown in Figure 3a,b during the second loading sequence, Figure 4a,b show corresponding jumps in AE energy. Such AE energy accumulations are observed in subsequent loading cycles, but they are orders of magnitude lower. Note that the observed pattern in Figure 4a,b is strikingly different from the results shown in Figure 4c,d. Accordingly, the recorded datasets were classified as significant vs. progressive in terms of the corresponding spar behavior. Specifically, Figure 4a,b is characterized by a sudden, dominant, discontinuous jump in AE energy, while Figure 4c,d is characterized by a gradual AE energy accumulation that appeared to be proportional to the load increase. Hence, the gradual cumulative AE energy profile observed in Figure 4c,d was classified as "progressive" (damage) behavior, in contrast to the large spikes ($>10^3$) displayed in Figure 4a,b, which were classified as "significant" behavior. Note that composite materials are known to exhibit progressive damage behavior and, therefore, test items that depart from this expected behavior (e.g., high intensity AE activity early in the loading cycle) may indicate design, production, or loading conditions that could lead to an early onset of catastrophic failure (i.e., lower ultimate load) when compared with similar parts.

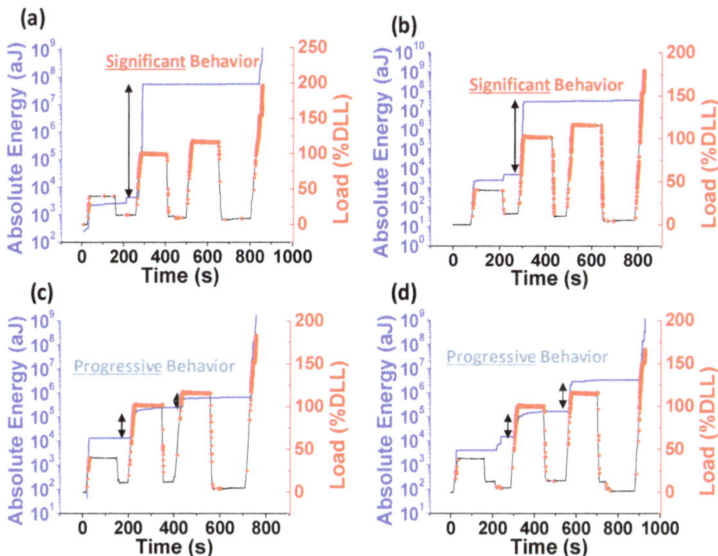

Figure 4. AE energy evolutions for the same tests shown in Figure 1. (**a**) Test 14, (**b**) Test 8, (**c**) Test 10, and (**d**) Test 4.

Similar trends were observed in all tested spars, and the behavior was classified as progressive or significant based on the cumulative absolute energy patterns noted in Figure 4. For example, Figure 5 shows four additional spar tests to further demonstrate the identified data trends using the same 10^3 aJ energy increase to denote the behavior as significant.

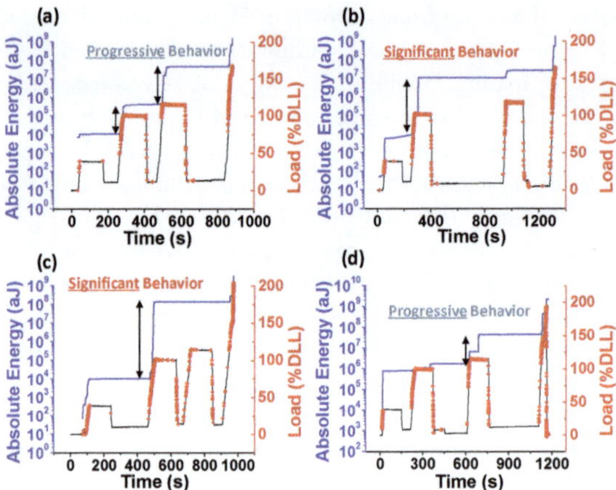

Figure 5. Additional examples of AE activity during four tests with some progressive and some significant behavior observed. (**a**) Test 5, (**b**) Test 3, (**c**) Test 15, and (**d**) Test 13.

Felicity ratio values for all spars at each loading step are given in Table 2. Most test specimens resulted in FR values below 1.0 and above 0.9. Noted exceptions are Test 8 and Test 9, both of which displayed the lowest FR values and further exhibited significant behavior during LL2. Interestingly, the spars that had the lowest FR values still resulted in near-average ultimate load. It should be also noted that some high FR spars failed at lower loads than those with lower FR values, potentially suggesting that these spars may enable damage distribution that leads to higher ultimate loads. In contrast, spars with higher FR values may potentially have less damage (which could not have been confirmed during testing as tests were not interrupted for post mortem inspection), however given that some of these spars failed at lower loads, damage in them is expected to be more localized and near the actual failure zone.

It can be inferred from the parametric AE data trends presented herein that, generally, appreciable damage initiates once the design-limit-load (LL2) is reached, which further progresses until ultimate failure. In conclusion, the AE parametric analysis presented provided additional insight into damage progression and sample-to-sample variation that may otherwise be undiscernible using traditional test methodologies. For example, note the AE energy profile in Figure 4d, Test 4. This particular spar showed progressive damage, with a higher comparable FR value averaging 1.00 over all three loads. It may appear that something about this spar e.g., its design, production, or loading parameters, provided an advantage in controlling damage progression that resulted in achieving a higher ultimate load. By correlating spar-to-spar differences in damage behavior as indicated by AE data with structural testing, production or design data, additional value can be mined from such nondestructive evaluation, which may support design or process improvements.

3.2. Identification of Probable Damage Regions

As determined during pre-testing, AE wave propagation velocity fluctuated across the length and among the different spar faces due to structural and material variations; therefore, a zonal technique based on AE hit amplitude was used to identify regions along the spars related to onset of damage. The AE sensors were distributed along the length of the spar, on the top surface, in order to achieve sufficient "zonal" coverage from root to tip (recall Table 1). The authors examined high amplitude AE hits recorded by certain sensors for which the locations are known. If a sensor had a relatively high concentration of high amplitude AE hits, the sensor "zone" was considered the "critical region" as a first order approximation to damage location. For the preponderance of the test cases, high

AE amplitude behavior was observed in sensors near the spar root. As noted in Table 3, the most commonly identified was Sensor 2 (S2), which was located approximately 6–13 inches from the closest root pin. Failure near the root was anticipated, due to the loading and design of the spar. Prior testing experience indicated that common failures may be in the form of cracking or buckling of the shear web, and/or bending of the cap near the spar root. These types of damage matched the failure modes observed in similar cantilevered wind turbine blade tests [39,41].

Table 3. Results summary.

File Name	Ultimate Crane Load [%DLL]	Target Load [%DLL] o First Damage Signal	Estimated Load [lbf] @ 1st Damage Signal	Cumulative AE Behavior	Critical Region
1	N/A	115	N/A	Significant	S1 (N/A)
2	151	100	90	Significant	S3, S5, S6, S19 (15,23.5,48,66.5 in)
3	165	100	103	Significant	S2 (6 in)
4	167	150	133	Progressive	S2 (6 in)
5	169	115	117	Progressive	S2 (7 in)
6	170	100	69	Significant	S2 (7 in)
7	178	100	101	Significant	S2 (6 in)
8	180	100	85	Significant	S2 (6.5 in)
9	182	100	97	Significant	S2 (6 in)
10	184	150	132	Progressive	S2 (6.5 in)
11	185	100	96	Significant	S2 (6 in)
12	192	100	96	Significant	S2, S4 (6.5&40.5 in)
13	197	115	115	Progressive	S2 (6.75 in)
14	197	100	100	Significant	S2 (6.5 in)
15	205	100	81	Significant	S4 (36 in)
16	216	150	134	Progressive	S2 (7.25)

3.3. Ultimate Failure Load Prognosis

In an effort to perform ultimate failure load prognosis, the ultimate failure crane-load for each spar was correlated with AE activity. Extracted features from the recorded AE signals were used to compare the progressive failure of the composite spars, and to infer similarities and differences among them. The results for all spars were tabulated in Table 3 and ranked according to the ultimate load value. The target crane-load for the first damage signal, the estimated load at first AE damage signal, the AE energy behavior (classified as either progressive or significant behavior based on the cumulative AE energy profile), and the critical region defined by AE sensor locations with high-amplitude AE hit concentrations are given for each spar. All load values are defined as a percentage of the design-limit-load (DLL). Significant AE energy behavior was defined as a jump in the cumulative absolute AE energy distribution of at least 10^3 aJ.

Although variability is observed in all listed parameters, some patterns emerged in the information presented in Table 3. As mentioned previously and supported by the analysis of multiple AE parameters, the majority (10 out of 16) of the first damage signals occurred during the second target load of 100% DLL. It appeared reasonable to observe the damage initiation and progression once the spar limit load was reached. Notably, three of the spars did not show the first damage signal until the final loading sequence reached the target load of 150% DLL. All three of these spars displayed progressive damage behavior based on their cumulative AE energy profiles. Despite the apparently controlled damage progression, this did not result in the three highest ultimate loads. The highest recorded ultimate load resulted from delayed damage initiation and progressive damage behavior, but the next two highest showed earlier damage initiation and significant damage behavior. This uncertainty suggests that there are other factors that may additionally influence the ultimate load prediction.

The estimated crane load applied at the time of the first AE damage signal and the target crane load value were given for comparison, showing that initial damage was often observed prior to reaching the maximum load for that cycle. Moreover, 11 out of the 16 spars tested exhibited significant AE energy behavior, rather than a progressive behavior, making the appearance of the first damage signal more critical than is the case for the progressive energy accumulation. Critical region examination by AE zonal location revealed that 12 out of 16 spars experienced the majority of high-amplitude damage signals near Sensor 2, which was located approximately 6–13 inches from the spar root—as expected given the cantilever load profile. Only a few spars (four out of 16) showed a critical region other than Sensor 2, and one case (Test 2) recorded multiple critical regions, indicating widespread damage, which led to the lowest recorded ultimate load.

3.4. Statistical Static Strength Assessment

To qualify for use on specific aircraft, spars were required to support 115% of the DLL with no permanent deformation, while being capable of supporting 150% of the DLL with no failure. For this test, failure was defined as a 30% drop from the maximum load. Assessing a strength criteria beyond this unidimensional pass/fail condition may provide additional feedback that can be used to evaluate spar behavior across a sample set. For example, the traditional pass/fail test only indicates that a group of spars meets or does not meet a threshold. In this case, the pass/fail classification may not allow for sufficient discretization to determine that one of those spars may have exceeded the criteria significantly. Hypothetically, that spar may have displayed certain favorable progressive damage characteristics that would predict its higher strength. Knowing that extra information could allow for the part production parameters to be studied, and lessons learned to be applied to the next production lot, thus continuously improving part quality.

In order to use the proposed statistical tests, the AE data distribution was examined and verified to follow a normal distribution using the Lilliefors test, a method employed when the population mean and standard deviation is unknown [61]. The Lilliefors analysis was executed using MATLAB R2017a (The MathWorks, Natick, MA, USA), and the results of the normality plot are shown in Figure 6. Both the first damage signal load and spar ultimate load followed a normal distribution (failed to reject the null hypothesis at a 1% significance level).

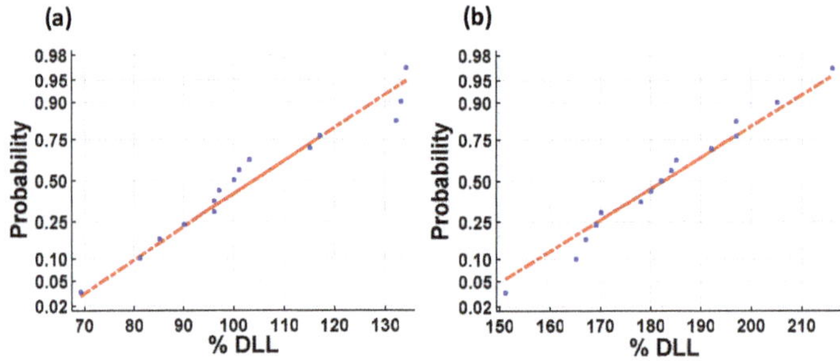

Figure 6. Data normality test for: (**a**) the load at first damage signal, and (**b**) the ultimate load datasets.

Probabilistic assessment (e.g., hypothesis testing) can give additional part performance insight, and the multi-dimensional AE datasets used in this research enable the use of statistical analysis methods. As an example of how AE data could be used to evaluate composite test article performance, a one-sided hypothesis test and lower confidence limit (LCL) was applied to assess both spar strength requirements. Summary statistics for both datasets are listed in Table 4, and are representatively displayed in Figure 7 for comparison.

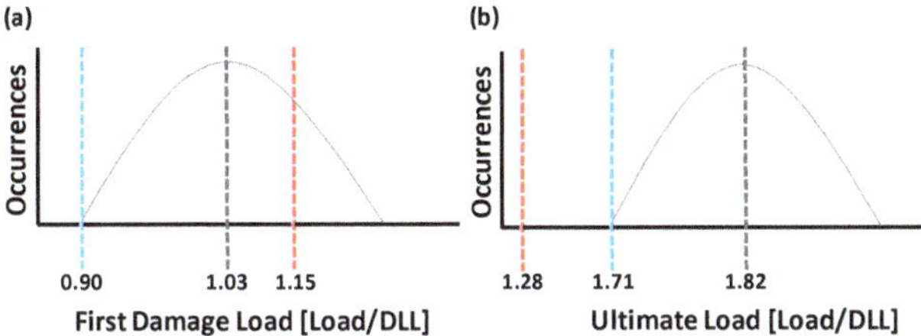

Figure 7. A schematic representation of the probabilistic assessment. (**a**) First damage AE signal distribution, (**b**) ultimate load distribution.

Table 4. Summary statistics for the response variables of interest.

	Crane-Load @ 1st Damage Signal	Ultimate Crane-Load
Mean [Load/DLL]	1.03	1.82
StDev [Load/DLL]	0.19	0.17
Spread [Load/DLL]	0.65	0.65
99% LCL [Load/DLL]	0.90	1.71

The lower confidence limits serve as average predictions for the lowest probable load where damage may initiate and the lowest probable ultimate load, respectively. They could potentially be used as a bound in statistical process control, or as quick indications of part lot performance from a "weakest link" perspective. Alternatively, upper confidence limits (not shown) could also be used to identify spars that perform exceptionally well during testing.

To evaluate spar strength requirements explicitly, the hypothesized population mean was first determined based on the requirement descriptions. Estimated crane-loads at the first AE damage signal were used as an interpretation of the "support 115% design limit load (DLL) with no permanent deformation" requirement, and ultimate loads were used to evaluate "support 150% DLL with no failure". Therefore, the null hypothesis for the first strength requirement was that the first damage signal population mean is equal to 1.15 DLL, and the alternative was that the first damage signal population mean is less than 1.15 DLL. In other words, the analysis was run to examine whether if on average, the first damage signals occurred at or below 1.15 DLL for the total population of spars, given the tested sample population. Figure 7a shows the requirement distance from the sample mean and the sample LCL, schematically. The analysis was run in MATLAB, and the null hypothesis was rejected at a 5% significance level ($p = 0.0174$). According to the results, the probability of the average first damage signal occurring below a crane-load of 115% DLL (the third test load increment and the first strength requirement) was 98%. This suggests that most spars will exhibit observable damage signals at or below the second loading level (100% DLL), a conclusion confirmed by the analysis of multiple AE parameters discussed earlier in this work.

Similarly, the ultimate load distribution can be examined (Figure 7b). The null hypothesis for the second strength requirement was that ultimate crane load population mean is equal to 1.28 DLL (a 30% reduction from the sample population average ultimate load) and the alternative was that the ultimate crane load population mean is less than 1.28 DLL. The analysis was run to examine whether on average, the average ultimate crane load could occur below a "30% drop from the maximum load" in accordance with the requirement. The MATLAB analysis failed to reject the null hypothesis at a 5% significance level ($p = 1$); there was not enough data to support the alternative hypothesis. Based on the results presented, the probability that the average spar ultimate crane-load was less than 128% DLL (30% less than the average ultimate load) was less than 1%. Therefore, almost all spars can withstand the maximum load without failure. Again, this was confirmed by the summary of results presented in

Table 3, to which the probabilistic assessment presented herein adds confidence; the lowest recorded individual spar ultimate load was 151% DLL.

4. Conclusions

Acoustic emission (AE) was successfully applied to full-scale, composite, fixed-wing aircraft spars during structural strength qualification testing. Correlations between AE activity and applied load were made that allow for the in-depth structural integrity assessment beyond the data available from pass/fail, legacy aircraft component testing. In addition to examining spar progressive failure, a criterion based on cumulative AE energy was introduced to characterize the recorded AE profile as indicating either significant or progressive behavior. Moreover, the criterion agreed well with particular instances during loading, and has potential for use during real-time AE data assessment, as there appears to be a link between AE data trends and ultimate failure loads. Using the knowledge of sensor placement and recorded AE activity at each sensor, damage critical regions were identified and determined to be generally concentrated at the spar root. Finally, statistical calculation methodologies for evaluating test performance using AE data were proposed, and feasibility was demonstrated. In summary, the presented investigation demonstrates that AE monitoring may assist in quantifying the effect of manufacturing or design differences on structural component performance parameters, which could lead to refined assessment criteria that are over and above the available legacy test methodologies.

Author Contributions: The authors contributed to the following tasks. Conceptualization, A.K. and J.G.; Methodology, A.K., P.A.V. and S.E.; Validation, S.E., B.J.W. and A.K.; Formal Analysis, P.A.V., S.E., B.J.W. and A.K.; Investigation, P.A.V., S.E. and A.K.; Resources, A.K. and J.G.; Data Curation, S.E., B.J.W. and A.K.; Writing-Original Draft Preparation, B.J.W., S.E. and A.K.; Writing-Review & Editing, B.J.W., S.E. and A.K.; Visualization, B.J.W. and S.E.; Supervision, A.K. and J.G.; Project Administration, A.K.

Funding: This investigation was supported financially by General Atomics Aeronautical Systems.

Acknowledgments: The authors acknowledge and thank Trilion Quality Systems, King of Prussia, PA, USA, MISTRAS Group Inc., Princeton Junction, NJ, USA, and General Atomics Aeronautical Systems, Adelanto, CA, USA for collaborating on this investigation.

Conflicts of Interest: The authors declare no conflict of interest.

References

1. U.S. Naval Air Systems Command. Airplane Strength and Rigidity: Ground Tests. In *MIL-A-8867C(AS)*; U.S. Department of Defense: Washington, DC, USA, 1987.
2. U.S. Federal Aviation Administration. Subchapter C—AIRCRAFT. In *U.S. Code of Federal Regulation*; Volume 14—Aeronautics and Space; Office of the Federal Register: Washington, DC, USA, 2018.
3. SAE International. Polymer Matrix Composites: Materials, Usage, Design, and Analysis. In *Composite Materials Handbook*; Volume 3 Revision G R-424; SAE International: Warrendale, PA, USA, 2012.
4. SAE International. Polymer Matrix Composites: Guidelines for Characterization of Structural Materials. In *Composite Materials Handbook*; Volume 1 Revision G R-422; SAE International: Warrendale, PA, USA, 2012.
5. U.S. Fire Administration. *Airframe Guide for Certification of Part 23 Airplanes*; U.S. Fire Administration: Frederick County, MD, USA, 2007.
6. U.S. Fire Administration. *Damage Tolerance and Fatigue Evaluation of Structure*; U.S. Fire Administration: Frederick County, MD, USA, 2011.
7. U.S. Fire Administration. *Proof of Structure*; U.S. Fire Administration: Frederick County, MD, USA, 2014.
8. U.S. Fire Administration. *Quality System for the Manufacture of Composite Structures*; U.S. Fire Administration: Frederick County, MD, USA, 2010.
9. U.S. Fire Administration. *Acceptance Guidance on Material Procurement and Process Specifications for Polymer Matrix Composite Systems*; U.S. Fire Administration: Frederick County, MD, USA, 2003.
10. Hamstad, M. Acceptance Testing Of Graphite/Epoxy Composite Parts Using An Acoustic Emission Monitoring Technique. *NDT Int.* **1982**, *15*, 307–314. [CrossRef]

11. Oster, R. Non-Destructive Testing Methodologies on Helicopter Fiber Composite Components Challenges Today and in the Future. In Proceedings of the 18th World Conference on Nondestructive Testing, Durban, South Africa, 16–20 April 2012; pp. 16–20.

12. Swindell, P.; Doyle, J.; Roach, D. Integration Of Structural Health Monitoring Solutions Onto Commercial Aircraft via The Federal Aviation Administration Structural Health Monitoring Research Program. In Proceedings of the AIP Conference, Atlanta, GA, USA, 17–22 July 2016; Volume 1806, p. 070001.

13. Shull, P.J. *Nondestructive Evaluation: Theory, Techniques, and Applications*; CRC Press: Boca Raton, FL, USA, 2016.

14. Pollock, A.A. Acoustic Emission from Solids Undergoing Deformation. Ph.D. Thesis, University of London, London, UK, 1970.

15. Ono, K. Acoustic Emission in Materials Research—A Review. *J. Acoust. Emiss.* **2011**, *29*, 284–308.

16. Moore, P.O. (Ed.) Fundamentals of Acoustic Emission Testing. In *Nondestructive Testing Handbook*, 3rd ed.; Volume 6 Acoustic Emission Testing; American Society for Nondestructive Testing: Columbus, OH, USA, 2005.

17. Mo, C.; Wisner, B.; Cabal, M.; Hazeli, K.; Ramesh, K.; El Kadiri, H.; Al-Samman, T.; Molodov, K.D.; Molodov, D.A.; Kontsos, A. Acoustic emission of deformation twinning in magnesium. *Materials* **2016**, *9*, 662. [CrossRef] [PubMed]

18. Wisner, B.; Cabal, M.; Vanniamparambil, P.; Hochhalter, J.; Leser, W.; Kontsos, A. In situ microscopic investigation to validate acoustic emission monitoring. *Exp. Mech.* **2015**, *55*, 1705–1715. [CrossRef]

19. Wisner, B.; Kontsos, A. In situ monitoring of particle fracture in aluminium alloys. *Fatigue Fract. Eng. Mater. Struct.* **2017**, *41*, 581–596. [CrossRef]

20. Wisner, B.; Kontsos, A. Investigation of particle fracture during fatigue of aluminum 2024. *Int. J. Fatigue* **2018**, *111*, 33–43. [CrossRef]

21. Castaneda, N.; Wisner, B.; Cuadra, J.; Amini, S.; Kontsos, A. Investigation of the Z-binder role in progressive damage of 3D woven composites. *Compos. Part A Appl. Sci. Manuf.* **2017**, *98*, 76–89. [CrossRef]

22. Cuadra, J.; Vanniamparambil, P.A.; Hazeli, K.; Bartoli, I.; Kontsos, A. Damage quantification in polymer composites using a hybrid NDT approach. *Compos. Sci. Technol.* **2013**, *83*, 11–21. [CrossRef]

23. Vanniamparambil, P.A.; Bartoli, I.; Hazeli, K.; Cuadra, J.; Schwartz, E.; Saralaya, R.; Kontsos, A. An integrated structural health monitoring approach for crack growth monitoring. *J. Intell. Mater. Syst. Struct.* **2012**, *23*, 1563–1573. [CrossRef]

24. Vanniamparambil, P.A.; Khan, F.; Hazeli, K.; Cuadra, J.; Schwartz, E.; Kontsos, A.; Bartoli, I. Novel optico-acoustic nondestructive testing for wire break detection in cables. *Struct. Control. Health Monit.* **2013**, *20*, 1339–1350. [CrossRef]

25. Khan, F.; Rajaram, S.; Vanniamparambil, P.A.; Bolhassani, M.; Hamid, A.; Kontsos, A.; Bartoli, I. Multi-sensing NDT for damage assessment of concrete masonry walls. *Struct. Control. Health Monit.* **2015**, *22*, 449–462. [CrossRef]

26. Salamone, S.; Bartoli, I.; Phillips, R.; Nucera, C.; di Scalea, F. Health monitoring of prestressing tendons in posttensioned concrete bridges. *Transp. Res. Rec. J. Transp. Res. Board* **2011**, *2220*, 21–27. [CrossRef]

27. Vanniamparambil, P.A.; Bolhassani, M.; Carmi, R.; Khan, F.; Bartoli, I.; Moon, F.L.; Hamid, A.; Kontsos, A. A data fusion approach for progressive damage quantification in reinforced concrete masonry walls. *Smart Mater. Struct.* **2013**, *23*, 015007. [CrossRef]

28. Baxevanakis, K.P.; Wisner, B.; Schlenker, S.; Baid, H.; Kontsos, A. Data-Driven Damage Model Based on Nondestructive Evaluation. *J. Nondestruct. Eval. Diagn. Progn. Eng. Syst.* **2018**, *1*, 031007. [CrossRef]

29. Cuadra, J.; Vanniamparambil, P.; Servansky, D.; Bartoli, I.; Kontsos, A. Acoustic emission source modeling using a data-driven approach. *J. Sound Vib.* **2015**, *341*, 222–236. [CrossRef]

30. Mazur, K.; Wisner, B.; Kontsos, A. Fatigue Damage Assessment Leveraging Nondestructive Evaluation Data. *JOM* **2018**, *70*, 1182–1189. [CrossRef]

31. Wisner, B.; Mazur, K.; Perumal, V.; Baxevanakis, K.; An, L.; Feng, G.; Kontsos, A. Acoustic emission signal processing framework to identify fracture in aluminum alloys. *Eng. Fract. Mech.* **2018**, in press. [CrossRef]

32. Zárate, B.A.; Caicedo, J.M.; Yu, J.; Ziehl, P. Probabilistic prognosis of fatigue crack growth using acoustic emission data. *J. Eng. Mech.* **2012**, *138*, 1101–1111. [CrossRef]

33. Ono, K. Application of acoustic emission for structure diagnosis. *Diagnostyka* **2011**, *2*, 3–18.

34. Anay, R.; Cortez, T.M.; Jáuregui, D.V.; ElBatanouny, M.K.; Ziehl, P. On-site acoustic-emission monitoring for assessment of a prestressed concrete double-tee-beam bridge without plans. *J. Perform. Constr. Facil.* **2015**, *30*, 04015062. [CrossRef]

35. De Santis, S.; Tomor, A.K. Laboratory and field studies on the use of acoustic emission for masonry bridges. *NDT E Int.* **2013**, *55*, 64–74. [CrossRef]

36. Golaski, L.; Gebski, P.; Ono, K. Diagnostics of reinforced concrete bridges by acoustic emission. *J. Acoust. Emiss.* **2002**, *20*, 83–89.

37. Pollock, A.; Smith, B. Stress-wave-emission monitoring of a military bridge. *Non-Destr. Test.* **1972**, *5*, 348–353. [CrossRef]

38. Blanch, M.; Dutton, A. Acoustic emission monitoring of field tests of an operating wind turbine. In *Key Engineering Materials*; Trans Tech Publ: Princeton, NJ, USA, 2003; pp. 475–482.

39. Han, B.-H.; Yoon, D.-J.; Huh, Y.-H.; Lee, Y.-S. Damage assessment of wind turbine blade under static loading test using acoustic emission. *J. Intell. Mater. Syst. Struct.* **2014**, *25*, 621–630. [CrossRef]

40. Paquette, J.; Van Dam, J.; Hughes, S. Structural testing of 9m carbon fiber wind turbine research blades. In Proceedings of the 45th AIAA Aerospace Sciences Meeting and Exhibit, Reno, NV, USA, 8–11 January 2007; p. 816.

41. Sutherland, H.; Beattie, A.; Hansche, B.; Musial, W.; Allread, J.; Johnson, J.; Summers, M. *The Application of Non-Destructive Techniques to the Testing of a Wind Turbine Blade*; Sandia National Labs.: Albuquerque, NM, USA, 1994.

42. Moore, P.O. (Ed.) Aerospace Applications of Acoustic Emission Testing. In *Nondestructive Testing Handbook*, 3rd ed.; Volume 6 Acoustic Emission Testing; American Society for Nondestructive Testing: Columbus, OH, USA, 2005.

43. Holford, K.M.; Pullin, R.; Evans, S.L.; Eaton, M.J.; Hensman, J.; Worden, K. Acoustic emission for monitoring aircraft structures. *Proc. Inst. Mech. Eng. Part G J. Aerosp. Eng.* **2009**, *223*, 525–532. [CrossRef]

44. Vanniamparambil, P.A.; Carmi, R.; Khan, F.; Cuadra, J.; Bartoli, I.; Kontsos, A. An active–passive acoustics approach for bond-line condition monitoring in aerospace skin stiffener panels. *Aerosp. Sci. Technol.* **2015**, *43*, 289–300. [CrossRef]

45. Prosser, W.H.; Gorman, M.R.; Madaras, E.I. *Acoustic Emission Detection of Impact Damage on Space Shuttle Structures*; NASA: Washington, DC, USA, 2004.

46. Staszewski, W.; Mahzan, S.; Traynor, R. Health monitoring of aerospace composite structures—Active and passive approach. *Compos. Sci. Technol.* **2009**, *69*, 1678–1685. [CrossRef]

47. Awerbuch, J.; Leone, F.A., Jr.; Ozevin, D.; Tan, T.-M. On the applicability of acoustic emission to identify modes of damage in full-scale composite fuselage structures. *J. Compos. Mater.* **2016**, *50*, 447–469. [CrossRef]

48. Weis, M.; Cejpek, J.; Juračka, J. Acoustic Emission Localization in Testing of Composite Structures. In *Applied Mechanics and Materials*; Trans Tech Publ: Princeton, NJ, USA, 2016; pp. 405–411.

49. Haile, M.A.; Bordick, N.E.; Riddick, J.C. Distributed acoustic emission sensing for large complex air structures. *Struct. Health Monit.* **2018**, *17*, 624–634. [CrossRef]

50. Cawley, P. Structural health monitoring: Closing the gap between research and industrial deployment. *Struct. Health Monit.* **2018**. [CrossRef]

51. Brunner, A.J. Identification of damage mechanisms in fiber-reinforced polymer-matrix composites with Acoustic Emission and the challenge of assessing structural integrity and service-life. *Constr. Build. Mater.* **2018**, *173*, 629–637. [CrossRef]

52. Moore, P.O. (Ed.) Acoustic Emission Signal Processing. In *Nondestructive Testing Handbook*, 3rd ed.; Volume 6 Acoustic Emission Testing; American Society for Nondestructive Testing: Columbus, OH, USA, 2005.

53. Tensi, H.M. The Kaiser-effect and its scientific background. *J. Acoust. Emiss.* **2004**, *22*, s1–s16.

54. Waller, J.; Andrade, E.; Saulsberry, R. Use of acoustic emission to monitor progressive damage accumulation in Kevlar® 49 composites. In Proceedings of the AIP Conference, Kingston, RI, USA, 26–31 July 2009; Volume 1211, p. 1111.

55. Fowler, T.J. Acoustic emission of fiber reinforced plastics. *J. Tech. Counc. ASCE* **1979**, *105*, 281–289.

56. Behnia, A.; Chai, H.K.; Shiotani, T. Advanced structural health monitoring of concrete structures with the aid of acoustic emission. *Constr. Build. Mater.* **2014**, *65*, 282–302. [CrossRef]

57. Ono, K.; Gallego, A. Research and applications of AE on advanced composites. *J. Acoust. Emiss* **2012**, *30*, 180–229.

58. Loutas, T.; Kostopoulos, V. Health monitoring of carbon/carbon, woven reinforced composites. Damage assessment by using advanced signal processing techniques. Part I: Acoustic emission monitoring and damage mechanisms evolution. *Compos. Sci. Technol.* **2009**, *69*, 265–272.

59. Djabali, A.; Toubal, L.; Zitoune, R.; Rechak, S. An experimental investigation of the mechanical behavior and damage of thick laminated carbon/epoxy composite. *Compos. Struct.* **2018**, *184*, 178–190. [CrossRef]
60. Joosse, P.; Blanch, M.; Dutton, A.; Kouroussis, D.; Philippidis, T.; Vionis, P. Acoustic emission monitoring of small wind turbine blades. *J. Sol. Energy Eng.* **2002**, *124*, 446–454. [CrossRef]
61. Lilliefors, H.W. On the Kolmogorov-Smirnov test for normality with mean and variance unknown. *J. Am. Stat. Assoc.* **1967**, *62*, 399–402. [CrossRef]

Article

Remote Monitoring and Evaluation of Damage at a Decommissioned Nuclear Facility Using Acoustic Emission

Marwa Abdelrahman [1], Mohamed ElBatanouny [1,*], Kenneth Dixon [2], Michael Serrato [2] and Paul Ziehl [3]

[1] Janney Technical Center, Wiss, Janney, Elstner Associates, Inc., Northbrook, IL 60062, USA; mabdelrahman@wje.com

[2] Savannah River National Laboratory, Aiken, SC 29808, USA; kenneth.dixon@srnl.doe.gov (K.D.); michael.serrato@srnl.doe.gov (M.S.)

[3] College of Engineering and Computing, University of South Carolina, Columbia, SC 29208, USA; ZIEHL@cec.sc.edu

* Correspondence: melbatanouny@wje.com

Received: 18 July 2018; Accepted: 12 September 2018; Published: 14 September 2018

Abstract: Reinforced concrete systems used in the construction of nuclear reactor buildings, spent fuel pools, and related nuclear facilities are subject to degradation over time. Corrosion of steel reinforcement and thermal cracking are potential degradation mechanisms that adversely affect durability. Remote monitoring of such degradation can be used to enable informed decision making for facility maintenance operations and projecting remaining service life. Acoustic emission (AE) monitoring has been successfully employed for the detection and evaluation of damage related to cracking and material degradation in laboratory settings. This paper describes the use of AE sensing systems for remote monitoring of active corrosion regions in a decommissioned reactor facility for a period of approximately one year. In parallel, a representative block was cut from a wall at a similar nuclear facility and monitored during an accelerated corrosion test in the laboratory. Electrochemical measurements were recorded periodically during the test to correlate AE activity to quantifiable corrosion measurements. The results of both investigations demonstrate the feasibility of using AE for corrosion damage detection and classification as well as its potential as a remote monitoring technique for structural condition assessment and prognosis of aging structures.

Keywords: acoustic emission; remote monitoring; corrosion; nuclear facilities

---

1. Introduction

The vast presence of aging infrastructure throughout the nation including transportation and energy-related infrastructure has raised concerns regarding the level of service, reliability, and vulnerability to natural disasters. The American Society of Civil Engineers (ASCE) 2013 Report Card stated a grade of "D$^+$" for US infrastructure and an estimated investment of \$3.6 trillion needed by 2020 for upkeep. One of the major challenges facing decision makers is resource allocation, which is dependent on available information related to the current state of each structure. Reliable monitoring techniques that can effectively assess structural conditions are needed to evaluate the robustness of structures and the urgency of any repair, replacement, or maintenance activities.

Monitoring nuclear facilities, in particular, is of special interest due to safety considerations and the relatively long half-life of nuclear waste products. Reinforced concrete elements are used to construct several portions of nuclear facilities. Potential degradation mechanisms of reinforced concrete [1] include corrosion of reinforcement [2–5], alkali-silica reaction [5–7], freeze-thaw cycling, sulfate attack, deformation mechanisms including creep and shrinkage, stresses due to structural constraint combined with seasonal effects such as thermal cycling and precipitation, and extreme events [8–10].

Advances in computing and data transfer over the last several decades have allowed for the development of wireless systems and remote monitoring. Acoustic emission (AE) is one emerging monitoring method that has proven to have the potential for early damage detection through laboratory and field applications [11,12]. As a passive piezoelectric sensing technique, acoustic emission is able to detect stress waves (in the kHz range) emitted from sudden releases of energy such as cracking of the concrete matrix [13,14]. The method is suitable for real-time monitoring over the long term and its high sensitivity enables it to detect active cracks long before they become visible (micro-cracking).

Corrosion of reinforcing steel is a degradation mechanism that affects the durability of concrete structures. The cracking of the concrete matrix associated with corrosion damage makes acoustic emission a well-suited method for monitoring its progression. Early investigations related to acoustic emission monitoring of corrosion damage in reinforced concrete date back to the 1980s [15–17]. Several investigations demonstrated the potential of utilizing AE for this degradation mechanism [18–20]. However, the quantification of damage was not fully resolved. Quantification of corrosion damage in reinforced concrete structures has been more recently addressed in a series of publications using accelerated corrosion results in laboratory settings [2,3,21] as well as natural corrosion tests [3,22,23].

This study investigates the applicability of deploying acoustic emission for the remote monitoring of selected areas at the Savannah River Site (SRS) 105-C Reactor Facility in Aiken, South Carolina (Figure 1). This is an inactive nuclear facility under surveillance and maintenance operations as well as deactivation and decommissioning operations. AE monitoring was conducted at areas known to have active corrosion damage and/or visible cracking. This allows the examination of the applicability of previously developed AE methods for corrosion damage detection and classification.

To aid in the development of damage algorithms and to provide a more controlled study, an aged reinforced concrete block specimen cut from a similar reactor facility was maintained and monitored in the University of South Carolina Structures and Materials Laboratory for the majority of the project duration. This specimen was subjected to wet/dry cycling to accelerate the corrosion process. Electrochemical measurements were periodically recorded while acoustic emission was monitored continuously.

The activities undertaken and reported in this study represent a step toward the development of an acoustic emission based approach for the assessment of reinforced concrete structural systems through remote monitoring. To the authors' knowledge, this research is the first of its kind to study the suitability of acoustic emission to monitor corrosion damage in the field while comparing the results to a test block from a similar structure tested in a controlled laboratory environment.

Figure 1. Reactor building 105-C at the Savannah River Site.

The study is divided into two main activities: (1) Remote acoustic emission monitoring and analysis of data collected at the 105-C Reactor Facility and (2) Accelerated corrosion testing to assess corrosion damage within an aged and reinforced concrete block supplied by SRNS at the University of South Carolina Structures and Materials laboratory.

2. AE Monitoring at the 105-C Reactor Facility

2.1. Acoustic Emission Sensing Systems

Two separate AE systems were utilized for remote monitoring at the 105-C Reactor Facility. These systems are referred to as a 'wired' AE system and a 'wireless' AE system. All acoustic emission system components and software were manufactured by Mistras Group, Inc. of Princeton Junction, New Jersey. The wired system utilized both R6I (peak resonance near 55 kHz) and WDI (relatively broadband) sensor types (calibration sheets for both sensor types are available in the manufacturer's website). Both sensor types utilize integral pre-amplifiers within the stainless steel sensor housing. The resonant sensors are more sensitive to damage in reinforced concrete structures than the broadband sensors. However, the broadband sensors provide higher fidelity frequency data, which can be useful for data reduction and interpretation. The sensors were connected to a 16-channel DiSP acoustic emission data acquisition system that utilizes four high speed data acquisition boards specifically designed and manufactured for the acquisition and processing of acoustic emission data as well as specialized software (AEWin).

The wireless acoustic emission system (type 1284) includes 4 channels and utilizes low power PK6I acoustic emission sensors. These sensors are resonant in the vicinity of 55 kHz and utilize integrated preamplifiers within stainless steel housing. The sensors were connected to the 1284 system where preliminary processing of the data is performed. The data is transmitted through an antenna and is received through a base station module that is connected to a conventional laptop computer. Specialized wireless acoustic emission software (AEWin for Wireless) is used for controlling the data acquisition. This system was powered through solar panels connected to 12V DC batteries.

2.2. Installation of Acoustic Emission Systems

Prior to the installation at the Savannah River Site (SRS), the consistency of the sensor readings was checked using pencil lead breaks on an acrylic rod [24,25]. Six pencil lead breaks were performed for each sensor. An appropriate sensor response was demonstrated as the average amplitude response of a sensor type that was within ± 6 dB of the average amplitude of the sensor group. A threshold of 40 dB was used for data collection. An analog filter was used to collect signals with a frequency between 1 kHz and 1 MHz. The waveform sampling rate was 1 million samples per second (MSPS) with 256 micro-seconds pre-trigger and 1 kilobyte length. Peak definition time (PDT), hit definition time (HDT), and hit lock-out time (HLT) were set to 200, 400, and 200 micro-seconds, respectively. Each AE system was connected to a cellular modem to allow for remote monitoring and each system was remotely controlled through appropriate software. This allowed for altering system settings and saving data at the University of South Carolina.

2.2.1. Crane Maintenance Area

The wired AE system was installed to monitor the activity in this area of building 105-C with ten sensors including five resonant sensors (type R6I) and five broadband sensors (type WDI). The sensors were installed at three different locations. The first location was near a column to roof interface (referred to as the 'vertical column to roof interface location'). This area had been visually assessed by the Savannah River Nuclear Solutions/Savannah River National Laboratory (SRNS/SRNL) personnel and is known to have deteriorated in comparison to the majority of the structural system comprising the 105-C reactor building. Spalling occurred in this area in the recent past and ongoing corrosion activity was suspected. The area has undergone at least one repair activity in the past. A total of six sensors (three resonant and three broadband) were installed at this location, which is shown in Figure 2. The locations of the sensors were chosen to be near the exposed reinforcing bars showing visual signs of corrosion damage. The locations of the sensors with respect to the red dot shown in Figure 2 are provided in Table 1.

The second location was chosen on a horizontal beam that forms the connection with the previously described column (referred to as the 'horizontal beam location'). Two sensors (one resonant and one broadband) were installed at a distance of 12 inches below the beam to roof interface where signs of deterioration were visually observed (Figure 3a). The spacing between the sensors was 6 inches. The third location was chosen at an area where no signs of damage were observed (referred to as the 'control location'). Two sensors (one resonant and one broadband) were installed at this location, which is shown in Figure 3b. The horizontal distance between the two sensors is 6 inches. The data collected from the control location was used to evaluate the effectiveness of data reduction approaches.

Table 1. Location of sensors shown in Figure 2.

Sensor Type-Channel	Horizontal Dimension (Inches)	Vertical Dimension (Inches)
WDI-9	0	−10.5 *
R6I-10	8.5	16
R6I-11	5.5	9
WDI-12	11	5.5
R6I-13	14.5	16
WDI-14	15.5	8.5

* Positive dimension indicates below the red dot are shown in Figure 2. Negative dimension indicates above the red dot are shown in Figure 2.

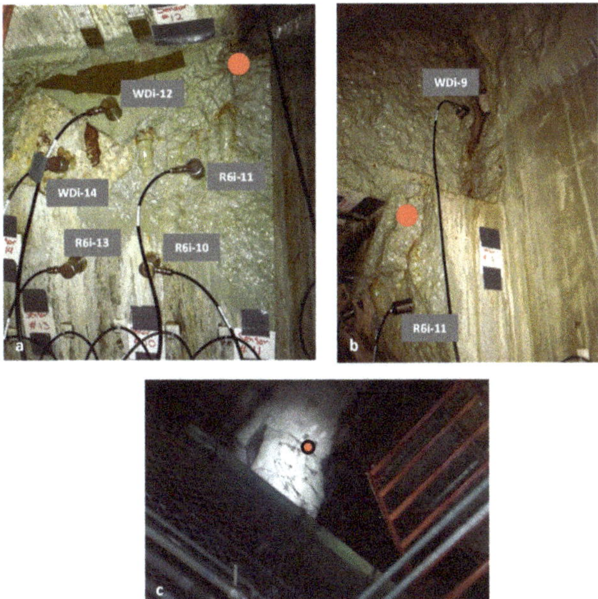

Figure 2. Photographs of the crane maintenance area: (**a**) Main sensor grid, (**b**) close-up of sensor on side of column, and (**c**) view of main grid from the floor level (red dot is at corner).

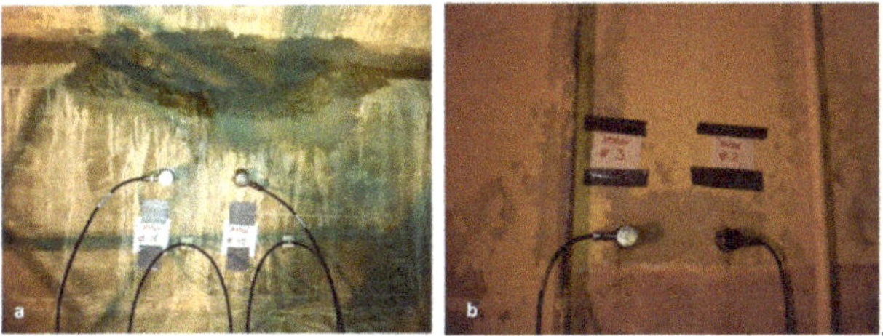

Figure 3. Photograph of: (**a**) Horizontal beam location and (**b**) control location.

2.2.2. Top Floor Level

The wireless AE system was installed at the top floor level (also known as +48 level) to monitor a vertical crack that may penetrate an exterior wall, which is shown in Figure 4, using four resonant sensors (type PK6I). Sensor layout and spacing is also shown in the figure.

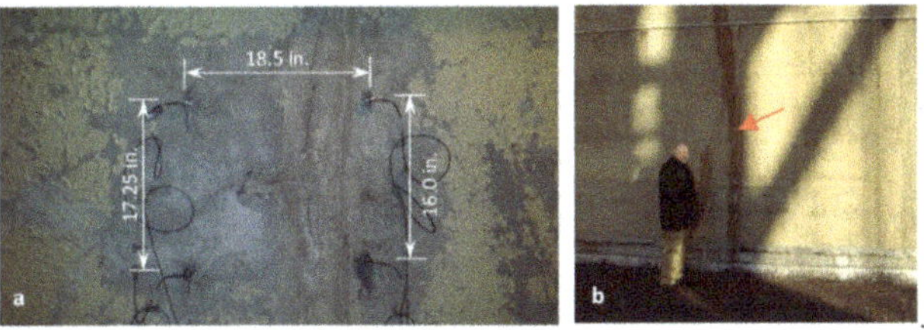

Figure 4. Photographs at top floor level (+48 level): (**a**) Sensor grid from interior and (**b**) vertical crack from exterior.

2.3. Results and Discussion

2.3.1. Remote Monitoring at the Crane Maintenance Area

Remote monitoring at the Crane Maintenance Location was performed from 10 September 2014 (commencement of test) through 25 August 2015. A cellular connection was utilized to remotely operate the wired system. Data loss due to a power outage at the system occurred between 18 December 2014 and 20 January 2015. The raw data was analyzed and appropriate filters were used to reject data arising from signals not related to the initiation or the growth of cracks such as wave reflections. The filters are primarily parameter-based filters that were developed based on visual inspection of AE waveforms, which is similar to those described in ElBatanouny et al. (2014a). The first is a Duration-Amplitude filter (D-A), which is also referred to as a Swansong II type filter, while the second is a Rise Time-Amplitude filter (R-A), which is described in Table 2. Additional filters known as Duration and RMS filters were developed during this study to minimize electrical noise. The additional filters were developed based on data collected from the concrete block discussed in the following sections.

Table 2. Data rejection limits.

D-A Filter		R-A Filter		Duration (μs)	RMS Filter (V)
Amplitude (dB)	Duration (μs)	Amplitude (dB)	Rise Time (μs)		
45–50	>500	45–50	>40	≤100	0.0019–0.0041
51–55	>1000	51–65	>100	—	—
56–65	>2000	66–100	>150	—	—
66–75	>3000	—	—	—	—
76–100	>4000	—	—	—	—

Figures 5 and 6 show the AE activity detected at the three monitored locations in the crane maintenance area for the resonant and broadband sensors, respectively. As shown in the figures, AE activity at the locations associated with visually observable damage ('vertical column to roof interface location' and 'horizontal beam location') was significantly higher than the AE activity at the control location. This indicates that the filters used were suitable for this application and also that intrinsic noise such as that potentially caused by electro-magnetic interference is not an obstacle for this application. The relatively high levels of AE activity indicate that damage (corrosion and related cracking) associated with the aging of reinforced concrete is progressing at the vertical column to roof interface and the horizontal beam locations.

Rain and temperature data were provided by SRNL to investigate the effect of environmental conditions on AE activity. Seasonal temperature fluctuations affected the data more significantly than daily temperature fluctuations. This may be attributed to the low coefficient of thermal expansion of concrete, which causes volumetric changes to be associated with prolonged exposure to temperature differentials. As a general statement, increased AE activity was recorded when temperatures decreased during the winter months. Rain events were not as closely correlated to AE activity as were temperature fluctuations. However, associated moisture and repeated wet/dry cycling from rain events may lead to acceleration of the degradation process. During one of the site visits, remnants of a crack sealing material were found on the floor of the 105-C building, which indicates one potential source of moisture intrusion in this area.

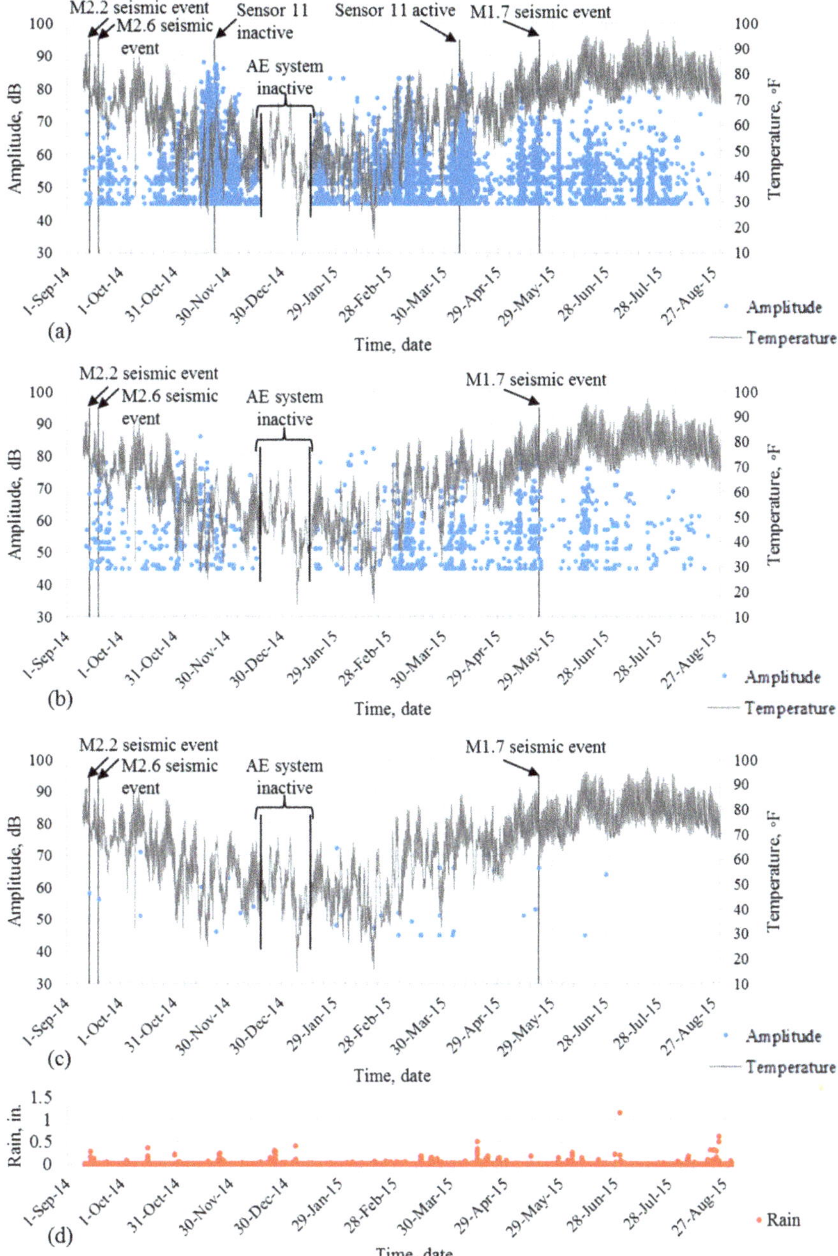

Figure 5. Amplitude and temperature versus time for resonant sensors: (**a**) Vertical column to roof interface location, (**b**) horizontal beam location, (**c**) control location, and (**d**) rain versus time.

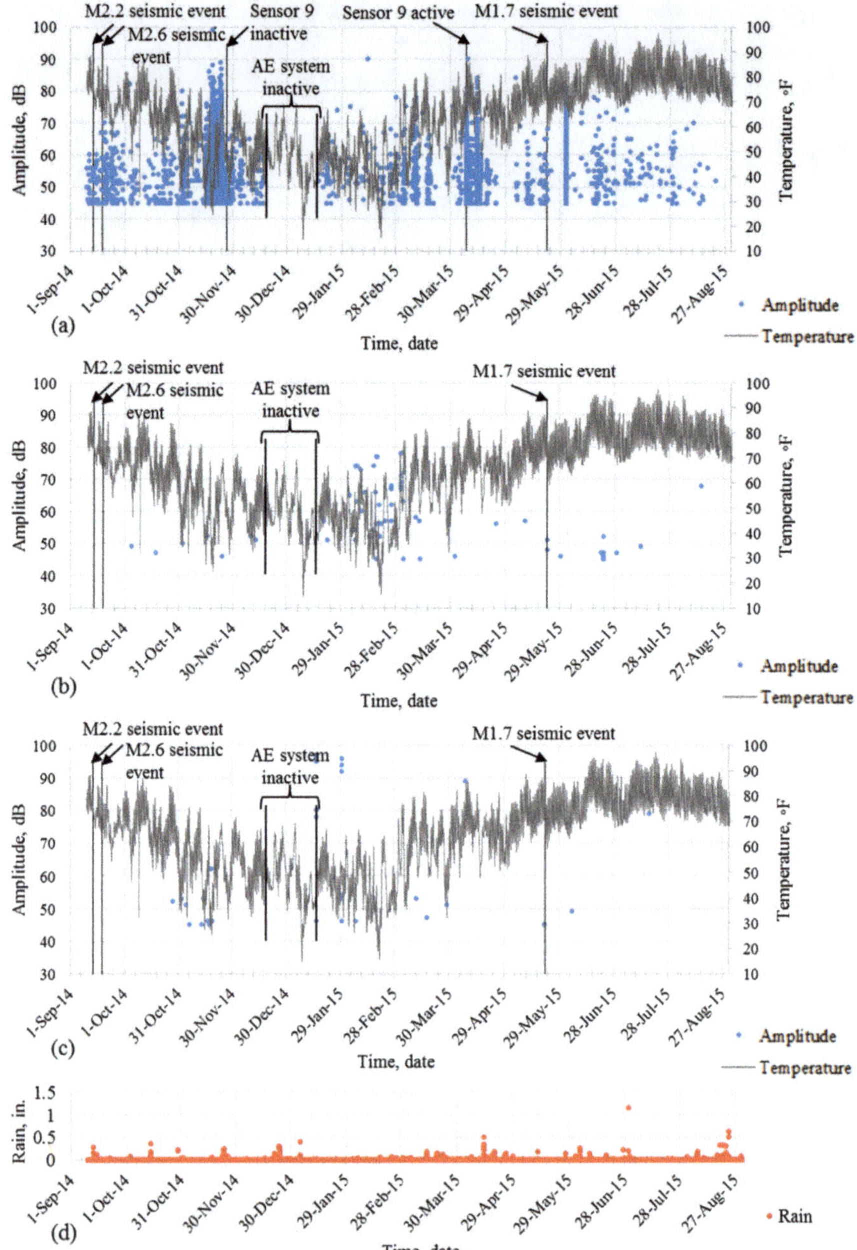

Figure 6. Amplitude and temperature versus time for broadband sensors: (**a**) Vertical column to roof interface location, (**b**) horizontal beam location, (**c**) control location, and (**d**) rain versus time.

The wired AE system was inactive between 18 December 2014 and 20 January 2015 due to a moisture-related event that adversely affected the laptop. Sensors corresponding to channels 9 and 11 (both at the vertical column to roof interface location) detached from the concrete surface on 27 November 2014 and 23 November 2014, respectively. Localized spalling that occurred at these locations during this time period is presumed to be the cause of the detachment. Both sensors were reattached on 8 April 2015.

Three seismic events occurred during the monitoring period: 14 September 2014 (M2.2), 19 September 2014 (M2.6), and 22 May 2015 (M1.96). Close inspection of data collected during this period did not reveal a correlation between these events and the AE data. Referring to the definition of acoustic emission (transient stress waves caused by a rapid release of energy within a material, [13], AE sensors would potentially be capable of detecting crack growth events caused by a seismic event provided the crack growth event or events that occurred within the range of sensitivity of the sensors. In the application at 105-C, the range of sensitivity for minor crack growth events (similar in energy to that caused by a pencil lead break) is in the range of 3 to 10 feet from each sensor. Due to the frequency range of AE sensors (30 kHz to 300 kHz), the sensors are not sensitive to global structural vibrations such as those potentially related to seismic activity.

2.3.2. Evaluation of Data at the Crane Maintenance Area

Figure 7a,b show the cumulative signal strength (abbreviated as CSS) at each monitored location for resonant and broadband sensors, respectively. Signal strength of an AE hit is a measure of the area under the recorded signal envelope (sometimes referred to as MARSE, Measured Area under the Rectified Signal Envelope) [13]. Higher levels of signal strength are associated with higher levels of energy release due to crack growth events.

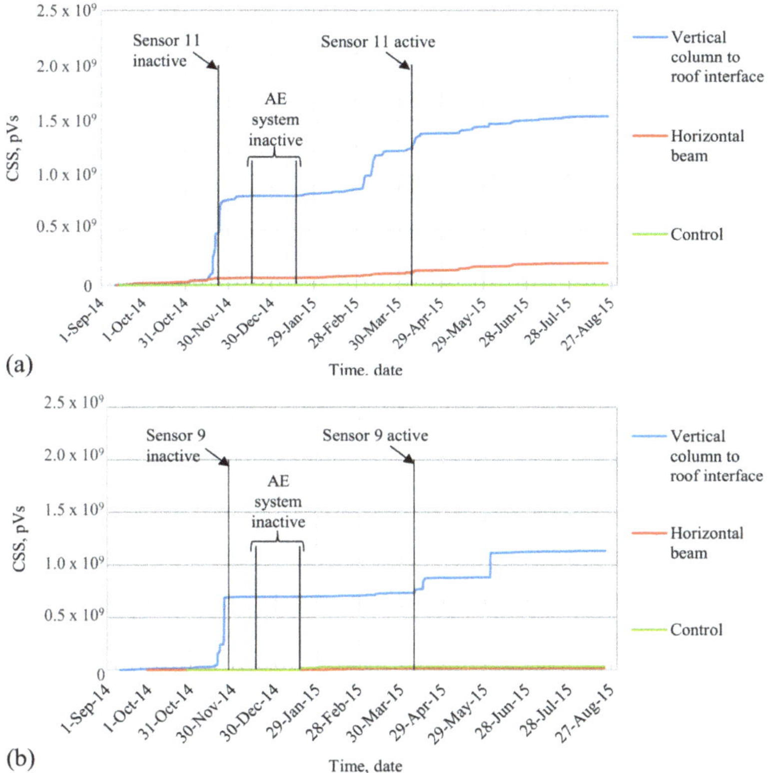

Figure 7. Cumulative signal strength (CSS) of: (**a**) Resonant sensor and (**b**) broadband sensors.

While the signal strength of an AE hit is related to the intensity of damage growth at a particular instant in time, cumulative signal strength is related to increases in damage growth rates over a particular testing period. Rapid increases in the cumulative signal strength curve are related to rapid increases in damage growth. The relationship between rapid changes in the cumulative signal strength curve and damage growth has been utilized to assess damage in different structural systems [26] including reinforced concrete bridges [12] and corrosion damage in reinforced concrete laboratory specimens [2,3].

In both Figure 7a,b, it is apparent that sharp changes in the slope of cumulative signal strength that indicate sharp increases in damage progression related to the vertical column to roof interface location occurred in several different instances. For example, a sharp increase in damage growth is noticed at the end of November, between 4 March 2015 and 13 March 2015, and between 8 April 2015 and 15 April 2015. These sharp increases were noticed for both the resonant and broadband sensor types. As expected, the broadband sensors exhibit slightly lower values of cumulative signal strength due to the relatively low sensitivity of this sensor type.

The highest change of slope for resonant and broadband sensors at the vertical column to roof interface occurred at the end of November in 2014. This sudden increase in cumulative signal strength was accompanied by localized spalling of concrete, which may have caused the detachment of two sensors previously mentioned. This spalling supports the findings that significant damage occurred during this time period.

To allow for the comparison of AE activity from each sensor, the response of broadband sensors was normalized to that of resonant sensors. The normalization was determined based on the application of a simulated source [25] applied at both resonant and broadband sensor locations on the reactor concrete block. Pencil lead breaks (PLBs) were applied at different angles around a resonant sensor (0, 45, 90, 135, and 180 degrees) at distances of 3 inches and 6 inches in each direction. Three PLBs were applied at each distance. The CSS recorded from PLBs applied at each distance was calculated separately. The same procedure was repeated for a broadband sensor. The ratio of CSS detected from the resonant sensor to the CSS from the broadband sensor was calculated for the cases of 3 inches and 6 inches from the sensor. The average of the ratios achieved at the two distances was found to be approximately equal to 10. Thus, cumulative signal strength detected from WDI sensors was normalized using a factor of 10.

Figure 8a is a visual representation of the intensity of damage at each sensor location using a contour plot. The plot is based on cumulative signal strength results (units of pico-Volt seconds) where high cumulative signal strength is plotted in red, which indicates high damage while low cumulative signal strength is plotted in blue. This indicates lower damage. The contour plots show relative intensity of AE activity.

As seen in the plot, the highest normalized cumulative signal strength values were detected at the top left of the elevation face sensors and at sensor 9 at the side of the vertical column. The 2D source location results (for the data detected from the five sensors at the same plane) show that most AE events were also detected at the top left of the sensor grid, which suggests that damage is progressing at this location (Figure 8b). Figure 8c likewise indicates very high damage progression in the vicinity of sensor 9 with the highest value of normalized CSS detected at sensor 9.

Figure 9 shows similar contour plots at the horizontal beam and control locations. Similar to the vertical column to roof interface location, normalized data was used to generate the plot. The same contour scale seen in Figure 8 was used to generate the plots. As seen in Figure 9, lower damage occurred at the horizontal beam location (Figure 9a) and the control location (Figure 9b) when compared to vertical column-to-roof interface location.

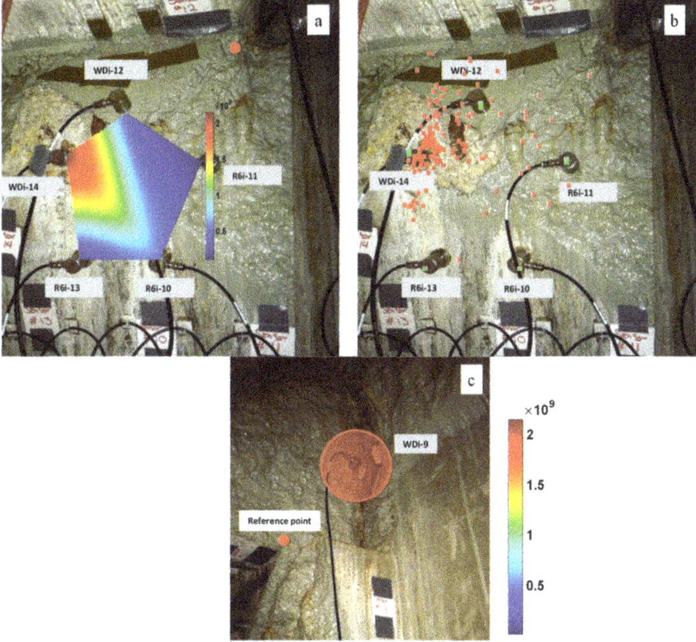

Figure 8. Vertical column to roof interface: (**a**) Signal strength contour plot at elevation face sensors, (**b**) source location at elevation face sensors, and (**c**) signal strength contour plot at the side face sensor.

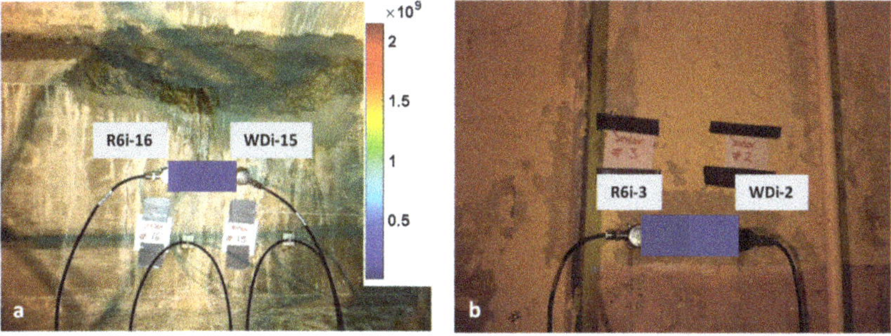

Figure 9. Signal strength contour plot: (**a**) Horizontal beam location and (**b**) control location.

2.3.3. Damage Classification Using Acoustic Emission

To provide a means for interpretation of the data, Intensity Analysis graphs were developed at each AE monitoring location. The method was first introduced by Fowler and others [26] for the evaluation of fiber reinforced polymer vessels and is based entirely on signal strength. Intensity Analysis is a graphical method that differs from many other forms of acoustic emission assessments in the sense that it is focused on trends in the AE data as opposed to individual events. Intensity Analysis uses two parameters, which are both based on signal strength: (a) historic index (plotted on the horizontal axis) and (b) severity (plotted on the vertical axis).

Historic index and severity can be calculated using Equations (1) and (2) where N is the number of hits up to time (t), S_{oi} is the signal strength of the i-th event, and K is an empirically derived factor that varies with the number of hits. The value of K has been previously selected in one case as:

(a) N/A if $N \leq 50$, (b) $K = N - 30$ if $51 \leq N \leq 200$, (c) $K = 0.85N$ if $201 \leq N \leq 500$, and (d) $K = N - 75$ if $N \geq 501$ [4].

$$H(t) = \frac{N}{N-K} \frac{\sum_{i=K+1}^{N} S_{oi}}{\sum_{i=1}^{N} S_{oi}} \tag{1}$$

$$S_r = \frac{1}{50} \sum_{i=1}^{i=50} S_{oi} \tag{2}$$

Historic index, $H(t)$, is a form of trend analysis that incorporates historical data in the current measurement. It is sensitive to changes of slope in cumulative signal strength versus time and compares the signal strength of the most recent hits to a value of cumulative hits. Severity, S_r, is defined as the average signal strength for the 50 hits with the highest numerical value of signal strength. The intensity analysis method has been widely used for the assessment of structural systems during load testing including reinforced concrete systems [12,27,28] and has been extended to the case of corrosion damage in pre-stressed concrete specimens [2–4,22].

By plotting the maximum historic index and severity values obtained over the duration of the test, an Intensity Analysis plot is generated. Due to the relationship between AE signal strength and damage growth, points that plot upward and to the right are associated with higher levels of damage.

Because Intensity Analysis (IA) uses historical information, an initial point on the Intensity Analysis plot must be chosen. This may be approached through visual inspection, numerical modeling, electrochemical measurements (in the case of corrosion damage), coring and petrographic examination, and other methods including suitable nondestructive evaluation techniques. Only a visual inspection was practical for the monitored locations within 105-C. Therefore, the initial point was chosen based on visual inspection.

The values of historic index and severity for the initial point were considered to account for pre-existing damage such that the historic index value at any time cannot be less than that for the initial point. For the severity, the distribution of the highest 50 signal strength values collected during the monitoring period in terms of their scattering from the mean value was used to develop the other 50 signal strength values with the same distribution but have a mean value equal to the severity of the initial point. Then the highest 50 numerical values from the collective set of 100 signal strengths, 50 from the monitoring period, and 50 developed from the initial point are used to calculate an updated severity value that takes into account the pre-existing condition.

Figures 10 and 11 are plots of Intensity Analysis results from the period beginning 10 September 2014 and ending 25 August 2015 for data recorded from resonant and broadband sensors, respectively. For the majority of field applications, only resonant sensors would be utilized due to the increased sensitivity of this sensor type in comparison to broadband sensors. The use of resonant sensors, therefore, reduces the number of sensors needed for a given application. However, resonant sensors do not provide high fidelity representations of the frequency content in comparison to broadband sensors. One purpose of using the two different sensor types is to investigate the associated differences in the results. The limits of the Intensity Analysis chart were developed based on data from resonant sensors [2]. Thus, it is expected that data collected from broadband sensors may yield underestimated damage classification if the same limits are used.

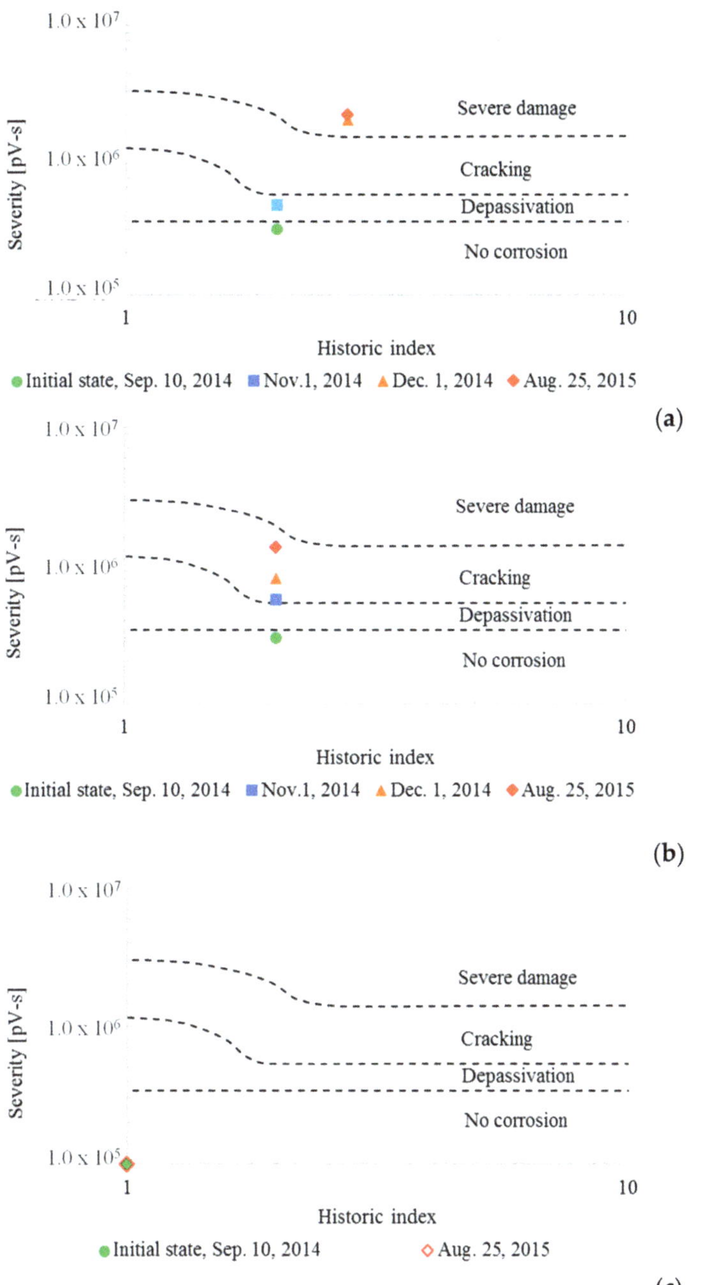

Figure 10. Intensity Analysis results for resonant sensors: (**a**) roof to column interface, (**b**) horizontal beam location, and (**c**) control location.

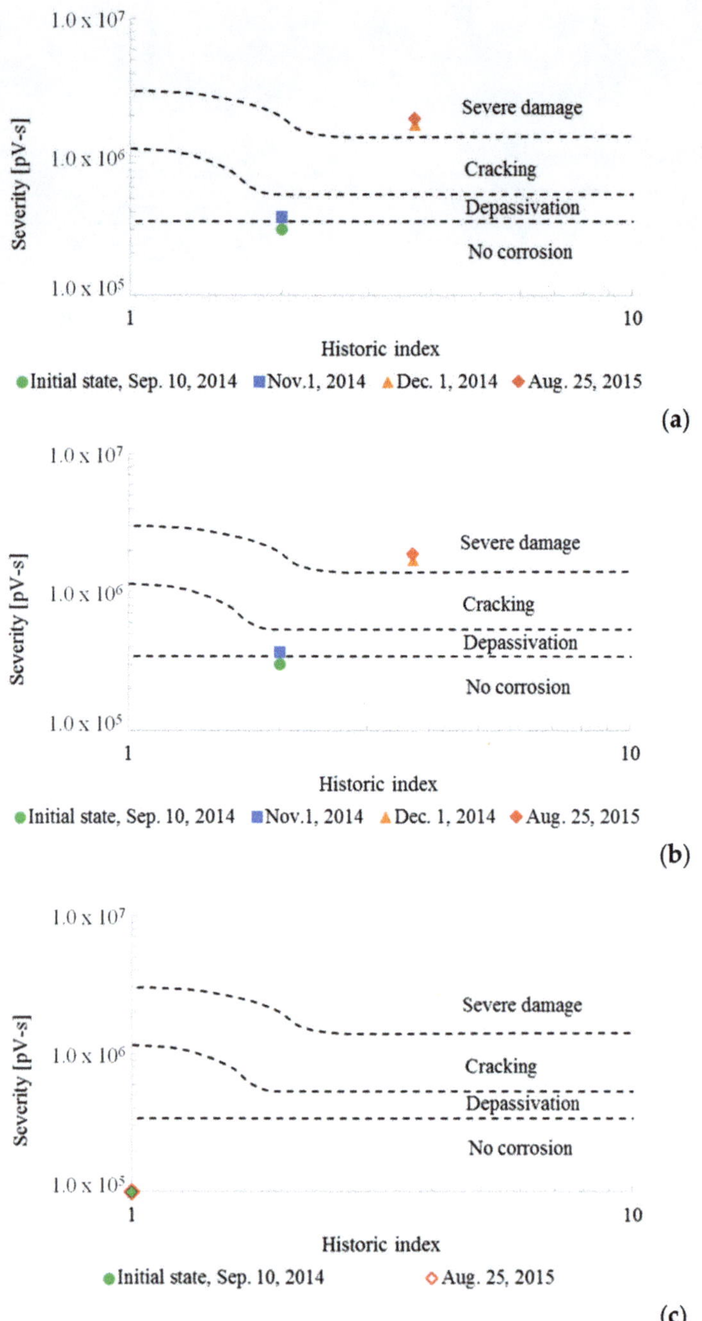

Figure 11. Intensity Analysis results for broadband sensors: (**a**) roof to column interface, (**b**) horizontal beam location, and (**c**) control location.

The preliminary estimation of damage was based on visual inspection during the initial visit to 105-C and was located near the border between the 'no damage' and the 'depassivation' regions of the chart (severity = 300,000 and historic index = 2.0) for both the vertical column to roof interface location and the horizontal beam location. This assumed level of damage underestimated the actual condition of the structures since these areas are known to have ongoing corrosion damage. Ideally, this initial point would be established through a combination of methods including visual inspection and electrochemical methods. Electrochemical methods, however, were not collected during this part of the study. For the control location, no damage was assumed and, therefore, the initial point was located at the left corner in the 'no corrosion' region of the chart.

Acoustic emission activity during the monitoring period (approximately one year) at the vertical column to the roof interface location indicated a progression from the initial state to the severe damage state for both sensor types. It is noted that the results of IA after approximately 2 months of monitoring (1 November 2014) showed that corrosion is ongoing at this location. On 1 December 2014, Intensity Analysis results indicated that severe damage occurred. For monitoring over this relatively short duration, such a progression from the initial state to the 'severe' damage state is indicative of a relatively high level of ongoing damage growth in the monitored areas. For this plot, the term 'cracking' refers to micro-cracking that is generally non-visible while 'severe damage' refers to visible cracking that may be accompanied by spalling. This result is supported by the spalling that occurred at this location during the monitoring period.

Acoustic emission data from the resonant sensor and the broadband sensor at the horizontal beam location progressed from the initial state to the cracking state over the duration of the monitoring period. This result is also an indication of ongoing damage growth at this location when the relatively short monitoring period is considered. The broadband sensor results (Figure 11b) indicated less damage than the resonant sensor results (Figure 10b) especially during the first 3 months of monitoring. This can be attributed to the lower sensitivity of the broadband sensors.

In contrast to the roof interface location and the horizontal beam location, the intensity analysis results for the control location indicate no damage progression during the monitoring period and, therefore, the initial state and final state coincide (plot on top of one another) for the control location.

2.3.4. Remote Monitoring at +48 Level

A cellular connection was used to remotely operate the wireless acoustic emission data acquisition system. Data from the wireless system was collected between 9 September, 2014 (commencement of test) and 13 November 2014. Due to the loss of power from the solar power/battery system, 10 days of data were lost starting from 11 September 2014. The power was reconnected and the system continued to monitor the connection until 15 October 2014 when a thunderstorm caused a power outage and data was lost for another 13 days. The system continued to collect data afterwards until the data acquisition laptop was damaged on 13 November 2014. The damage was most likely due to moisture and was not repairable.

As described for the wired system data, the raw data was analyzed and appropriate data filters were used to separate meaningful data from spurious emissions. The limits of the data filters are shown in Table 2. Figure 12 shows acoustic emission activity in terms of amplitude versus time (showing both rain and temperature data) collected between 9 September 2014 and 13 November 2014 from the wireless acoustic emission system. This data set contains a significant number of hits with amplitude exceeding 80 dB. These hits are of relatively high amplitude and may be correlated to ongoing damage.

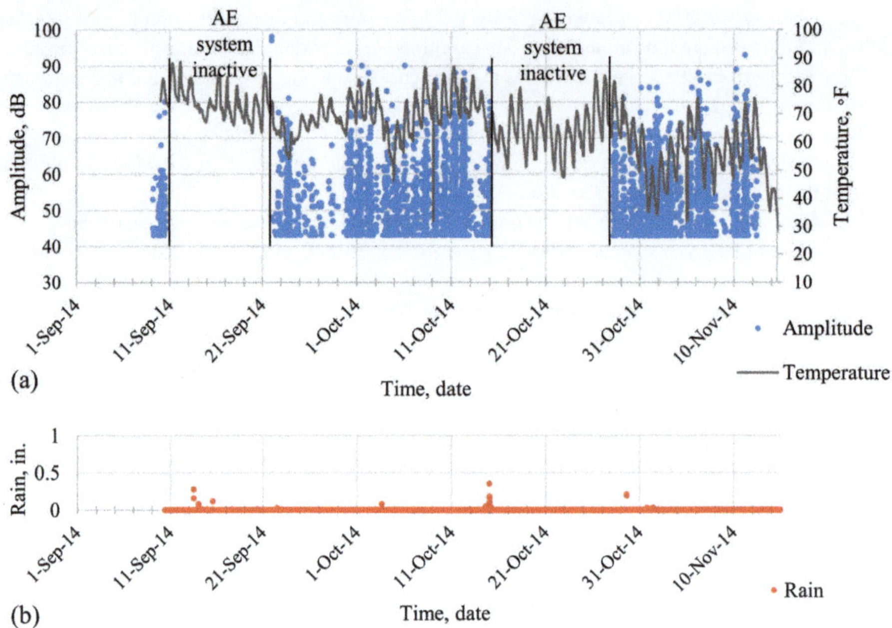

Figure 12. (**a**) Amplitude and temperature versus time for four wireless sensors at the +48 level and (**b**) rain versus time.

One objective of monitoring this location was to assess whether the large vertical crack in the wall is still active. This vertical crack has a width between 0.125 and 0.25 inches with several small hairline cracks extending from it in the horizontal direction. Figure 13 plots the cumulative signal strength (units of pico-Volt seconds) versus time (days) for the collected signals over the monitoring period. An increasing trend in the acoustic emission activity is observed in the figures, which indicates that damage may be progressing at this location.

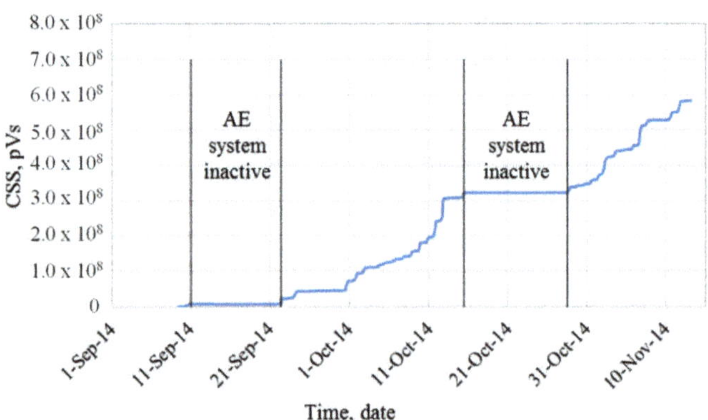

Figure 13. Cumulative signal strength (pVs) versus time (days) for four wireless sensors at the +48 level.

To further investigate the trends in this data set, triangulation algorithms were used to investigate if AE events were generated from crack growth. Figure 14 shows the source location results from filtered acoustic emission data. In this figure, each red dot indicates a located acoustic emission event, which means that all four sensors received data with a specified time increment. The time increment was determined based on the characteristic wave speed of the structure, which was experimentally determined during the installation site visit, and the geometry of the sensor grid. Source location from raw data was inconclusive since it showed acoustic emission activity throughout the monitored area. The 6 acoustic emission events from the filtered data set were located in the vicinity of the vertical crack. These results imply that crack growth or friction between crack surfaces was ongoing in this area during the monitoring period.

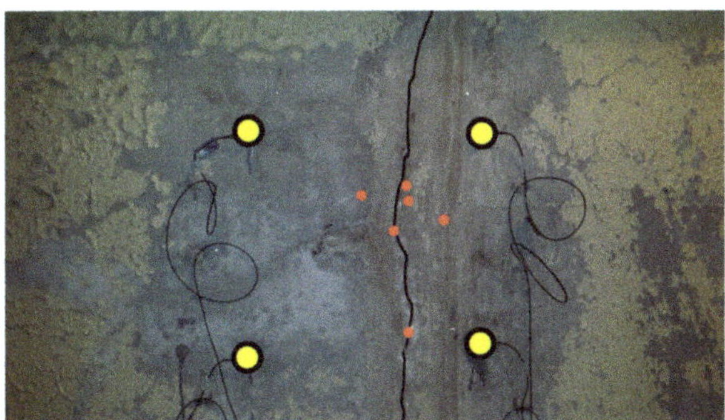

Figure 14. Source location results at the +48 level. Red dots indicate located AE events.

3. Accelerated Corrosion Testing of the Reactor Concrete Block

3.1. Experimental Program

A reinforced concrete block was cut from the reactor facility with a length, width, and depth of 7 feet 4 inches, 3 feet, and 3 feet 4 inches, respectively. An accelerated corrosion test was conducted to corrode three different areas over the course of this study. Three concrete cores were drilled (3 inches in diameter and 9 inches in length) at three locations to create different concrete cover thickness for three vertical steel reinforcing bars adjacent to the cores (Figure 15). During the coring process, a transverse reinforcing bar was unavoidably cut at a depth of approximately 6 inches from the surface of the concrete test block specimen.

The test was initiated by placing 3% NaCl solution in the drilled holes to a depth of 3 inches on 2 December 2014. The solution was maintained in the drilled holes for two months to ensure that chloride concentration reached a needed level for corrosion initiation [29,30]. Wet/dry cycles were then initiated (three days wet and four days dry) on 19 February 2015 to accelerate the corrosion process. A galvanic cell was created during the wet days by inserting a copper plate in the cored locations. Figure 15d shows the 'as measured' concrete cover after the cores were drilled.

The first location has a concrete cover of 0.25 inches and was monitored using three broadband sensors (WDI) and one resonant sensor (R15I) while the second location has a cover of 1.0 inch and was monitored using four resonant sensors (R6I). The third location has a cover of 0.125 inches and was monitored using eight resonant sensors (R6I). On 22 May 2015 one of the sensors at the 1.0 inch cover location was removed from the test block and, on 27 May 2015, it was placed on a small concrete specimen (control specimen) with dimensions of 3.0 inches × 3.0 inches × 11.25 inches. The control specimen is not reinforced and, therefore, is known not to have corrosion activity. Data collected from the control specimen was used to verify the efficiency of the data filters developed during the course of the project. Acoustic emission activity was recorded continuously throughout the test period.

Half-cell potential (HCP) and linear polarization resistance (LPR) measurements were recorded once a week with the objective of providing insight related to the corrosion process of targeted reinforcement locations. HCP method is described in ASTM C876 [31] and is traditionally employed to determine the likelihood of corrosion activity, which is described in Table 3. Linear polarization resistance (LPR) is a method used to measure polarization resistance (Rp), which can be used to calculate corrosion current (*Icorr*) and corrosion current density (*icorr*). These parameters can be used to estimate the corrosion rate (*CR*). Figure 16 shows a schematic of the test setup and acoustic emission sensor layout to monitor the corrosion process of the reinforcing bars. A schematic of the aged concrete block control specimen is also shown in this figure.

Table 3. ASTM corrosion for Cu-CuSO4 reference electrode [31].

Potential Against Cu-CuSO4 Electrode	Corrosion Condition
>-200 mV	Low risk (10% probability of corrosion)
-200 to -350 mV	Intermediate corrosion risk
<-350 mV	High corrosion risk (90% probability)
<-500 mV	Severe corrosion damage

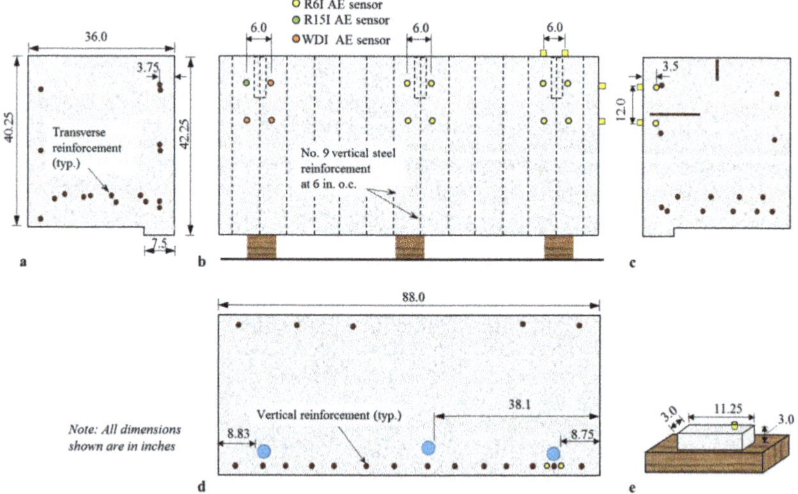

Figure 15. Aged concrete block specimen: (**a**) Left side view, (**b**) front view, (**c**) right side view, (**d**) top view, and (**e**) control location.

Figure 16. Schematic of aged reactor concrete test block: (**a**) Left side view, (**b**) front view, (**c**) right side view, (**d**) top view, and (**e**) control location.

3.2. Results and Discussion

3.2.1. Electrochemical Measurements

Initial electrochemical measurements known as the half-cell potential were taken prior to the initiation of the conditioning period. These measurements indicated a passive state of steel reinforcement. The NaCl solution was then placed in the cored areas on 2 December 2015 and electrochemical readings were recorded weekly thereafter. As shown in Figure 17, three weeks after conditioning, half-cell potential values were observed to be more negative than -350 mV (referred to as the corrosion threshold) at all three locations. At the conclusion of the wet/dry cycles, half-cell potential readings indicated high corrosion risk in one of the three locations (0.25 inch cover location) and severe corrosion damage (more negative than -500 mV) in the other two locations (0.125 inch and 1.0 inch cover locations). The 1.0 inch cover location is known to have leakage associated with it since the NaCl solution drained continuously from this location from the commencement of the testing. While chloride diffusion is often assumed to be the primary initiator of corrosion damage, the presence of cracking in the concrete matrix may have a more profound effect on corrosion in some instances. The 0.125 inch cover and the 0.25 inch cover locations did not experience similar issues with leakage. The bottom of the hole at the 1.0 inch cover location was sealed with epoxy in the first week of April in 2015.

Figure 18 shows linear polarization resistance results at the three locations with a logarithmic fit of the data points. The x-axis in Figure 18 represents the number of days after the solution was placed in the cored areas (initiated on 2 December 2014). The results indicate that all locations had relatively high corrosion rates since the polarization resistance was less than 100 ohms [4]. As seen in the figure, data was not collected between 24 December 2014 and 19 February 2015 (between 22 and 79 days) due to a malfunction with the potentiostat/galvanostat cable over that time period. This was addressed and the testing was resumed after 29 February, 2015. Since these readings are taken weekly over a time span of 300 days and due to the instantaneous nature of the readings, trends in the data set are more important than readings taken on a particular day. Therefore, trend lines with both upper and lower estimates are shown in the figures. A statistical method was used to eliminate outliers with low values to obtain the upper estimate and eliminate outliers with high values to obtain the lower estimate.

3.2.2. Detection of Damage Using Acoustic Emission

Figure 19 shows the acoustic emission activity, in terms of amplitude versus time, recorded at locations monitored with resonant sensors (the 1.0 inch concrete cover location, the 0.125 inch concrete cover location, and the control location which was initiated on 27 May 2015). Figure 20 shows the acoustic emission activity recorded using broadband sensors at the 0.25 inch concrete cover location. The data shown in Figures 19 and 20 was filtered using the data filters discussed in Table 2. An unusual amount of data that had characteristics related to electromagnetic interference was continually collected at the control location potentially due to damage in the sensor or cable during the removal and re-installation process. RMS and Duration data rejection limits were developed and were able to delete the majority of the false data without affecting data collected from other locations.

As seen in Figures 19 and 20, acoustic emission activity at the 1.0 inch concrete cover location and the 0.125 inch concrete cover location was higher than the acoustic emission activity at the 0.25 inch concrete cover location. This is attributed to the inherently higher sensitivity of the resonant sensors. It is noted that the rate of activity recorded at 1.0 inch concrete cover locations decreased during wet days after sealing the bottom of the hole.

To reduce the possibility of contaminating the acoustic emission data set with unrelated data generated from ongoing work in the University of South Carolina Structures and Materials Laboratory, the acoustic emission data acquisition system was intentionally paused on several occasions. Significant pauses in data acquisition are shown in the figures. A video camera monitoring the system was utilized to cross-verify and to aid in the development of data filters that are specific to ongoing work in the laboratory environment.

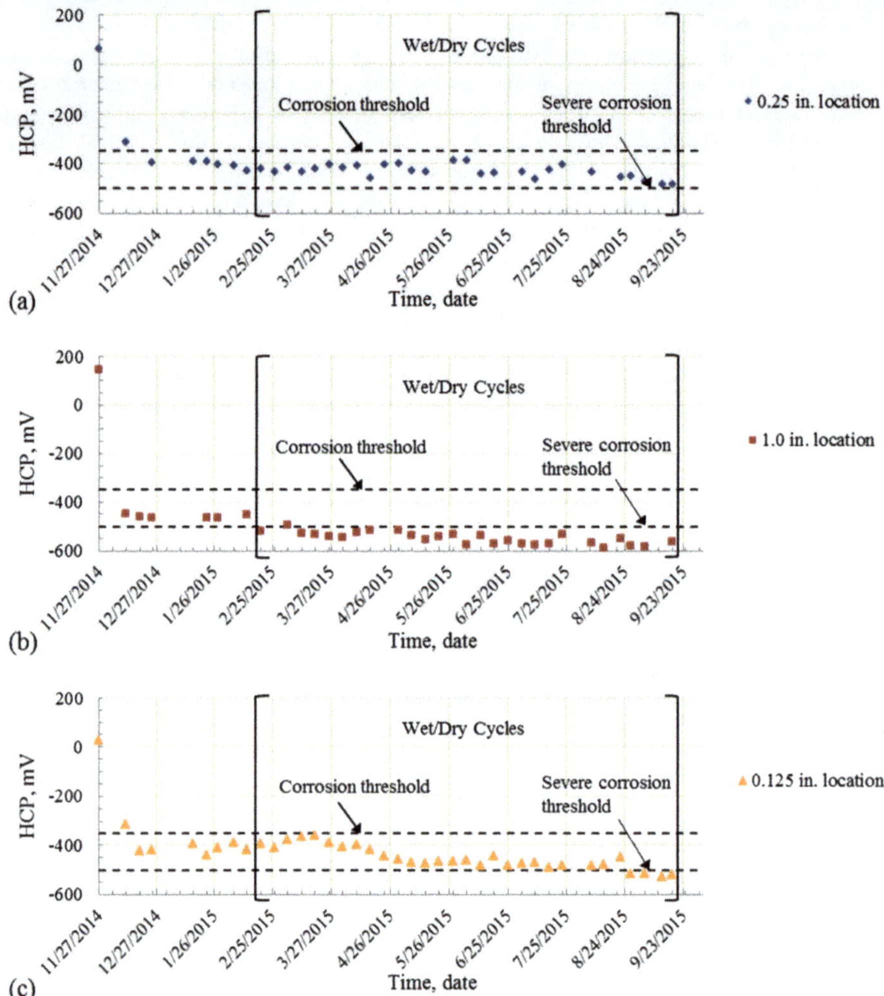

Figure 17. Half-cell potential measurements at: (**a**) 0.25-inch concrete cover location, (**b**) 1.0-inch concrete cover location, and (**c**) 0.125-inch concrete cover locations.

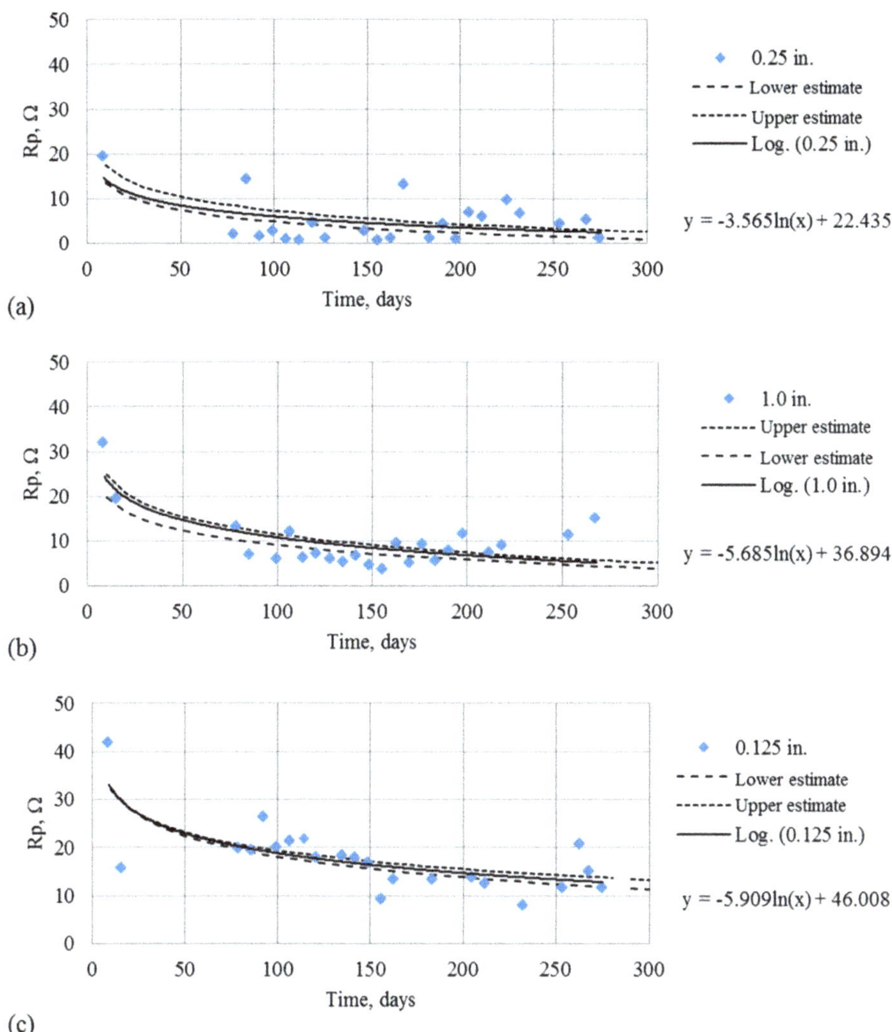

Figure 18. Linear polarization resistance (LPR) measurements at: (**a**) 0.25-inch concrete cover location, (**b**) 1.0-inch concrete cover location, and (**c**) 0.125-inch concrete cover location.

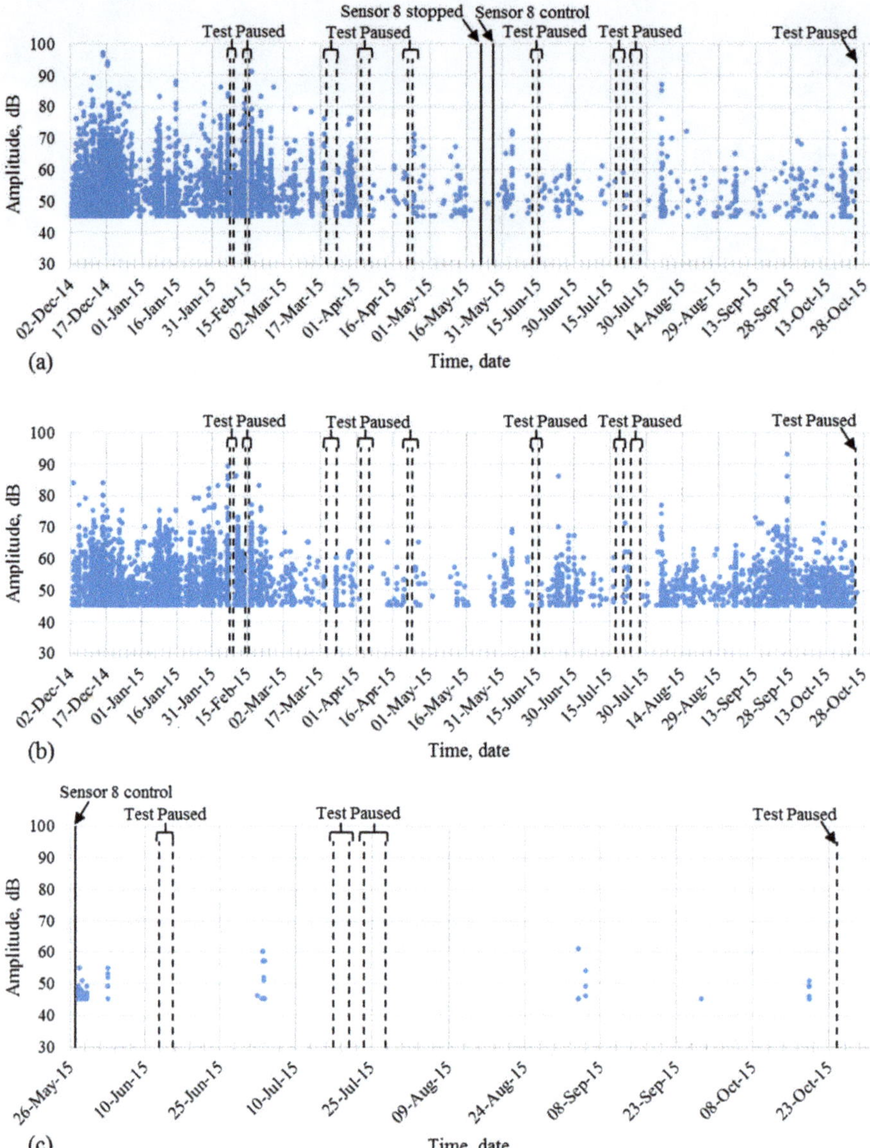

Figure 19. AE data recorded from resonant sensors on the reactor concrete block specimen: (**a**) 1.0-inch concrete cover location, (**b**) 0.125-inch concrete cover location, and (**c**) control location.

Appl. Sci. **2018**, *8*, 1663

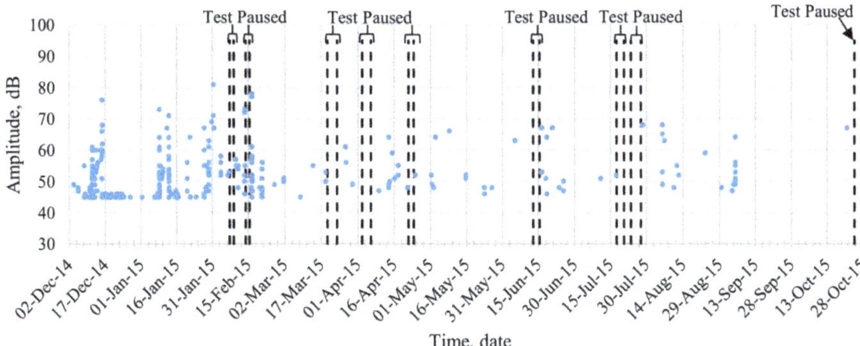

Figure 20. AE data recorded from broadband sensors on the reactor concrete block specimen at the 0.25-inch concrete cover location.

Figure 21 shows cumulative signal strength versus time at locations monitored using resonant sensors. It can be seen from this figure that cumulative signal strength increases rapidly at the beginning of the test, which corresponds to a period of rapid damage growth associated with corrosion initiation, enters a dormant period, and then increases slightly near the end of the testing period for the 1.0 inch and 0.125 inch locations. This trend in the data mirrors a trend noticed in the linear polarization resistance plots. The magnitude of the cumulative signal strength is greater for the 1.0-inch location when compared to the 0.125-inch location, which indicates increased acoustic emission activity and, therefore, increased damage growth at the 1.0-inch location. This is consistent with the electrochemical readings at this location and may be attributable to the presence of cracking in this location. The control location has minimal cumulative signal strength magnitude as would be expected. The relatively low cumulative signal strength magnitude at the control location demonstrates that unwanted acoustic emission data caused by ongoing laboratory activities in the vicinity of the test block specimen were minimized in the data sets.

The broadband sensor data shows a similar trend of rapidly increasing damage early in the testing period, which is followed by a relatively dormant period at the 0.25 inch location. This is shown in Figure 22. The magnitude of cumulative signal strength from the broadband sensors is lower in comparison to the resonant sensors, which is expected due to the lower sensitivity of the broadband sensors.

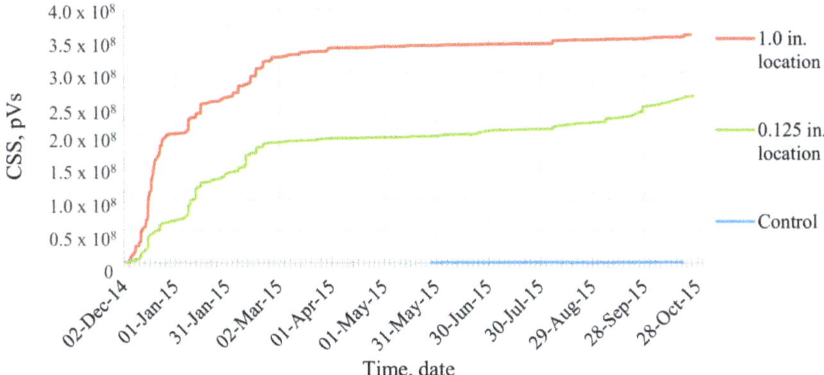

Figure 21. Cumulative signal strength from resonant sensors on the aged concrete block specimen.

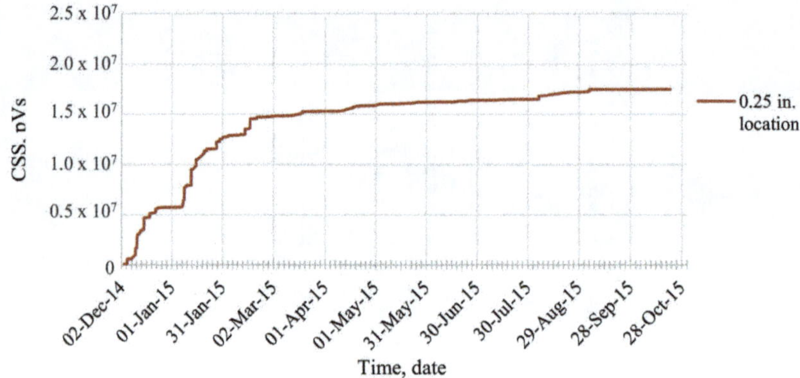

Figure 22. Cumulative signal strength versus time from broadband sensors on the aged concrete block specimen.

Figures 23 and 24 show the Intensity Analysis results calculated at each location. The estimation of initial damage for the aged concrete block specimen based on visual inspection and electrochemical results was located near the border between the 'no damage' region and the 'depassivation' region of the chart. For the control location, a lower initial damage state was used since no corrosion damage is expected in this specimen. AE activity from the resonant sensors at the 1.0-inch concrete cover location progressed from the initial state to the severe damage zone over the duration of the monitoring period. AE activity from the resonant sensors at the 0.125-inchconcrete cover location progressed from the initial state to the border of the cracking and severe damage zones. For the broadband sensors at the 0.25-inch concrete cover location, acoustic emission activity progressed from the initial state to the border of the cracking and severe damage zones.

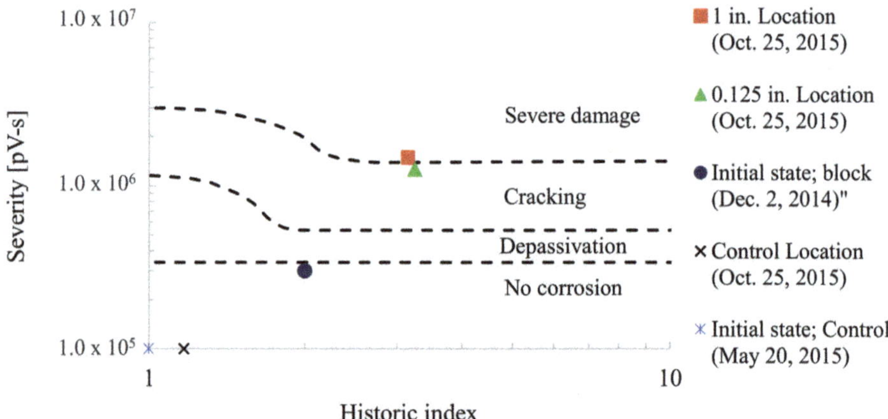

Figure 23. Intensity Analysis for resonant sensors on the reactor concrete block specimen.

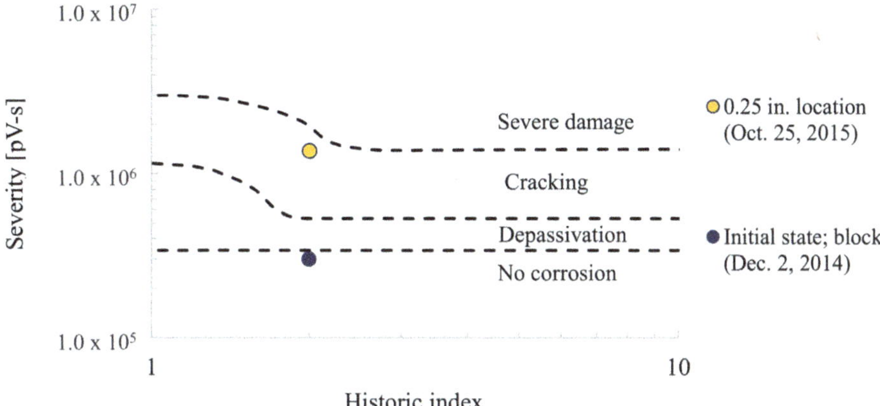

Figure 24. Intensity Analysis for broadband sensors on the reactor concrete block specimen.

The above results are indicative of cracking in the concrete matrix due to corrosion activity at all three locations. As with the electrochemical measurements, the acoustic emission activity indicated that the most severe damage occurred at the 1.0-inch concrete cover location. As mentioned above, this location is affected by cracking as noticed through leakage of the NaCl solution at this location. While many degradation models for reinforced concrete are based on diffusion and, therefore, do not directly address the presence of cracking in the matrix. The effect of cracking in the matrix may, nonetheless, be significant. Similarly, many models assume a homogeneous concrete matrix. The lack of homogeneity in the concrete matrix for actual structures such as the concrete test block may also play a significant role in the results.

4. Summary and Conclusions

This investigation explores the implementation of acoustic emission monitoring as a remote structural assessment method. Acoustic emission systems were used to monitor corrosion damage and cracking in a decommissioned nuclear reactor facility as well as to monitor corrosion damage in a concrete block cut from the nuclear facility in laboratory conditions. The monitoring period in this study extended to approximately one year.

The study showed that long-term remote monitoring of ongoing damage in large scale existing structures is feasible using acoustic emission systems. For the wired system, AC power and cellular network connection are required for successful operation of the system. No major issues were encountered in terms of electromagnetic interference with the sensors, external noise and remote monitoring, and data transfer. The wireless system used has the potential to be used with solar power paired with cellular connections for the remote monitoring, which makes this approach well suited for long-term monitoring efforts. However, adequate protection to the electrical components is required especially in humid environments, which is illustrated by the failure of the data acquisition laptop due to moisture damage.

For the Reactor Building 105-C Crane Maintenance Area, the acoustic emission activity recorded at the 'vertical column to roof interface location' and 'horizontal beam location' varied throughout the monitoring period and tended to be associated with seasonal temperature fluctuations. The acoustic emission activity recorded at the 'control location' was significantly less when compared to the activity from the other two locations. Intensity Analysis was used to quantify the damage progression over the course of the monitoring period for both the broadband and resonant sensor types. The results of this method were in agreement with visually observed distress in the monitored locations. The assessed condition of the actively corroding areas progressed from the assumed condition of 'no corrosion/approaching depassivation' to 'severe damage' over the monitoring period while no change was observed in the state of the control location. It is noted that the assessed condition based on Intensity Analysis progressed to 'cracking/severe damage' within the first two months of monitoring. This shows the feasibility of this technique to successfully qualify active corrosion damage in structures in relatively small monitoring periods.

For the Reactor Building 105-C +48 level, the acoustic emission activity at the +48 location also varied with seasonal temperature fluctuations. This area contained a vertical crack in the exterior wall and it is possible that crack growth or friction between surfaces of this crack was the cause of much of the acoustic emission activity. Source location was carried out at this location and events were located in the vicinity of the vertical crack, which shows the feasibility of acoustic emission to detect and locate ongoing damage from cracking given that appropriate data filters are used.

For the Aged Concrete Test Block, both electrochemical results and acoustic emission cumulative signal strength versus time indicated that the corrosion activity occurred primarily during the first three to four months of conditioning and then continued at a reduced rate. Intensity Analysis based on the acquired data indicated that damage progressed from the assumed initial condition of 'no corrosion/approaching depassivation', determined based on electrochemical results upon arrival at the laboratory to 'cracking/severe damage' over the monitoring period for all three locations and for both sensor types. This Intensity Analysis result is similar to the one reported for the 'vertical column-to-roof interface location'.

One of the main areas that hinder wide implementation of structural health monitoring systems is the large amounts of data that is collected and the subsequent effort needed to interpret and analyze this data in order to produce meaningful assessment of the condition of the structures. An important contribution of this study is that it proved the ability of well-developed data reduction and damage assessment algorithms to provide accurate evaluation of the condition of the structures. The results of the study showed that the developed filtering techniques along with the Intensity Analysis chart used for corrosion damage classification were able to successfully qualify the damage in the monitored areas. A method to account for pre-existing damage in the AE Intensity Analysis chart was also developed. These methods can be easily programmed and used to provide meaningful information to facility managers without the need for further assessment of large data sets. This can subsequently help in maintenance planning and prioritization especially in large scale and complex infrastructure systems.

Future research is needed to correlate acoustic emission data to electrochemical measurements especially in early corrosion stages, which will further enhance the outcome of monitoring. This will provide more insight regarding the progression of corrosion damage and can ultimately enable estimation of remaining service life.

Author Contributions: Conceptualization, M.A., M.E., K.D., M.S. and P.Z.; Methodology, P.Z., M.E. and M.A.; Validation, M.A. and M.E.; Formal Analysis, M.A.; Writing-Original Draft Preparation, M.A.; Writing-Review & Editing, M.E., P.Z., K.D. and M.S.; Supervision, P.Z.; Project Administration, P.Z. and M.S.; Funding Acquisition, P.Z., M.S., K.D. and M.E.

Acknowledgments: The authors would like to thank the support of Savannah River National Laboratory. Portions of this work are supported by the U.S. Department of Energy and Savannah River National Laboratory through SCUREF Funding. Portions of this work are supported by the U.S. Department of Commerce, National Institute of Standards and Technology (NIST), Technology Innovation Program (TIP), Cooperative Agreement 70NANB9H9007.

Conflicts of Interest: The authors declare no conflict of interest.

References

1. Clifton, J.F. *Predicting the Remaining Service Life of Concrete*; National Institute of Standards and Technology: Gaithersburg, MD, USA, 1991.
2. Mangual, J.; ElBatanouny, M.; Ziehl, P.; Matta, F. Acoustic-emission-based characterization of corrosion damage in cracked concrete with prestressing strand. *ACI Mater. J.* **2013**, *110*, 89–98.
3. Mangual, J.; ElBatanouny, M.; Ziehl, P.; Matta, F. Corrosion damage quantification of prestressing strands using acoustic emission. *J. Mater. Civ. Eng.* **2013**, *25*, 1326–1334. [CrossRef]

4. ElBatanouny, M.; Mangual, J.; Ziehl, P.; Matta, F. Early corrosion detection in prestressed concrete girders using acoustic emission. *J. Mater. Civ. Eng.* **2014**, *26*, 504–511. [CrossRef]

5. Abdelrahman, M.; ElBatanouny, M.; Serrato, M.; Dixon, K.; Larosche, C.; Ziehl, P. Classification of alkali-silica reaction and corrosion distress using acoustic emission. *AIP Conf. Proc.* **2016**, *1706*, 140001.

6. Fournier, B.; Berube, M.A.; Folliard, K.J.; Thomas, M. *Report on the Diagnosis, Prognosis, and Mitigation of Alkali-Silica Reaction (ASR) in Transportation Structures*; Transport Research International Documentation (TRID); Available online: https://trid.trb.org/view/1099043 (accessed on 30 June 2014).

7. Abdelrahman, M.; ElBatanouny, M.; Ziehl, P.; Fasl, J.; Larosche, C.; Fraczek, J. Classification of alkali-silica reaction damage using acoustic emission: A proof-of-concept study. Constr. *Build. Mater.* **2015**, *95*, 406–413. [CrossRef]

8. Braverman, J.I.; Xu, J.; Ellingwood, B.R.; Costantino, C.J.; Morante, R.J.; Hofmayer, C.H. *Evaluation of the Seismic Design Criteria in ASCE/SEI Standard 43-05 for Application to Nuclear Power Plants*; U.S. Nuclear Regulatory Commission: Washington, DC, USA, 2007; pp. 1–73.

9. Kojima, F. *Structural Health Monitoring of Nuclear Power Plants using Inverse Analysis in Measurements*; Kobe University: Hyogo, Japan, 2009; Available online: http://www.research.kobe-u.ac.jp/csi-applmath/proc/workshopKobe/final/Section10.pdf (accessed on 31 December 2011).

10. Abdelrahman, M.; ElBatanouny, M.; Ziehl, P. Acoustic emission based damage assessment method for prestressed concrete structures. *Eng. Struct.* **2014**, *60*, 258–264. [CrossRef]

11. Ono, K. *Application of Acoustic Emission for Structure Diagnosis*; BazTech; Available online: http://yadda.icm.edu.pl/baztech/element/bwmeta1.element.baztech-article-BAR0-0060-0020 (accessed on 31 December 2011).

12. Golaski, L.; Gebski, P.; Ono, K. Diagnostics of concrete bridges by acoustic emission. *J. Acoust. Emiss.* **2002**, *20*, 83–98.

13. ASTM E1316-16a. *Standard Terminology for Nondestructive Examinations*; ASTM International: West Conshohocken, PA, USA, 2016; pp. 1–38.

14. Pollock, A.A. Classical Wave Theory in Practical AE Testing. In Proceedings of the 8th International AE Symposium, Tokyo, Japan, 21–24 October 1986; pp. 708–721.

15. Weng, M.S.; Dunn, S.E.; Hartt, W.H.; Brown, R.P. Application of Acoustic Emission to Detection of Reinforcing Steel Corrosion in Concrete. *Corrosion* **1982**, *38*, 9–14. [CrossRef]

16. Dunn, S.E.; Young, J.D.; Hartt, W.H.; Brown, R.P. Acoustic emission characterization of corrosion induced damage in reinforced concrete. *Corrosion* **1984**, *40*, 339–343. [CrossRef]

17. Zdunek, A.D.; Prine, D.W.; Li, Z.; Landis, E.; Shah, S. *Early Detection of Steel Rebar Corrosion by Acoustic Emission Monitoring*; NACE International: Houston, TX, USA, 1995.

18. Li, Z.; Zudnek, A.; Landis, E.; Shah, S. Application of acoustic emission technique to detection of reinforcing steel corrosion in concrete. *ACI Mater. J.* **1998**, *95*, 68–76.

19. Assouli, B.; Simescu, F.; Debicki, G.; Idrissi, H. Detection and identification of concrete cracking during corrosion of reinforced concrete by acoustic emission coupled to the electrochemical techniques. *NDT E Int.* **2005**, *38*, 682–689. [CrossRef]

20. Ohtsu, M.; Tomoda, Y. Phenomenological model of corrosion on process in reinforced concrete identified by acoustic emission. *ACI Mater. J.* **2008**, *105*, 194–199.

21. Di Benedetti, M.; Loreto, G.; Matta, F.; Nanni, A. Acoustic emission monitoring of reinforced concrete under accelerated corrosion. *J. Mater. Civ. Eng.* **2013**, *25*, 1022–1029. [CrossRef]

22. Vélez, W.; Matta, F.; Ziehl, P. Acoustic emission monitoring of early corrosion in prestressed concrete piles. *Struct. Control Health Monit.* **2015**, *22*, 873–887. [CrossRef]

23. Appalla, A.; ElBatanouny, M.; Velez, W.; Ziehl, P. Assessing corrosion damage in post-tensioned concrete structures using acoustic Emission. *J. Mater. Civ. Eng.* **2015**, *28*, 04015128. [CrossRef]

24. ASTM E2075/E2075M-15. *Standard Practice for Verifying the Consistency of AE-Sensor Response Using an Acrylic Rod*; ASTM International: West Conshohocken, PA, USA, 2015; pp. 1–5.

25. ASTM E2374-15. *Standard Guide for Acoustic Emission System Performance Verification*; ASTM International: West Conshohocken, PA, USA, 2015; pp. 1–5.

26. Fowler, T.; Blessing, J.; Conlisk, P.; Swanson, T.L. The MONPAC system. *J. Acoust. Emiss.* **1989**, *8*, 1–8.

27. Nair, A.; Cai, C.S. Acoustic emission monitoring of bridges: Review and case studies. *Eng. Struct.* **2010**, *32*, 1704–1714. [CrossRef]

28. ElBatanouny, M.; Ziehl, P.; Larosche, A.; Mangual, J.; Matta, F.; Nanni, A. Acoustic emission monitoring for assessment of prestressed concrete beams. Constr. *Build. Mater.* **2014**, *58*, 46–53. [CrossRef]
29. Nilsson, L.O.; Sandberg, P.; Poulsen, E.; Tang, L.; Andersen, A.; Frederiksen, J.M. *HETEK, a System for Estimation of Chloride Ingress into Concrete Theoretical Background*; Danish Technological Institute: Taastrup, Danmark, 2011.
30. Vélez, W.; ElBatanouny, M.; Matta, F.; Ziehl, P.H. *Assessment of Corrosion in Prestressed Concrete Piles in Marine Environment with Acoustic Emission*; NACE International: Salt Lake City, UT, USA, 2012.
31. ASTM C876-09. *Standard Test Method for Half-Cell Potentials of Uncoated Reinforcing Steel in Concrete*; ASTM International: West Conshohocken, PA, USA, 2009; pp. 1–8.

applied
sciences

Article

Damage Mechanism Evaluation of Large-Scale Concrete Structures Affected by Alkali-Silica Reaction Using Acoustic Emission

Vafa Soltangharaei [1,*], Rafal Anay [1], Nolan W. Hayes [2], Lateef Assi [1], Yann Le Pape [3], Zhongguo John Ma [2] and Paul Ziehl [1]

[1] Department of Civil and Environmental Engineering, University of South Carolina, Columbia, SC 29208, USA; ranay@email.sc.edu (R.A.); lassi@email.sc.edu (L.A.); ziehl@cec.sc.edu (P.Z.)

[2] Department of Civil and Environmental Engineering, University of Tennessee, Knoxville, TN 37916, USA; nhayes3@vols.utk.edu (N.W.H.); zma2@utk.edu (Z.J.M.)

[3] Oak Ridge National Laboratory, Oak Ridge, TN 37831, USA; lepapeym@ornl.gov

[*] Correspondence: vafa@email.sc.edu

Received: 25 September 2018; Accepted: 31 October 2018; Published: 3 November 2018

Abstract: Alkali-silica reaction has caused damage to concrete structures, endangering structural serviceability and integrity. This is of concern in sensitive structures such as nuclear power plants. In this study, acoustic emission (AE) was employed as a structural health monitoring strategy in large-scale, reinforced concrete specimens affected by alkali-silica reaction with differing boundary conditions resembling the common conditions found in nuclear containments. An agglomerative hierarchical algorithm was utilized to classify the AE data based on energy-frequency based features. The AE signals were transferred into the frequency domain and the energies in several frequency bands were calculated and normalized to the total energy of signals. Principle component analysis was used to reduce feature redundancy. Then the selected principal components were considered as features in an input of the pattern recognition algorithm. The sensor located in the center of the confined specimen registered the largest portion of AE energy release, while in the unconfined specimen the energy is distributed more uniformly. This confirms the results of the volumetric strain, which shows that the expansion in the confined specimen is oriented along the thickness of the specimen.

Keywords: alkali-silica reaction; acoustic emission; pattern recognition; confinement; damage evaluation

---

1. Introduction

Alkali-silica reaction (ASR) is a chemical processes that has caused damage in concrete structures such as bridges [1–4], nuclear power plants [5–9], and concrete dams [5,10]. This reaction usually occurs between alkali hydroxides in the pore solution and the reactive silica in some aggregates. The result is an alkali-silica gel, which is hygroscopic and imbibes water and humidity. The gel expands in humidity exceeding 80% [6]. This expansion induces pressure on the concrete matrix and aggregates and causes micro-cracks and cracks when the pressure exceeds the tensile strength of the concrete [11,12]. Several traditional methods such as visual inspection, coring, and petrographic analysis have been utilized for monitoring and identifying the behavior of damage caused by ASR. Visual inspection is very subjective and depends on experience [13]. In addition, damage can only be detected by using this method, once it has reached the surface. Coring is a destructive method, which is discouraged in sensitive structures like nuclear power plants. Moreover, it is difficult to generalize the global condition of structures through evaluation of a few cores. Petrographic analysis is also mostly dependent on the experience of the examiner, is likewise a destructive method, and is costly and time-consuming. Linking the micro-state of the material examined to the condition of the entire structure using petrographic analysis is difficult [11], although this is an active research area (RILEM TC-259). Therefore, to monitor structures affected by ASR, non-destructive and global/semi-global methods are attractive. Acoustic emission (AE) is a health monitoring approach which acts as a passive receiver to record internal activities in structures. This method is capable of continuous monitoring which is not the case with most traditional methods. In addition, AE sensors are very sensitive and can capture signals due to micro-scale defect formations coming from the internal regions of structures rather than only those at the surface [14,15]. Several studies have been conducted to evaluate the use of AE in detecting the damage in concrete caused by ASR in the laboratory [11,16–19].

Lokajíček et al. [16] employed ultrasonic sounding and AE to monitor the microstructural changes in mortar bars with different levels of aggregate reactivity. The P-wave velocity increased in the first days of the experiments and then declined abruptly for the specimens with the reactive aggregates. Furthermore, AE cumulative energy indicated the clear difference between the specimens with the reactive aggregates compared to the specimens without any ASR reactivity. Farnam et al. [17] characterized the AE signature for crack formations in aggregates, cement matrix, and interfacial transition zones (ITZ) during alkali-silica reaction using peak frequency and frequency of centroid of the AE signals. The high-frequency signals were assigned to the crack formation in the aggregate, while low-frequency signals were attributed to the crack formation in ITZ and cement matrix. Abdelrahman et al. [11], conducted an accelerated ASR test on concrete prism specimens. A correlation between cumulative signal strength and length change of the prism specimens was identified, and an AE based intensity analysis chart for the classification of ASR damage conditions in correlation with petrographic damage rating indices was developed.

All previously mentioned, research has been focused on ASR-induced damage detection using AE data in small-scale specimens (laboratory scale), which is far from the conditions found in real structures in the field. Furthermore, the stress boundary condition has generally not been taken into consideration in previous studies, while the stress boundary condition has proven to have a large effect on damage distribution in concrete structures affected by ASR [20,21].

In this study, acoustic emission is utilized to monitor the effects of ASR on large-scale, reinforced concrete structures while considering stress induced boundary conditions in completely plane-confined and partially plane-confined specimens. There are several differences between damage detection of small-scale and large scale structures using acoustic emission. The main challenge with monitoring of small-scale structures using AE is reflections. An example of an ASR study using AE on small-scale specimens was conducted by Abdelrahman et al. [11]. For large-scale structures, attenuation and dispersion of waves are of primary concern. Many sensors with an appropriate layout are needed for damage detection and source localization.

The test specimens were assembled, cured, and monitored at the University of Tennessee, Knoxville, and are part of a test program sponsored by the US DOE Light Water Reactor Sustainability Program. The confined specimen has a complete confinement provided by a rigid steel frame and steel reinforcement meshes. The unconfined specimen has partial confinement provided by steel reinforcement meshes. The specimens without transverse reinforcement resemble construction reminiscent of nuclear power plant containments. An unsupervised pattern recognition algorithm was employed to classify the AE signals based on frequency-energy based features. Different damage mechanisms for the confined and unconfined specimens were identified using AE data.

2. Materials and Test Setup

2.1. Specimen and Chamber Preparation

In this study, three large-scale reinforced concrete blocks were cast. Two specimens were reactive and one was a control specimen (Figure 1). Reactive specimens are the specimens that experience expansion due to alkali-silica reaction. Two layers of steel reinforcement mesh were located at the top and bottom surfaces of the specimens. The reinforcement mesh includes US #11 Grade 60 with a nominal diameter of 36 mm (1.41 inches) at 25 cm (10 inches) spacing (Figure 2). The specimens have 7.62 cm (3 inches) cover at top and bottom of the specimens. Moreover, the square steel plates were installed on the ends of the rebar to achieve the full development length in a relatively short distance inside the specimens. For concrete mixture, 350 kg/m^3 cement, 175 kg/m^3 water, 1180 kg/m^3 coarse aggregate, and 728 kg/m^3 fine aggregate were used for both reactive and control specimens. The main difference between the specimens was using of NaOH or LiNO$_3$. In the reactive specimens, 9.8 kg/m^3 NaOH solution was added, while in the control specimen 1.9 kg/m^3 LiNO$_3$ was included in the mixture [22].

The coarse aggregates in both control and reactive specimens were greenschist coarse aggregates from North Carolina, which were highly-reactive. Non-reactive manufactured sand and low-alkali Portland cement Type II were also utilized in the concrete mixture. Water to cement weight ratio for both control and reactive specimens was 50%. To increase the alkali content of reactive specimens, 50% sodium hydroxide solution was added to the mixture for reactive specimens. Additionally, a 30% lithium nitrate solution was added to the control specimens to mitigate the ASR effect. More details regarding concrete mixture design was presented by Hayes et al. [22].

The specimens shown in Figure 1 have a cubic shape with dimensions of 3.50 m × 3.0 m × 1.0 m (136 inch × 116 inch × 40 inch). There are two reactive specimens; confined and unconfined specimens. A rigid steel brace was utilized to restrain the confined specimen from expanding in the lateral plane. The unconfined specimen does not have a steel brace in its plane. The two reactive specimens resemble the different stress boundary conditions found in real structures. A 1.5 mm layer of polyethylene was employed between the concrete and the steel frame in the confined specimen to reduce friction. More detailed information about the specimens and steel frame design were presented in Hayes, et al. [22]. In this experiment, the specimens stand on four short, steel columns (76 cm height) to provide access to the bottom of the specimens.

Each specimen's formwork was removed 12 days after casting which gave them time for final finishing and concrete setting. After casting, the specimen surfaces were sprayed with curing compound and covered with wet burlap and plastic sheets to reduce moisture loss and cracks from shrinkage during drying. The burlap was kept wet until chamber construction was complete [22].

A large chamber with the dimensions of 16.2 m long, 7.3 m wide, and 3.7 m high was constructed around the specimens to control the temperature and relative humidity of the environment. The chamber was equipped with systems that kept the specimens at the environmental temperature of 38 ± 1 °C and humidity of 95% ± 5%. The environmental chamber was initiated 26 days after casting and operated continuously except for shutdown for measurements and inspections [22].

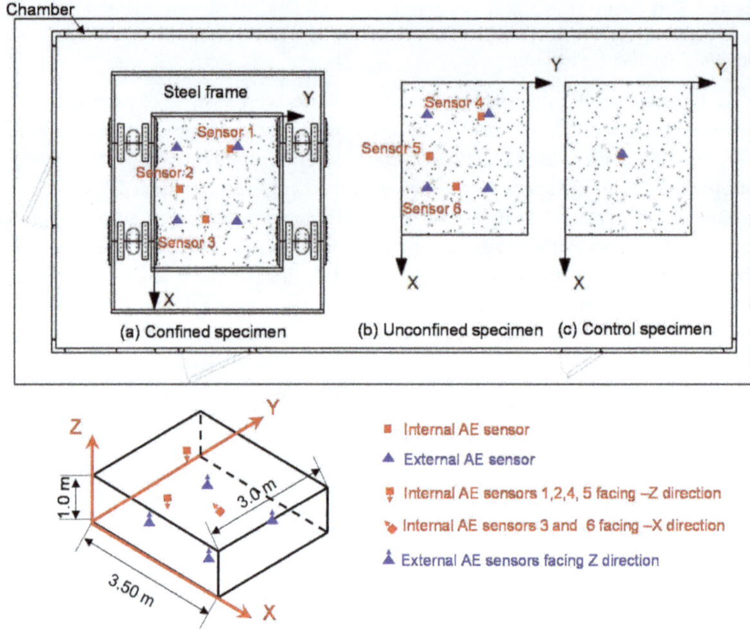

Figure 1. Test specimens.

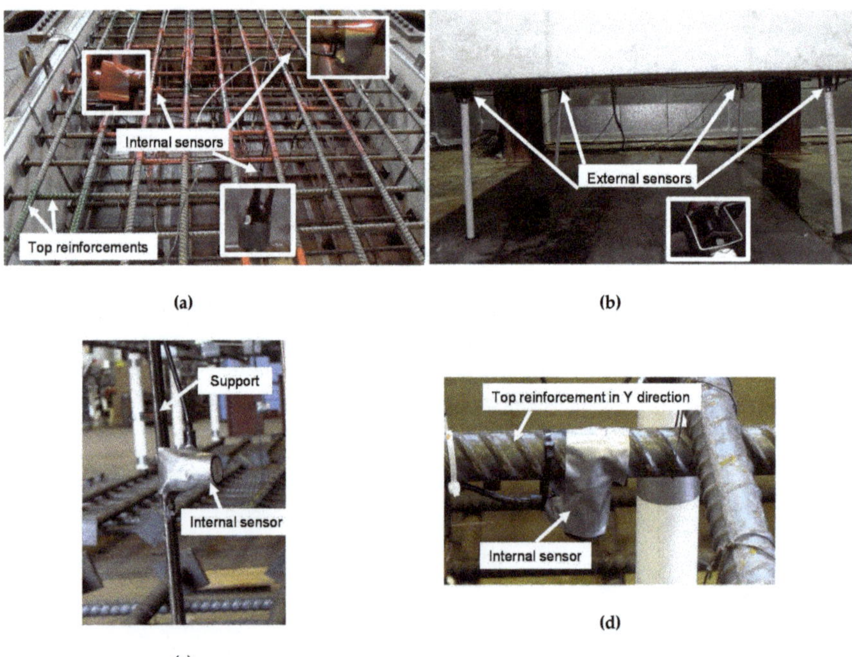

Figure 2. Acoustic emission (AE) sensor locations. (**a**) Internal AE sensors (broadband); (**b**) External AE sensors (resonant); (**c**) Internal sensors 3 and 6 installation; (**d**) Internal sensors 1, 2, 4 and 5 installation.

2.2. Measurement Equipment

To measure the strain inside the specimens, two kinds of strain sensors, alkali-tolerant strain transducers (KM-100B from Tokyo Sokki Kenkyujo) and long-gauge fiber optic extensometers (from SMARTECH/Roctest), were employed. In total, sixty-four 100 mm-gauge strain transducers were placed inside the specimens, along with the specimen dimensions. The sensors were fixed on the support fabricated by smooth steel bars with a diameter of 3 mm. Additionally, five high-precision, fiber optic extensometers were installed inside the specimens. The sensors in the X-Y plane had 1.5 m and 1.8 m gauge lengths and were located at the bottom of the specimens. Two fiber optic sensors were installed along the specimen thickness with a gauge length of 0.8 m [22].

Seven acoustic emission sensors were utilized in each of the confined and unconfined specimens (reactive specimens) as shown in Figures 1 and 2. Two sensors were employed for the control specimen. The internal AE sensors were broadband WDIUC-AST (manufactured by MISTRAS Group, Inc., Princeton Junction, NJ, USA), with an operating frequency range of 200–900 kHz and an internal low-noise 40 dB preamplifier. The external sensors were resonant R6I-UC (manufactured by MISTRAS Group, Inc., Princeton Junction, NJ, USA), with an operating frequency range of 35–100 kHz and an internal 40 dB preamplifier. The special polymer coating on the sensors (internal and external) with an internal waterproof cable makes the sensors insulated, non-conductive, and capable of operating under water.

Three broadband sensors were installed inside the reinforcement cages prior to casting for reactive specimens and one sensor was installed in the control specimen. Sensors 1, 2, 4, and 5 were attached under the second layer of top reinforcement mesh (under reinforcement in the Y direction) with epoxy and fastened to the rebar by several cable ties and duct tape, with the sensing surface facing downward along the negative Z axis (Figures 1d and 2d). Sensors 3 and 6 were in the middle of the specimen thickness, with the sensing surface facing toward the negative X axis (Figure 1d). Sensors 3 and 6 were attached on the supports by epoxy and duct tape and cables were fastened on the support by cable ties (Figure 2c). The supports were fastened to the top and bottom reinforcement meshes. The locations of internal sensors were colored on the top reinforcements to avoid over-vibrating during casting and potentially decoupling the sensors.

Four resonant sensors were attached at the bottom of the reactive specimens and one to the control specimen (Figure 2b). The resonant sensors were attached with epoxy on the surface of the concrete and fixed by holders as shown in Figure 2b. Internal and external sensors are referred to as broadband and resonant sensors, respectively, in this paper. The holders were made of stainless steel and polyurethane pads were attached between the concrete and the holders to avoid potential corrosion in the holders.

External and internal sensors pose differing technical challenges. Internal sensors are of scientific interest and are not well-suited for implementation in existing structures. One potential technical consideration is the formation of voids around the sensors due to concrete shrinkage. This may result in interruption of data or reduction of sensitivity. Another technical challenge is lack of access after casting. The external sensors are accessible, but decoupling may be a consideration.

A 16-channel Sensor Highway II (SHII), manufactured by MISTRAS Group, Inc. (Princeton Junction, NJ, USA), was utilized as a data acquisition system. The sensitivity of external sensors was checked by applying Hsu-Nielsen sources [23].

The sampling rate was set to 1000 kHz. Pre-trigger time is the period required for data acquisition to save a signal prior to threshold intersection, which was set to 256 μs. HDT (hit definition time) is the time for which if a signal crosses the threshold the signal will continue until the end of HDT without stopping. HDT was set to 400 μs. HLT (hit lockout time) is the time defined at the end of a signal for neglecting any reflected signal that exceeds the threshold, which was set to 200 μs. PDT (peak definition time) is the time which ensures correct identification of the signal peak for rise time and peak amplitude measurements [24]. In this study, PDT was set to 200 μs. The initial threshold was 32 dB.

3. Analysis Method

3.1. Pattern Recognition Algorithm

To identify the damage mechanisms in the reactive specimens a pattern recognition algorithm was employed for data classification. Pattern recognition is under the machine learning field and has two major types: unsupervised and supervised. When a background history for a data set is available or the data set has labeled classes, supervised pattern recognition is employed. If there are no labeled classes available, unsupervised pattern recognition can be utilized to identify the potential patterns in the data set based on the selected features.

An agglomerative hierarchical clustering algorithm [25] was employed for classifying the AE data into subsets. The clustering procedure is chronologically illustrated in Figure 3. The first step in the pattern recognition was deriving the frequency-based features for the signals. AE signals were transferred into the frequency domain using the Fast Fourier Transform (FFT). The FFT amplitude spectra were determined for each AE signal and the frequency domain was divided into ten frequency ranges. Then, signal energy in each frequency range was calculated and normalized to the total energy of the signal. These ten signal energies are the signal features. For example, in Figure 3a, the area of the hatched region is the energy of the signal in the frequency band from 200 kHz to 250 kHz. This value is normalized to the total energy of the signal, which is the entire area under the FFT spectrum. Principle component analysis (PCA) was used to reduce redundancy in the data. In this analysis, the original data is projected to the new orthogonal coordinates having high variation. An input for PCA is a matrix with the number of columns and rows equal to the feature numbers and the number of hits. Then a variance-covariance matrix for the features was calculated. The coefficients and variance of a specific principle component were calculated by eigenvalue analysis of the variance-covariance matrix. The principal components were selected in a way that represented more than 90% of the entire data. The principal components of the original features were selected as the input features for the pattern recognition algorithm. The algorithm initially calculated the Euclidian distance between the resulting data from PCA analysis. The result was a proximity matrix that contained distances between the original objects (data).

The objects were initially linked together according to calculated distances in the previous step and Ward's method. Ward's method is based on calculating the total within-cluster sum of squares of the data resulting from combining the clusters [25]. In each level, the data was merged into a binary linkage and the clusters were again merged into new clusters according to the Ward's method. This procedure was continued to form a single cluster which includes all data. The number of the cluster was determined according to the developed dendrogram and the height of each link with respect to the adjacent links [26].

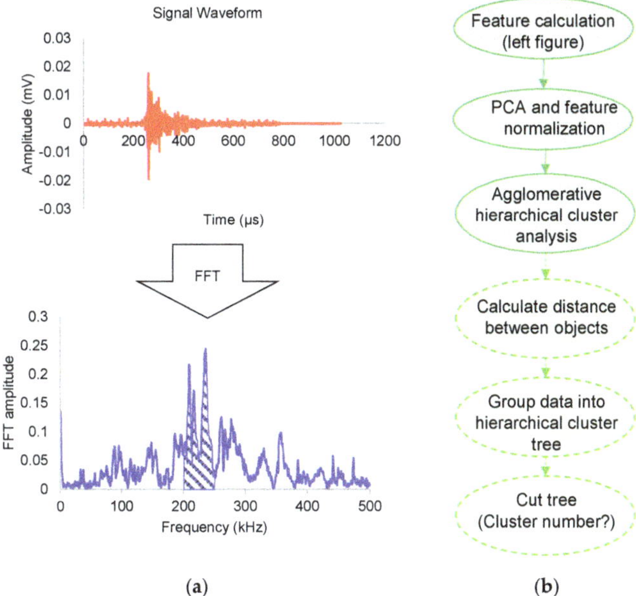

(**a**)　　　　　　　　　　(**b**)

Figure 3. Clustering procedure. (**a**) Energy-frequency based feature extraction; (**b**) Flow chart of data clustering steps.

4. Results and Discussions

The internal strains of the specimens in X, Y, and Z directions were recorded from the casting date. The results are presented up to 200 days in Figure 4. The data is presented from 51 days in the figures since external sensors were attached in this day and all the figures in this paper remain in a same time scale for a convenient comparison. Figure 4a shows the average expansion along the specimen dimensions for the reactive specimens and the average volumetric expansions for both reactive and control specimens are presented in Figure 4b. The results for volumetric strains show that reactive specimens were expanded. Conversely, the control specimens had no expansion and the internal strain was relatively constant, only showing small shrinkage over time. However, the volumetric strains for both confined and unconfined specimens were very close. This observation indicates that an imposed confinement in the specimens did not influence the expansion of the entire specimens (volumetric strain). Furthermore, the confinement changed the direction of expansion (Figure 4a).

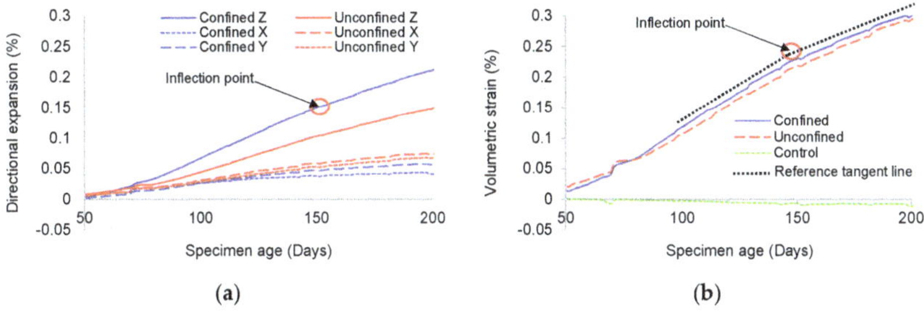

(**a**)　　　　　　　　　　(**b**)

Figure 4. Strain distribution of specimens versus concrete age. (**a**) Directional expansion; (**b**) Volumetric strain.

Generally, the expansion through the thickness of the reactive specimens (Z direction) is much larger than the expansion throughout the X-Y plane, due to partial or complete confinement in the plane and lack of steel reinforcement through the thickness. The maximum strains in the Z direction for confined and unconfined specimens are approximately 3.6 and 2.1 times the strain in the Y direction, respectively, at 200 days. The steel frame in the confined specimen causes a reduction of both in-plane strains. In the confined specimen, the strain along the X direction is less than the Y direction, while in the unconfined specimen, the strain along X and Y are almost the same. The strain along the X direction in the confined specimen is 58% of the strain along the X direction in the unconfined specimen at 200 days. However, the strain along the Y direction in the confined specimen is 84% of the strain along the Y direction in the unconfined specimen at 200 days. Moreover, the confinement in X-Y plane for the confined specimen caused an increased strain rate along the Z direction (thickness). The expansion strain rate in terms of time decreases at the point around 150 days, which is named inflection point as seen in Figure 4. The inflection point in volumetric strain is considered as one of the important point for ASR modeling, where the curvature of volumetric strain is changed [27,28]. The latency and characteristic times (two modeling parameters) are experimentally determined by knowing the location of inflection point [6,28]. The inflection point is shown in Figure 4b and marked in other figures in this paper.

Acoustic emission data was recorded through the internal sensors from the casting day and resonant sensors started recording from the concrete age of 51 days. Filtering AE data is an important step for reducing the amount of non-relevant data for post-processing. The possible sources of false AE data can be friction between the structural components, such as the steel frame and specimens, and water dripping from the chamber ceiling due to high humidity. Two different filtering procedures were developed to minimize the non-genuine data for the internal and external sensors, separately. The filtering is different for the internal and external sensors due to their differences in sensitivity and location. For instance, the internal sensors are much less sensitive than the external sensors and were located inside the specimens, thereby receiving less environmental noise. The AE data below 32 dB and 41 dB for the internal and external sensors was filtered from the data set. Then, the signals were further filtered using a Swansong filtering procedure [29,30]. This method is based on the observation that genuine AE signals with high amplitude should have long duration and vice versa [11]. Therefore, false signals in this method are categorized by long duration with low amplitude and short duration with high amplitude. The data is presented in terms of duration versus amplitude distribution. Signals which did not comply with the characteristics of genuine signals were deleted based on visual observation of the waveforms. The rejection limits for the internal and external sensors are presented in Table 1. In addition, suspicious signals were removed by inspection of waveforms and the chamber activity timetable provided by the University of Tennessee, Knoxville. The filtered AE data from 50 to 195 days after casting for all sensors is presented in Figure 5. The figure illustrates the amplitude and cumulative signal strength (CSS) in terms of specimen age. The time window of 50 to 195 days was selected because the highest strain change rate occurred during this period. The gap in the data from 65 to 100 days is missing data due to a difficulty with the data acquisition system caused by an unexpected energy surge. The amount of data in the control specimen is much lower than the data for the reactive specimens. The difference between AE data for the control and reactive specimens is also observable from plots of cumulative signal strength versus time (Figure 5). Therefore, the relatively high AE activity for the reactive specimens can likely be associated with expansion caused by the alkali-silica reaction.

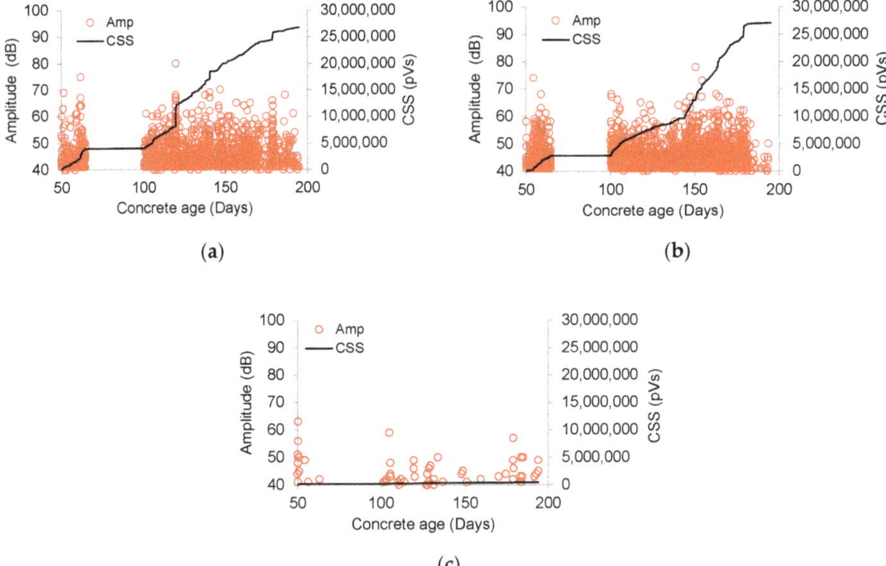

Figure 5. Amplitude and cumulative signal strength of AE data in terms of concrete age. (**a**) Confined specimen; (**b**) Unconfined specimen; (**c**) Control specimen.

Table 1. Duration-amplitude rejection limits.

External Sensor		Internal Sensor	
Amplitude (dB)	**Duration (μs)**	**Amplitude (dB)**	**Duration (μs)**
41–43	400<	32–35	155<
44–45	500<	36–42	260<
46–47	600<	43–100	330<
48–49	650<	-	-
50–53	820<	-	-
54–56	940<	-	-
57–65	1080<	-	-
66–100	1400<	-	-

4.1. Acoustic Emission Energy Release

To enable comparisons between the confined and unconfined specimens only AE data recorded by the internal sensors is discussed. These sensors are less sensitive than the resonant sensors, which results in lower volumes of data. The main reason for choosing the broadband sensors is that their broad frequency response makes them suitable for frequency analysis. The resonant sensors attached on the bottom surface of the specimens are more representative of what may be used in practice for optimized detection and source location. Three dimensional source location of large-scale specimens requires a method to accurately calculate time of arrival for very weak signals, which is a future step of this research. This study involves frequency analysis and therefore focuses on broad band sensors. Analysis of the resonant sensor data is a future consideration.

In Figure 1, the red marks and corresponding labels in red font denotes the schematic sensor locations for both confined and unconfined specimens. Sensor coordinates are presented in Table 2. The last column of the table represents the coordinates of normal vectors, which are perpendicular to the sensing surfaces of sensors and the directions of vectors are toward the outside of the sensors. The orientation of sensors is shown in Figure 1d.

Table 2. Sensor coordinates.

Sensor No.	X (m)	Y (m)	Z (m)	$\vec{n}$
Sensor 1	1.08	1.91	0.86	(0,0,−1)
Sensor 2	1.85	0.78	0.86	(0,0,−1)
Sensor 3	2.34	1.46	0.5	(−1,0,0)
Sensor 4	1.1	1.59	0.86	(0,0,−1)
Sensor 5	1.83	0.88	0.86	(0,0,−1)
Sensor 6	2.36	1.59	0.5	(−1,0,0)

The AE cumulative signal strengths of the internal sensors for the reactive specimens are presented in Figure 6a,b. The total CSS versus time for the confined specimen is much higher than the CSS for the unconfined specimen at 195 days (the CSS for the confined specimen is 2.35 times the value for the unconfined specimen). Moreover, in Figure 5a,c, it can be seen that a significant portion of released AE energy is attributed to sensor 3, which is located at mid-height of the confined specimen. The CSS rate for sensor 3 (5805 pVs/Day) is much larger than for sensors 1 and 2 (1545 and 1212 pVs/Day), thereby increasing the difference in the CSS between the sensors. On the other hand, this trend in the confined specimen is not observed in the unconfined specimen where AE energy release is not concentrated in one specific sensor. The distribution of energy was approximately uniform among the sensors. However, the CSS for the sensors at the top reinforcement mesh (sensors 4 and 5) is larger than in sensor 6.

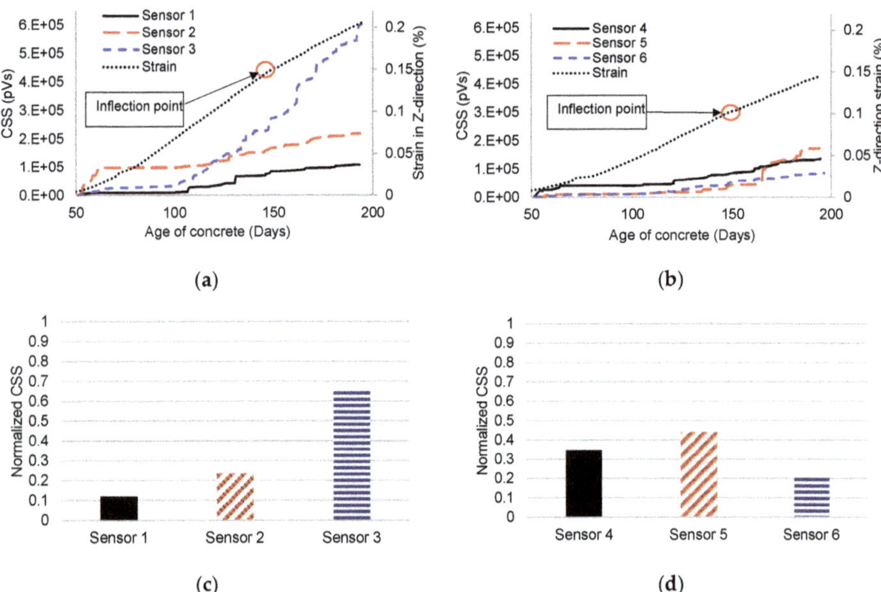

Figure 6. Cumulative signal strength (CSS) contribution for sensors. (**a**) CSS vs. age of concrete for confined specimen; (**b**) CSS vs. age of concrete for unconfined specimen; (**c**) Normalized CSS for confined specimen; (**d**) Normalized CSS for unconfined specimen.

These observations illustrate that the confined specimen has larger AE energy release in the middle layers of the specimen, which is increasing with the progression of the ASR reaction, than the unconfined specimen. This may be due to a larger expansion strain through the thickness of the confined specimen. This large expansion is expected to cause more damage through the thickness of specimen and consequently more AE energy release at middle of the thickness. In the unconfined specimen the crack distribution is expected to be less anisotropic than in the confined specimen. Therefore, it is expected that the AE energy was more uniformly distributed through the thickness of this specimen.

4.2. Pattern Recognition of AE Data

The AE signals were classified according to the agglomerative hierarchical algorithm as explained in Section 3.1. Dendrograms resulting from the analysis are presented in Figure 7. Dashed red lines in the subfigures indicate the desired height of links for clustering. The results of the pattern recognition show three clusters for each reactive specimen (both confined and unconfined). The clusters of confined specimens are indicated by Cluster-1, Cluster-2, and Cluster-3. Accordingly, the clusters of the unconfined specimen are labeled Cluster-4, Cluster-5, and Cluster-6. The horizontal axis in Figure 7 gives the data labels, which show either labels of the original data sets (signal number) or the label number of the clusters that resulted from merging the original data. The height of each link shows the distance between the two objects. Each link between two objects is shown by an upside-down U-shaped line in the figure. The data is shown in terms of the first three principal components (PC) to visualize the distribution of clusters with respect to each other in Figure 8. Although some signals in the cluster 5 and 6 were not ideally separated, in general, the clusters indicate a reasonable separation in the PC space (Figure 8). The reason for an overlap between clusters 5 and 6 is that some signals in cluster 5 have a similar energy contribution in a specific frequency range to some signals in cluster 6. As seen in Figure 9b, clusters 5 and 6 have similar average energy contribution between 250 to 300 kHz.

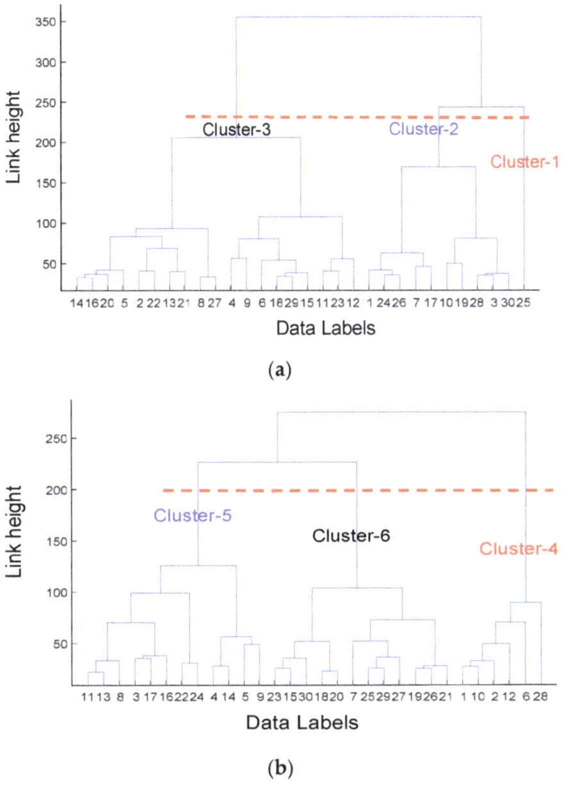

Figure 7. Clustering dendrograms. (**a**) Confined specimen; (**b**) Unconfined specimen.

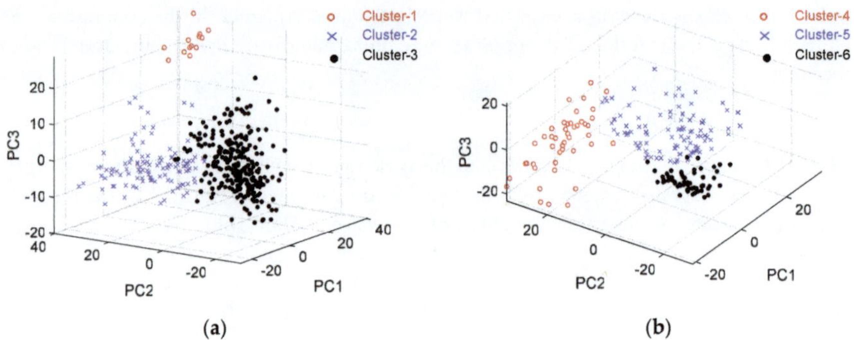

Figure 8. Clusters in principle component dimensions. (a) Confined specimen; (b) Unconfined specimen.

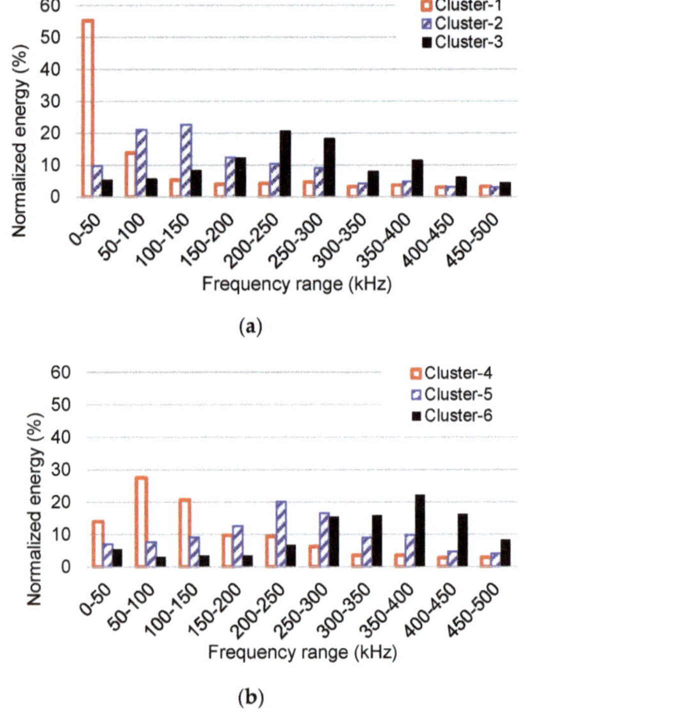

Figure 9. Average normalized signal energy in frequency domain. (a) Confined specimen; (b) Unconfined specimen.

The average energy of signals in terms of frequency ranges is shown in Figure 9 for the reactive specimens. These values, as previously mentioned, were calculated using the FFT amplitude spectrum. Afterward, the calculated values were normalized by the total energy of the signal. The average values for each cluster were then calculated. The energy shift to the higher frequency components for the unconfined specimen is apparent when viewed alongside the confined specimen. The clusters in each specimen can be separated based on the frequency content. In the confined specimen, the low-frequency cluster (Cluster-1) has approximately 69% of its energy in a frequency range of 0–100 kHz. The medium-frequency cluster (Cluster-2) has 42% of its energy concentrated in a frequency range of 50–150, while the high-frequency cluster (Cluster-3) has 51% of its energy between 150–300 kHz.

In the unconfined specimen, the low-frequency cluster (Cluster-4) has 62% of its energy in a frequency range of 0–150 kHz and the medium-frequency cluster (Cluster-5) has 49% of its signal energy in the frequency range of 150–300 kHz. The high-frequency cluster (Cluster-6) has 54% of its signal energy concentrated between the frequencies of 300–450 kHz. The Cluster-3 and Cluster-5 share similar frequency content. More signal features for the clusters are illustrated in Figure 10. The average feature values for each cluster were normalized by the maximum feature values for each cluster. In the confined specimen, Cluster-1 initiated at a higher concrete age (data was analyzed through a concrete age of 195 days) than clusters with higher frequencies (Cluster-2 and Cluster-3). The average amplitude of the signals in Cluster-3 (the highest frequency) is higher than the other signals. Average signal strength for Cluster-1 is lower than the values for Cluster-2 and Cluster-3, and duration is higher for the cluster with the lowest frequency content, for example, Cluster-1.

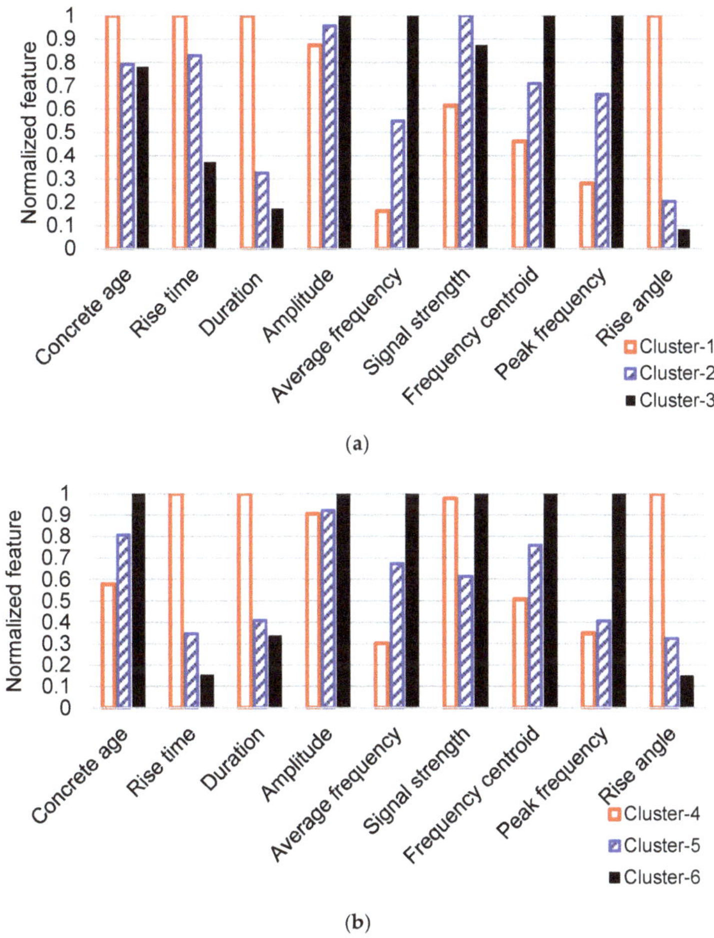

Figure 10. Normalized signal features. (**a**) Confined specimen; (**b**) Unconfined specimen.

A clear correlation is present between the frequency content of the signal clusters and the rise angle values (rise time over amplitude ratio) as has been observed by other researchers [31,32]. The higher the frequency components are in a signal, the lower rise angle value the signal possesses. In the unconfined specimen, Cluster-5 and Cluster-6 exhibit higher hit rates at the higher concrete age compared to Cluster-4, and the average amplitude of signals for the clusters with higher frequency components is slightly higher than for signals in Cluster-4. However, the average duration for the signals in the cluster with the low-frequency components (Cluster-4) is much longer than the duration for Cluster-5 and Cluster-6.

In Figure 11, the variation of cumulative signal strength in terms of the age of the concrete for each cluster is presented. The cumulative signal strengths were normalized by the maximum value for each specimen. In the confined specimen, the signals with the highest frequency components (Cluster-3) have dominant CSS from the early age. However, the CSS of Cluster-2 is very close to the CSS of Cluster-3 up to the concrete age of 150 days. After 150 days, the CSS rate for Cluster-3 increases, while the CSS rate of Cluster-2 continues with approximately the same rate. The signals in cluster Cluster-1 have negligible signal strength compared to Cluster-2 and Cluster-3 and initiate primarily after 120 days. In the unconfined specimen, the AE energy is primarily attributed to cluster Cluster-4 up to approximately 150 days. After 150 days, the clusters with the higher frequency components (Cluster-5 and Cluster-6) become prominent in terms of AE energy release.

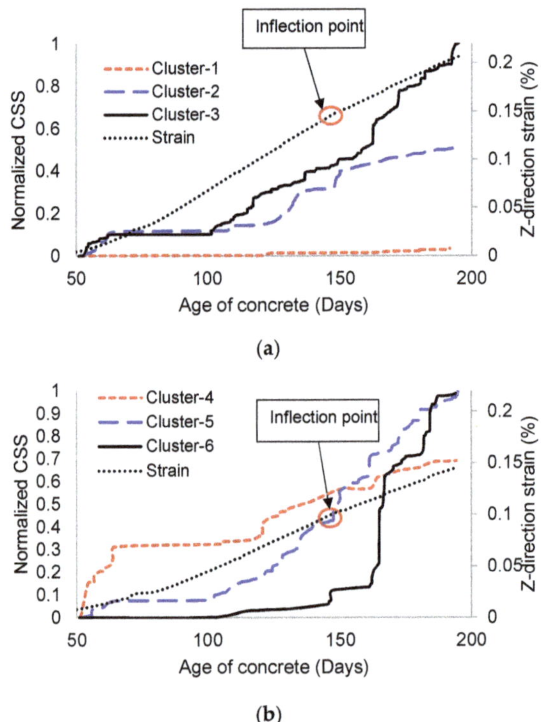

Figure 11. Normalized cumulative signal strength. (**a**) Confined specimen; (**b**) Unconfined specimen.

The distribution of total AE signal strength for the classified clusters and sensors at different ages of the concrete (66, 150, 195 days) are illustrated in Figure 12. These distributions are referred to as signal strength contribution factors (SSCF). The 66th day and 195th day were selected to illustrate the trend of data at the beginning and end of the evaluated time window. The 150th day was selected because in both reactive specimens there was an obvious change in the rate of CSS of the clusters with high-frequency components in comparison to the lower frequency components. The figures on the left show results of the confined specimen and the figures to the right show data from the unconfined specimen.

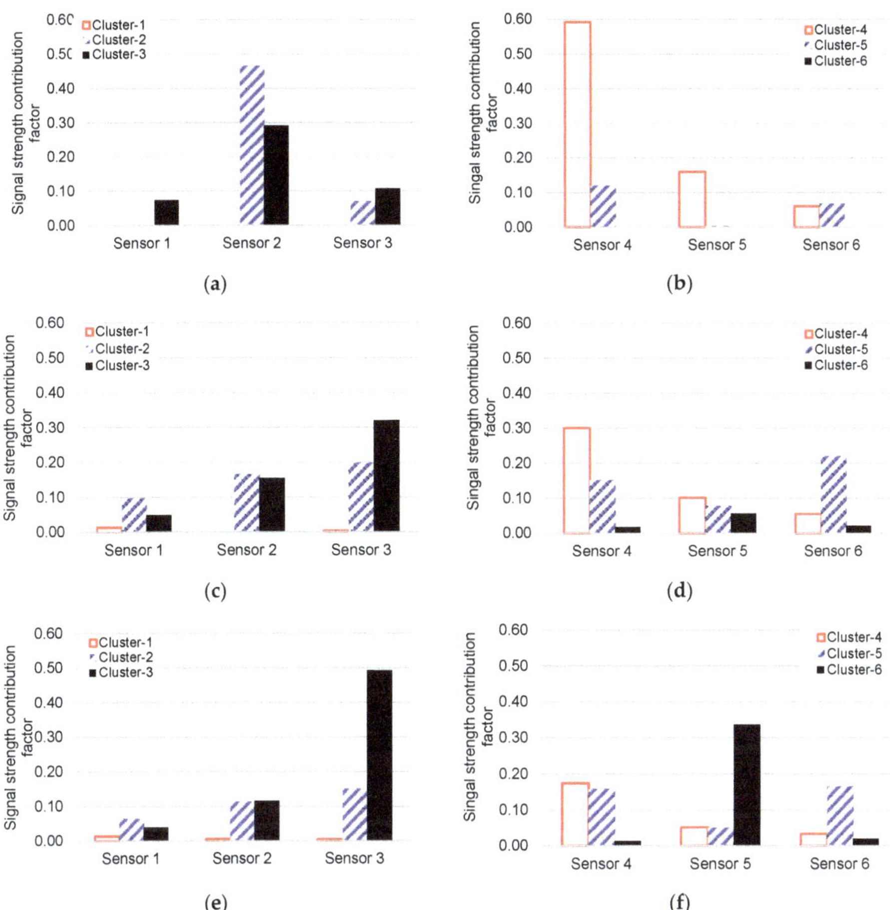

Figure 12. Distribution of total AE signal strength in terms of clusters and sensors. (**a**) Confined specimen at age of 66 days; (**b**) Unconfined specimen at age of 66 days; (**c**) Confined specimen at age of 150 days; (**d**) Unconfined specimen at age of 150 days; (**e**) Confined specimen at age of 195 days; (**f**) Unconfined specimen at age of 195 days.

In the confined specimen, most of the energy contribution is related to Cluster-2 (energy concentration in 50–150 kHz) and Cluster-3 (energy concentration in 150–300 kHz). The SSCF for Cluster-3 is increases with time, particularly after 150 days. Most of the AE energy for Cluster-3 is concentrated in sensor 3 (mid-thickness of the specimen) after 100 days. Cluster-2 and Cluster-3 both have prominent AE energy at sensor 2 before the 66th day. Then the highest AE energy portion moves to sensor 3, whereas SSCF of Cluster-2 is much lower than Cluster-3, especially after 150 days. The SSCF for cluster Cluster-1 is negligible compared to other clusters. In the unconfined specimen, the highest SSCF is for cluster Cluster-4 (energy concentration between 0–150 kHz) at sensor 4 (attached to top reinforcement) at 66 days. However, this energy contribution decreases with time, and the SSCF of the clusters with high-frequency components (Cluster-5 and Cluster-6) increases with time. There is no obvious energy concentration in the sensor located at mid-thickness of the specimen (sensor 6), which is different from what is observed in the confined specimen. In both specimens, the SSCF declines in low-frequency signals and increases in high-frequency signals with time. This trend in signal frequency from low to high in the confined specimen is not pronounced before 150 days. The SSCF for cluster Cluster-3 is slightly greater than the Cluster-2 at 150 days. The signal frequency trend initiates primarily after 150 days in this specimen. On the other hand, in the unconfined specimen, the frequency content evolution of AE signals is obvious from an earlier stage of ASR reaction (66 days) and is more significant after 150 days.

The confined specimen has a higher extensional strain along the Z direction than the unconfined specimen (approximately 42% more at 195 days). This expansion leads to tension concentration through the thickness of the confined specimen. Since there is no confinement through the specimen thickness it is susceptible to crack formation. In the unconfined specimen, the expansion strain is more evenly distributed between the X-Y plane and the thickness. Therefore, the tension is more uniformly distributed in the entire specimen in comparison to the confined specimen. This is also observable from the AE data, where sensor 3, located at the mid-height of the confined specimen, has a larger SSCF than the unconfined specimen (e.g., 65% for the confined specimen versus 35% for the unconfined specimen at 195 days).

As mentioned previously, the frequency of AE signals progresses from low to high as the concrete ages. This may be attributable to the formation of cracking through the coarse aggregate due to ASR progression. The crack formation inside the aggregate is expected to have higher frequency components than the cement matrix and interfacial transition zone (ITZ) as mentioned by Farnam et al. [17]. The transition from low-frequency signals to the high-frequency signals in the confined specimen initiated later than for the unconfined specimen (after 150 days). However, there are different contradictory hypotheses relating to formation of cracks in concrete due to ASR [33]. For instance, osmotic pressure theory was proposed by Hanson to describe the mechanism of expansion [34]. In this theory, the cement paste surrounding to reactive aggregates acts as a semi-permeable membrane, which water solution can pass inside the region around the reactive aggregates, but alkali-silica ions are enclosed in the reactive regions. This causes osmotic pressure and alkali-silica gel swells and exerts pressure to the cement paste. This pressure leads to crack formation in the cement paste [34]. McGowan and Vivian also proposed a similar theory as osmotic theory, which transforming a solid alkali-silica layer on a reactive aggregate to a gel by absorbing moisture from the pore solution was explained as a main reason of cracking in the cement paste due to ASR [35]. Bazant and Steffens suggested that the cracking is caused in the cement paste and interfacial transition zone due to accumulation of alkali-silica gel in the interfacial transition zone and resulting gel pressure [36]. On the other hand, Dron and Brivot assumed that crack formation occurred far away from reactive aggregates due to diffusing dissolved silica away from aggregate into the pores in the cement paste [37]. Some researchers observed that ASR gel initially forms inside a reactive aggregate and causes the pressure and crack formation inside the aggregate and surrounding cement paste [12,38–41]. Ponce and Batic [42] related the cracking pattern of concrete due to ASR to the types of reactive aggregate. ASR cracks start to form inside aggregate or in the cement matrix depending on the aggregate type [42].

In the confined specimen, the CSS rate of Cluster-3 (energy concentration 150–300 kHz), started to increase at the age of 150 days. In the unconfined specimen, the CSS rate for the cluster with the higher frequency components also increases around that time. 150 days is close to the inflection point of the volumetric strain curve, after which point expansion rates decrease. In addition, the first visible cracks were observed at the age of 150 days on the sides of the unconfined specimen, but no cracks were visible on the top surface of the unconfined specimen. Cracks could not be traced in the confined specimen on the sides due to the steel confinement frame. From the above observations, 150 days is a significant time period for ASR in the specimens, which generally agrees with trends in the AE data.

5. Conclusions

Acoustic emission was utilized for monitoring the activities caused by ASR in large-scale reinforced concrete specimens. The specimens resemble common nuclear power plant containments with no shear reinforcement. An agglomerative hierarchical algorithm was used to classify the AE data based on the energy-frequency dependent features to study and identify the damage mechanisms in the specimens with different stress boundary conditions. The conclusions of this study are summarized as follows:

- A significant portion of the AE data (in terms of cumulative signal strength) in the confined specimen was recorded by sensor 3, which was located at mid-thickness. However, the portion of cumulative signal strength for the corresponding sensor at mid-thickness of the unconfined specimen was less in comparison to the other two sensors in that specimen. This agrees with expectations, as the confined specimen exhibited increased out-of-plane expansion in comparison to the unconfined specimen, meaning that the crack distribution is expected to be more concentrated near mid-thickness of the confined specimen than near the reinforcement layers.

- The frequency contents of signals in the confined and unconfined specimens evolved from low to high frequency with the age of concrete although this evolution started later in the confined specimen than the unconfined specimen. Since the high-frequency AE signals have been associated to the cracking in aggregates [17], different crack mechanisms in aggregate for the confined and unconfined specimens are expectable. However, there are different contradictory ASR cracking hypotheses that have been proposed by the researchers [33–38,41–43].

- There is a coincident point observed in the strain curves and the CSS of Cluster-3 in the confined specimen and Cluster-6 in the unconfined specimen. The CSS rate in terms of concrete age increases at the time (around 150 days), when the strain rate is decreasing. This point is named as inflection point of strain curve, where the curvature of strain curve changes from positive to negative. The inflection point location in terms of concrete age depends on the kinetic of ASR reaction and diffusion process. Determining the inflection point, latency and characteristic time are experimentally estimated, which are the two important modeling parameters. According to the results of AE data and clustering, the inflection point location in terms of concrete age could be estimated from the variation in CSS rate change of clustered AE data.

- Monitoring of a structural system with acoustic emission can provide useful information regarding condition based maintenance and/or retrofit. For example, one potential time of action for treating affected structures is around the inflection point in the volumetric strain curve which can be approximated through acoustic emission data. This point coincided with observation of first visible surface cracking. After identifying the time of action, treatment methods may be implemented to mitigate the effects of ASR. Injection of lithium solution is a chemical alternative to mitigate ASR provided that enough solution penetration in the structure can be achieved. Another method is to remove moisture through coatings and sealers such as silane sealers and bituminous or elastomeric coatings. After conducting these methods, structures should be monitored for enough time to evaluate the efficiency of the method or methods.

Author Contributions: Conceptualization, P.Z., V.S., Y.L.P., and Z.J.M.; Data curation, V.S., R.A., and N.W.H.; Formal analysis, V.S.; Funding acquisition, Z.J.M., Y.L.P and P.Z.; Investigation, V.S., R.A., N.W.H. and L.A.; Project administration, Y.L.P., Z.J.M. and P.Z.; Resources, Y.L.P., Z.J.M. and P.Z.; Supervision, Y.L.P., Z.J.M. and P.Z.; Visualization, V.S.; Writing – original draft, V.S.; Writing – review & editing, V.S, N.W.H., Z.J.M., Y.L.P. and P.Z.

Acknowledgments: This material is based upon work partly supported by the U.S. Department of Energy, Office of Nuclear Energy, Light Water Reactor Sustainability Program, under contract number DE-AC05-00OR22725. This manuscript has been co-authored by UT-Battelle, LLC under Contract No. DE-AC05-00OR22725 with the U.S. Department of Energy. The United States Government retains and the publisher, by accepting the article for publication, acknowledges that the United States Government retains a non-exclusive, paid-up, irrevocable, worldwide license to publish or reproduce the published form of this manuscript, or allow others to do so, for United States Government purposes. The Department of Energy will provide public access to these results of federally sponsored research in accordance with the DOE Public Access Plan (https://www.energy.gov/downloads/doe-public-access-plan).

Conflicts of Interest: The authors declare no conflicts of interest.

References

1. Schmidt, J.W.; Hansen, S.G.; Barbosa, R.A.; Henriksen, A. Novel shear capacity testing of ASR damaged full scale concrete bridge. *Eng. Struct.* **2014**, *79*, 365–374. [CrossRef]
2. Clark, L. *Critical Review of the Structural Implications of the Alkali Silica Reaction in Concrete*; Contactor Report 169; Transportation and Road Research Laboratory, Department of Transport: Crowthone, UK, 1989.
3. Bach, F.; Thorsen, T.S.; Nielsen, M. Load-carrying capacity of structural members subjected to alkali-silica reactions. *Constr. Build. Mater.* **1993**, *7*, 109–115. [CrossRef]
4. Bakker, J. Control of ASR related risks in the Netherlands. In Proceedings of the 13th International Conference on Alkali-Aggregate Reaction in Concrete, Trondheim, Norway, 16–20 June 2008; pp. 21–31.
5. Charlwood, R.; Solymar, S.; Curtis, D. A review of alkali aggregate reactions in hydroelectric plants and DAMS. In Proceedings of the International Conference of Alkali-Aggregate Reactions in Hydroelectric Plants and Dams, Fredericton, NB, Canada, 28 September–2 October 1992; Volume 129.
6. Saouma, V.E.; Hariri-Ardebili, M.A. A proposed aging management program for alkali silica reactions in a nuclear power plant. *Nucl. Eng. Des.* **2014**, *277*, 248–264. [CrossRef]
7. Takakura, T.; Ishikawa, T.; Mitsuki, S.; Matsumoto, N.; Takiguchi, K.; Masuda, Y.; Nishiguchi, I. Investigation on the expansion value of turbine generator foundation affected by alkali-silica reaction. In Proceedings of the 18th International Conference on Structural Mechanics in Reactor Technology (SMiRT), Beijing, China, 7–12 August 2005; pp. 2061–2068.
8. Takakura, T.; Masuda, H.; Murazumi, Y.; Takiguchi, K.; Masuda, Y.; Nishiguchi, I. Structural soundness for turbine-generator foundation affected by alkali-silica reaction and its maintenance plans. In Proceedings of the 19th International Conference on Structural Mechanics in Reactor Technology (SMiRT), Toronto, ON, Canada, 12–17 August 2007.
9. Tcherner, J.; Aziz, T. Effects of AAR on seismic assessment of nuclear power plants for life extensions. In Proceedings of the 20th International Conference on Structural Mechanics in Reactor Technology (SMiRT), Espoo, Finland, 9–14 August 2009; pp. 1–8.
10. Charlwood, R. Predicting the long term behaviour and service life of concrete dams, Long Term Behaviour of Dams. In Proceedings of the 2nd International Conference, Graz, Austria, 12–13 October 2009.
11. Abdelrahman, M.; ElBatanouny, M.K.; Ziehl, P.; Fasl, J.; Larosche, C.J.; Fraczek, J. Classification of alkali–silica reaction damage using acoustic emission: A proof-of-concept study. *Constr. Build. Mater.* **2015**, *95*, 406–413. [CrossRef]
12. Garcia-Diaz, E.; Riche, J.; Bulteel, D.; Vernet, C. Mechanism of damage for the alkali–silica reaction. *Cem. Concr. Res.* **2006**, *36*, 395–400. [CrossRef]
13. Sargolzahi, M.; Kodjo, S.A.; Rivard, P.; Rhazi, J. Effectiveness of nondestructive testing for the evaluation of alkali–silica reaction in concrete. *Constr. Build. Mater.* **2010**, *24*, 1398–1403. [CrossRef]
14. Anay, R.; Soltangharaei, V.; Assi, L.; DeVol, T.; Ziehl, P. Identification of damage mechanisms in cement paste based on acoustic emission. *Constr. Build. Mater.* **2018**, *164*, 286–296. [CrossRef]
15. Soltangharaei, V.; Anay, R.; Assi, L.; Ziehl, P.; Matta, F. Damage identification in cement paste amended with carbon nanotubes. *AIP Conf. Proc.* **2018**, *1949*, 030006. [CrossRef]
16. Lokajíček, T.; Přikryl, R.; Šachlová, Š.; Kuchařová, A. Acoustic emission monitoring of crack formation during alkali silica reactivity accelerated mortar bar test. *Eng. Geol.* **2017**, *220*, 175–182. [CrossRef]
17. Farnam, Y.; Geiker, M.R.; Bentz, D.; Weiss, J. Acoustic emission waveform characterization of crack origin and mode in fractured and ASR damaged concrete. *Cem. Concr. Compos.* **2015**, *60*, 135–145. [CrossRef]
18. Rajabipour, F.; Giannini, E.; Dunant, C.; Ideker, J.H.; Thomas, M.D. Alkali–silica reaction: Current understanding of the reaction mechanisms and the knowledge gaps. *Cem. Concr. Res.* **2015**, *76*, 130–146. [CrossRef]
19. Weise, F.; Katja, V.; Pirskawetz, S.; Dietmar, M. Innovative measurement techniques for characterising internal damage processes in concrete due to ASR. In Proceedings of the International Conference on Alkali Aggregate Reaction (ICAAR), University of Texas, Austin, TX, USA, 20–25 May 2012.

20. Morenon, P.; Multon, S.; Sellier, A.; Grimal, E.; Hamon, F.; Bourdarot, E. Impact of stresses and restraints on ASR expansion. *Constr. Build. Mater.* **2017**, *140*, 58–74. [CrossRef]

21. Saouma, V.E.; Hariri-Ardebili, M.A.; Le Pape, Y.; Balaji, R. Effect of alkali–silica reaction on the shear strength of reinforced concrete structural members. A numerical and statistical study. *Nucl. Eng. Des.* **2016**, *310*, 295–310. [CrossRef]

22. Hayes, N.W.; Gui, Q.; Abd-Elssamd, A.; Le Pape, Y.; Giorla, A.B.; Le Pape, S.; Giannini, E.R.; Ma, Z.J. Monitoring Alkali-Silica Reaction Significance in Nuclear Concrete Structural Members. *J. Adv. Concr. Technol.* **2018**, *16*, 179–190. [CrossRef]

23. Hsu, N. Characterization and calibration of acoustic emission sensors. *Mater. Eval.* **1981**, *39*, 60–68.

24. *Express-8 AE System User Manual*; Mistras Group, Inc.: Princeton Junction, NJ, USA, 2014.

25. Murtagh, F.; Legendre, P. Ward's hierarchical agglomerative clustering method: Which algorithms implement Ward's criterion? *J. Classif.* **2014**, *31*, 274–295. [CrossRef]

26. Tan, P.N.; Steinbach, M.; Karpatne, A.; Kumar, V. *Introduction to Data Mining*, 2nd ed.; Pearson Education: New York, NY, USA, 2018; ISBN 978-013-312-890-1.

27. Saouma, V.; Perotti, L. Constitutive model for alkali-aggregate reactions. *ACI Mater. J.* **2006**, *103*, 194.

28. Ulm, F.J.; Coussy, O.; Kefei, L.; Larive, C. Thermo-chemo-mechanics of ASR expansion in concrete structures. *J. Eng. Mech.* **2000**, *126*, 233–242. [CrossRef]

29. Fowler, T.J.; Blessing, J.A.; Conlisk, P.J.; Swanson, T.L. The MONPAC system. *J. Acoust. Emiss.* **1989**, *8*, 1–8.

30. ElBatanouny, M.K.; Ziehl, P.H.; Larosche, A.; Mangual, J.; Matta, F.; Nanni, A. Acoustic emission monitoring for assessment of prestressed concrete beams. *Constr. Build. Mater.* **2014**, *58*, 46–53. [CrossRef]

31. Aggelis, D.; Soulioti, D.; Gatselou, E.; Barkoula, N.-M.; Matikas, T.E. Monitoring of the mechanical behavior of concrete with chemically treated steel fibers by acoustic emission. *Constr. Build. Mater.* **2013**, *48*, 1255–1260. [CrossRef]

32. Xargay, H.; Folino, P.; Nuñez, N.; Gómez, M.; Caggiano, A.; Martinelli, E. Acoustic Emission behavior of thermally damaged Self-Compacting High Strength Fiber Reinforced Concrete. *Constr. Build. Mater.* **2018**, *187*, 519–530. [CrossRef]

33. Pan, J.; Feng, Y.; Wang, J.; Sun, Q.; Zhang, C.; Owen, D.J. Modeling of alkali-silica reaction in concrete: A review. *Front. Struct. Civ. Eng.* **2012**, *6*, 1–18. [CrossRef]

34. Hanson, W. Studies Relating To the Mechanism by Which the Alkali-Aggregate Reaction Produces Expansion in Concrete. *J. Am. Concr. Inst.* **1944**, *40*, 213–228.

35. McGowan, J.K. Studies in Cement-Aggregate Reaction XX, the Correlation between Crack Development and Expansion of Mortar. *Aust. J. Appl. Sci.* **1952**, *3*, 228–232.

36. Bazant, Z.P.; Steffens, A. Mathematical model for kinetics of alkali–silica reaction in concrete. *Cem. Concr. Res.* **2000**, *30*, 419–428. [CrossRef]

37. Dron, R.; Brivot, F. Thermodynamic and kinetic approach to the alkali-silica reaction. Part 1: Concepts. *Cem. Concr. Res.* **1992**, *22*, 941–948. [CrossRef]

38. Idorn, G.M. A discussion of the paper "Mathematical model for kinetics of alkali-silica reaction in concrete" by Zdenek P. Bazant and Alexander Steffens. *Cem. Concr. Res.* **2001**, *7*, 1109–1110. [CrossRef]

39. Jun, S.S.; Jin, C.S. ASR products on the content of reactive aggregate. *J. Civ. Eng.* **2010**, *14*, 539–545. [CrossRef]

40. Goltermann, P. Mechanical predictions of concrete deterioration; Part 2: Classification of crack patterns. *Mater. J.* **1995**, *92*, 58–63.

41. Ichikawa, T.; Miura, M. Modified model of alkali-silica reaction. *Cem. Concr. Res.* **2007**, *37*, 1291–1297. [CrossRef]

42. Ponce, J.M.; Batic, O.R. Different manifestations of the alkali-silica reaction in concrete according to the reaction kinetics of the reactive aggregate. *Cem. Concr. Res.* **2006**, *36*, 1148–1156. [CrossRef]

43. Escalante-Garcıa, J.; Sharp, J. The microstructure and mechanical properties of blended cements hydrated at various temperatures. *Cem. Concr. Res.* **2001**, *31*, 695–702. [CrossRef]

Article

A New Closed-Form Solution for Acoustic Emission Source Location in the Presence of Outliers

Zilong Zhou [1], Yichao Rui [1,*], Jing Zhou [1], Longjun Dong [1], Lianjun Chen [2,*], Xin Cai [1] and Ruishan Cheng [1]

[1] School of Resources and Safety Engineering, Central South University, Changsha 410083, China; zlzhou@csu.edu.cn (Z.Z.); zhoujing205@csu.edu.cn (J.Z.); lj.dong@csu.edu.cn (L.D.); xincai@csu.edu.cn (X.C.); chengruishan@csu.edu.cn (R.C.)

[2] State Key Laboratory of Mining Disaster Prevention and Control Co-founded by Shandong Province and the Ministry of Science and Technology, Shandong University of Science and Technology, Qingdao 266590, China

* Correspondences: ruiyichao@csu.edu.cn (Y.R.); creejxk@163.com (L.C.); Tel.: +86-187-1106-6336 (Y.R.); +86-137-8987-0628 (L.C.)

Received: 21 April 2018; Accepted: 5 June 2018; Published: 8 June 2018

Abstract: The accuracy of an acoustic emission (AE) source location is always corrupted by outliers due to the complexity of engineering practice. To this end, a preconditioned closed-form solution based on weight estimation (PCFWE) is proposed in this study. Firstly, nonlinear equations are linearized, and initial source coordinates are obtained by using equal weights. Residuals, which are calculated by source coordinates, are divided into three regions according to normal distribution. Secondly, the weight estimation is developed by establishing the relationship between residuals and weights. Outliers are filtered by the iteration between the weight estimation and source location. Subsequently, linear equations are reconstructed with the remaining measurements containing no outliers, while they are ill-conditioned. Finally, the preconditioning method is applied to weaken the ill condition of the reconstructed linear equations, so as to improve the location accuracy. This new method is verified by a pencil-lead break experiment. Tests results show that the location accuracy and stability of the new method are superior to traditional methods. In addition, outlier tolerance and the velocity sensibility of the new method are investigated by simulating tests.

Keywords: acoustic emission; closed-form solution; outlier; time difference of arrival; weight estimation

---

1. Introduction

The acoustic emission (AE) source location method based on sensor array has been a hot research topic with widespread applications in many fields of structural health, underground tunneling, and deep mining [1–7]. Nevertheless, locating the source is not a simple task, because there are two major challenges, namely: measurement errors and real-time implementation [8–12].

The least squares principle, which is employed in most methods, can achieve a favorable location accuracy on the condition that all of the measurements are accurate or only have minor errors [13–18]. However, adding an outlier will cause the residual value to dramatically increase, which is enlarged due to the square nature of the least squares principle, and the location performance to significantly deteriorate [19,20]. Accurate estimation for the arrival times of acquired signals associated with each sensor is the underlying challenge for the most picking-based location method. Researchers have proposed varying algorithms, trying to pick arrivals accurately and in real-time, such as a short and long-time average ratio [21], higher order statistics method [22]. However, these methods only have good performances in clean or high signal-to-noise ratio environments. Niccolini et al. [23,24] proposed the joint autoregressive modeling of noise and the signal in windows using Akaike information criterion with an automatic procedure for signal data processing under a low signal-to-noise ratio environment. This method, which can eliminate certain false or doubtful arrivals before the source location, can be utilized to obtain the real-time arrivals and reduce the probability of an outlier. However, it does not mean that all of outliers will be eliminated, because there are other factors that can generate outliers besides the background noise. Practical applications of monitoring systems have reported that the sensors are easy to be triggered incorrectly by interference signals due to the complex engineering environment, which brings external outlier signals [25,26]. In addition, outliers can also occur in these cases, such as signal interference, sensor location errors, weak and ambiguous of arrivals, interchannel crosstalk, or simply hardware failure. Therefore, all the above cases may bring outliers in measurements, especially in the use of automated picking methods. Considering the difficult acquisition of an accurate measurement and the deterioration effect of the outliers on the source location, the error-tolerant location methods have been developed by scholars to reduce the influence of outliers [27,28].

A large amount of error-tolerant methods of time difference of arrival (TDOA) have been proposed, which can be classified into nonlinear methods and linear methods [29–33]. Nonlinear location methods, including maximum likelihood [34,35], the virtual field optimization method [36], and the least absolute deviation method [37], have been suggested to improve the error-tolerance. To gain the optimum solution of these nonlinear methods, iterative techniques should be used because of nonlinear TDOA equations. Alternative choices include the simplex method [38], Newton iterative method [39], and the Gauss–Newton method [40]. However, these methods not only require an appropriate guess of initial location near the true solution, which is difficult to obtain in practice, but also probably suffer from convergence problems and a large computational burden caused by the iterative nature. Compared with nonlinear methods, linear methods in the close form guarantee the convergence and real-time implementation. The weighted least squares algorithm, which gives different weights to different equations, is a generic form of the error-tolerant closed-form solution. Two-step weighted least squares (TSWLS), which linearizes the nonlinear TDOA equations by introducing an additional variable [41–44], is one of the prevalent weighted least squares methods. This method has been widely used in various fields as a benchmark for subsequent studies [9,31,45]. Firstly, source location and the additional variable are considered independent variables to conduct the estimation. Secondly, the estimation is developed by considering the relationship between the additional variable and the source location. However, the covariance matrix of measurement errors used in this location process is difficult to accurately estimate, which causes performance degradation at the source location [9]. Unlike the method of weight least squares, Dong et al. [46] recently proposed comprehensive closed-form solutions using a set of sensor networks and the logistic probability density function, so as to improve the location accuracy. However, location errors of the closed-form solution are still large, because only six sensors are considered in each calculation process, and the coefficient matrix of the linear equations is close to ill-conditioned. In addition, since many subsets of six sensors exist, it requires enormous computing power; thus, it has a poor performance in real-time implementation [47].

Aiming at improving the error-tolerance and real-time implementation of the AE source location, this paper proposes the preconditioned closed-form solution based on weight estimation (PCFWE) to seek the optimal source coordinates by filtering outliers and reducing the ill condition of linear equations. PCFWE is also compared with the TSWLS [41] and linear least squares (LLS) methods. The effectiveness and accuracy of the proposed PCFWE are verified by the pencil-lead break experiment and simulating tests. In Section 2, the ordinary closed-form method that fairly treats all of the measurements is stated. In Section 3, the theory of the proposed method is formulated mathematically. In Section 4, the location results of the pencil-lead break experiment are reported. Finally, outlier tolerance and the velocity sensibility of the proposed method are discussed and concluded in Section 5.

2. Ordinary Closed-Form Method

To achieve an accurate and real-time AE source location, many closed-form methods exploiting the LLS principle are proposed [15,16,48,49]. The process of the ordinary closed-form location method mainly contains two steps as follows. Firstly, nonlinear equations are transformed into linear forms to reduce the difficulty in solving nonlinear equations by iterative methods. Secondly, the LLS principle is incorporated to solve the overdetermined liner equations, which can use as many measurements as possible, and yield a better location accuracy than those that only use a minimum number of measurements [50,51], i.e., the number of equations is equal to the number of unknowns. From the viewpoint of error control, the dataset is statistically more reliable, and the array geometry is more reasonable when more measurements are used. The process of an ordinary closed-form location method with the LLS principle is stated as follows.

Assuming that an AE source is located at source θ (x, y, z), and the $n + 1$ sensors are located at S_i (x_i, y_i, z_i) $(i = 0, 1, \cdots, n)$, for an arbitrary AE event, the source coordinates satisfy the following control equations:

$$(x_i - x)^2 + (y_i - y)^2 + (z_i - z)^2 = v^2(\Delta t_i + t_0)^2 \tag{1}$$

where t_0 is the propagation time of the acoustic wave from the source to the nearest sensor. The sensors are numbered S_i $(i = 0, 1, \cdots, n)$ according to the arrival orders of the AE signal. The nearest sensor S_0 that receives the AE signal first is regarded as the reference sensor. Δt_i is the time difference between reference sensor S_0 and sensor S_i $(i = 1, 2, \cdots, n)$, namely the time difference of arrival (TDOA). Besides, when $i = 0$, $\Delta t_i = 0$, nonlinear equations can be linearized by subtracting the equation $i = 0$ from $i \geq 1$.

$$2(x_i - x_0)x + 2(y_i - y_0)y + 2(z_i - z_0)z + 2\Delta t_i v^2 t_0 = L_i, \ i = 1, 2, \cdots n \tag{2}$$

where $L_i = x_i{}^2 - x_0{}^2 + y_i{}^2 - y_0{}^2 + z_i{}^2 - z_0{}^2 - \Delta t_i^2 v^2$.

Equation (2) can be expressed in matrix form as:

$$A\theta = b \tag{3}$$

where $A = 2 \begin{bmatrix} x_1 - x_0 & y_1 - y_0 & z_1 - z_0 & \Delta t_1 v^2 \\ x_2 - x_0 & y_2 - y_0 & z_2 - z_0 & \Delta t_2 v^2 \\ & & \vdots & \\ x_n - x_0 & y_n - y_0 & z_n - z_0 & \Delta t_n v^2 \end{bmatrix}, \theta = \begin{bmatrix} x \\ y \\ z \\ t_0 \end{bmatrix}, b = \begin{bmatrix} L_1 \\ L_2 \\ \vdots \\ L_n \end{bmatrix}.$

To find the location parameters θ in Equation (3), which minimizes the sum of the residual square for the linear equation system:

$$\arg\min_{\theta}||b - A\theta|| = \arg\min_{\theta}\sum_{i=1}^{n}(b_i - A_i\theta)^2 \tag{4}$$

The closed-from solution with the least squares principle can be obtained by:

$$\theta = \left(A^T A\right)^{-1} A^T b \tag{5}$$

where the symbol T denotes the matrix transpose, and $A^T A$ is the Hessian matrix.

The solution of LLS is the best linear unbiased estimator for the AE source location when the measurement errors are independent and identically distributed. However, it may produce dramatic location errors due to the equal treatments of all of the measurements, even the outliers. Therefore, outliers are considered to be filtered by weight estimation first; then, linear equations are reconstructed with the remaining measurements that contain no outliers to obtain more accurate location results. However, the linear equation system (3) is always close to ill-conditioned, which can cause a large location error with a minor disturbance. Therefore, the preconditioning of linear equations is adopted to weaken the ill condition and further improve the location accuracy.

3. Theory of Proposed Method

To solve the above problems, the PCFWE method is proposed to seek the optimal location results. The proposed method mainly includes two processes: filtering the outliers by weight estimation and reconstructing the linear equations with remaining measurements.

3.1. Filtering Outliers by Weight Estimation

3.1.1. Weight Estimation

The main idea of weight estimation is that a small weight is given to the measurement with a large residual, which in turn will reduce the influence of the outlier. Therefore, it is crucial to establish the relationship between weights and residuals.

However, the residuals of the real model cannot be obtained directly, so the weights are also difficult to be determined. In this work, the AE source is located by the LLS method, and each measurement is treated with equal weights firstly. Then, the residuals of the fitting model can be obtained and used to replace the residuals of the true model:

$$\Delta_i = b_i - A_i\theta, \quad i = 1, 2, \cdots, n \tag{6}$$

where, Δ_i is the ith element of a residual vector of the fitting model.

After obtaining the residuals, the weight estimation is developed by establishing the relationship between weights and residuals. The simplest method is to exploit the reciprocal value of the residuals as the weights directly. However, when a residual tends to be zero, the corresponding measurement is assigned with an unreasonably high weight, resulting in overfitting. Therefore, the weight should have a maximum upper limit to avoid overfitting, which can be realized by setting a lower limit of λ_m. Second, residuals exceeding the threshold λ_L are considered outliers, and the relative measurements are also regarded as outliers and filtered by setting the weight to zero. Therefore, the weight estimation can be expressed as follows [52]:

$$w_i = \begin{cases} 1/\lambda_m, & if \ |\Delta_i| \le \lambda_m \\ 1/\Delta_i, & if \ \lambda_m < |\Delta_i| < \lambda_L \\ 0, & else \end{cases} \tag{7}$$

To determine the specific parameters of the weight estimation, it is assumed that residuals obey the normal distribution. Therefore, the threshold λ_m and λ_L in (7) can be defined by:

$$\lambda_m = k_m\sigma \tag{8}$$

$$\lambda_L = k_L\sigma \tag{9}$$

where k_m and k_L are the correlation factors, and σ is the estimated value of the standard deviation of residuals, with expression as follows:

$$\sigma = \sqrt{\frac{1}{n-m} \frac{\sum_{i=1}^{n} w_i(b_i - A_i\theta)^2}{\frac{1}{n}\sum_{i=1}^{n} w_i}} \tag{10}$$

where m is the number of the unknown parameters. The denominator $\frac{1}{n}\sum_{i=1}^{n} w_i$ is used to standardize the weights.

Figure 1 shows the regional division of residuals according to the normal distribution. The parameters k_m and k_L divide the complete residuals into three regions, including minor residuals, medium residuals, and outlier residuals. Besides, the corresponding weights transformed from these residuals are divided into three parts, including an equal weight part, a variable weight part, and a zero weight part, as shown in Figure 2. The first part has the equal and highest weights out of the consideration that at least half of the measurement data are accurate. Therefore, $k_m = 0.765$ is selected, so that 50% of the normal residuals fall in the equal weight area. To avoid the unbalance of weighting, λ_m should have a lowest limitation equal to $0.05\max(|\Delta_i|)$ when there are many minor residuals. The weights of the second part varies with the relative residuals according to Equation (7). The third part is the zero weight region. Any observations from the n residuals that lie inside the probability band ($|\Delta| > k_L\sigma$) can be considered outliers. Besides, the corresponding measurements are regarded as outlier measurements, which are filtered on the following calculation by setting the weights to zeros.

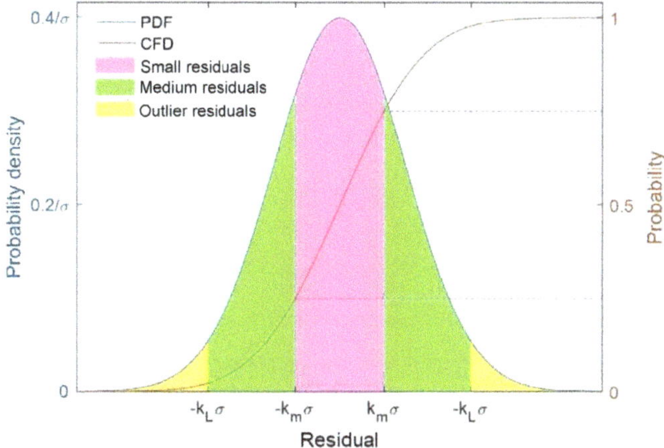

Figure 1. Residual division according to the normal distribution. In this figure, PDF is the normal probability density function and CDF means the cumulative distribution function.

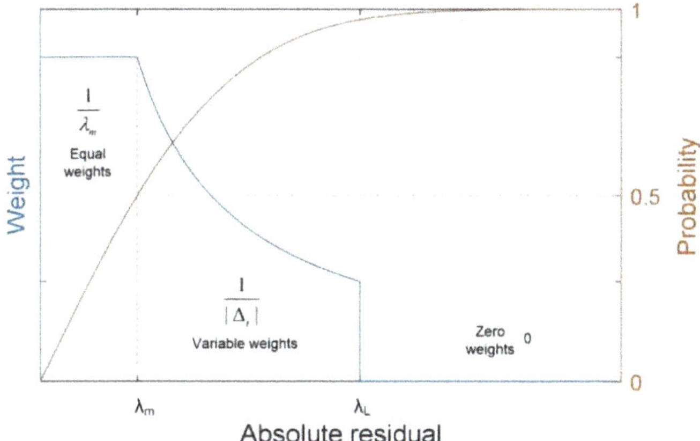

Figure 2. Change of weights with the increasing absolute residuals and the probability of the absolute residuals less than $k\sigma$ ($|\Delta| < k\sigma$).

To determinate the parameter k_L, half of the residuals are considered to be situated in the third region (i.e., outside of the probability band $|\triangle| \leq k_L \sigma$). Therefore, the probability band ($|\triangle| \leq k_L \sigma$) should only account for $n - \frac{1}{2}$ residuals [53]. The probability P_{in} equal to $n - \frac{1}{2}$ out of n residuals is looked for, which can be expressed as:

$$P_{in} = 1 - \frac{1}{2n} \tag{11}$$

where P_{in} is the probability band, and n is the number of residuals.

The possibility P_{in} for the residual $\triangle$ falling in the interval of $-k_L \sigma$ to $k_L \sigma$ can be obtained by the integral of the probability density function:

$$
\begin{aligned}
P_{in} &= \int_{-k_L\sigma}^{k_L\sigma} \frac{1}{\sqrt{2\pi}\sigma} e^{-\frac{(\Delta-\mu)^2}{2\sigma^2}} d\Delta \\
&= 2\int_0^{k_L\sigma} \frac{1}{\sqrt{2\pi}\sigma} e^{-\frac{(\Delta-\mu)^2}{2\sigma^2}} d\Delta \\
&= 2\left[\frac{1}{\sqrt{2\pi}\sigma}\sqrt{\frac{\pi}{2}}\sigma \cdot erf\left(\frac{\Delta-\mu}{\sqrt{2}\sigma}\right)\right]_0^{k_L\sigma} \\
&= erf\left(\frac{k}{\sqrt{2}}\right)
\end{aligned}
\tag{12}
$$

where erf is the error function with expression $\mathrm{erf}(t) = \frac{2}{\sqrt{\pi}}\int_0^{k_L\sigma} e^{-t^2} dt$.

Therefore, the threshold parameter k_L determined by n can be obtained from Equations (11) and (12).

$$k_L = \sqrt{2} \cdot \mathrm{erfinv}\left(1 - \frac{1}{2n}\right) \tag{13}$$

where *erfinv* is the inverse error function corresponding to the *erf* function.

3.1.2. Iteration between Weight Estimation and Source Location

Preliminary weights can be obtained after the initial location with equal weights, while they may not be the optimal. First, the location with equal weights is likely to be inaccurate or even incorrect, and the standard deviation σ of the residuals is always large, especially when outliers exists. In this case, it is difficult to distinguish the outliers and minor residuals from all of the observations, because both the thresholds λ_m and λ_L are high. Second, the residual $\triangle_i$ does not reflect the deviation of the real model; it only describes the deviation of the fitting model. The residuals and the estimated weights will be both incorrect, if the location result of the fitting model deviates from the true source greatly. Therefore, further iterations between weight estimations and source locations are required [54].

In the process of iterations, the source location at each step needs to solve a weighted least square solution:

$$\theta = \arg\min\sum_{i=1}^{n} w_i(b_i - A_i\theta)^2 = \arg\min\sum_{i=1}^{n} w_i\Delta_i^2 = (A^TWA)^{-1}A^TWb \tag{14}$$

where W is the diagonal matrix of weights, and all of the elements of w_i are updated according to Equation (7).

After several iterations, the optimal weight estimation is obtained, and all of the outliers are identified and filtered.

3.2. Reconstruct Linear Equations

After filtering outliers, the linear equations in Equation (3) are reconstructed by using the remaining m measurements as:

$$\dot{A}_{(m)}\theta_0 = \dot{b}_{(m)} \tag{15}$$

However, the linear equations are always close to ill-conditioned, because the elements in the coefficient matrix differ by orders of magnitude. Besides, the unreasonable sensor layout can also result in the linearly dependence of certain equations, in turn to the ill condition of linear equations. The ill condition of linear equations indicates a large difference in the final location results, with a minor change in the coefficient matrix. Moreover, an ill-conditioned linear equation system always has a large condition number of the coefficient matrix $\dot{A}$:

$$cond(A) = ||\dot{A}|| \cdot ||\dot{A}^{-1}|| \tag{16}$$

Besides, location results deviate largely with a minor disturbance in the coefficient. Therefore, a preconditioning method is applied to linear equations to reduce the condition number and give a more accurate closed-form solution, which is named the preconditioned closed-form solution.

The non-singular and diagonal matrix P is found, then the solution of Equation (15) is transformed into that of Equation (17) as:

$$P\dot{A}\theta_o = P\dot{b} \tag{17}$$

where $P = diag(\frac{1}{s_1}, \frac{1}{s_2}, \cdots, \frac{1}{s_n})$ and, $s_i = \max_{1 \le j \le m} |(A)_{ij}|, i = 1, 2, \cdots, n$.

Finally, a more accurate location result θ_o can be obtained by:

$$\theta_o = [(P\dot{A})^T P\dot{A}]^{-1} (P\dot{A})^T P\dot{b} \tag{18}$$

due to the improved condition number $cond(P\dot{A}) \ll cond(\dot{A})$.

3.3. Location Process of Proposed Method

The whole procedures of PCFWE are shown in Figure 3 as below:

1. Establish linear equations for the source location.
2. Set index $I = 0$ and $w_i^{(0)} = 1$ for the primary iteration and obtain the initial location result $\theta^{(0)}$.
3. Calculate the residuals $\Delta^{(0)} = A\theta^{(0)} - b$, which are used to generate the preliminary weights $w_i^{(1)}$ according to Equation (7).
4. Conduct the next iteration, $I = 1$, and solve the weighted least squares solution $\theta^{(1)}$ from Equation (14).
5. Update the weights $w_i^{(2)}$ according to the residuals $\Delta^{(1)}$.
6. Use the new weights in the next location, where $I = 2$ and $\theta^{(2)}$ is calculated.
7. Repeat the above calculations (steps 4–6) to optimize the weights and filter the outliers, until $\sum \left| \Delta^{(I)} - \Delta^{(I-1)} \right|_2 \le \varepsilon$.
8. Reconstruct linear equations with the remaining measurements.
9. Precondition linear equations to lower the condition number and eliminate the ill condition.
10. Obtain the optimal closed-form solution by solving the preconditioned linear equations with Equation (18).

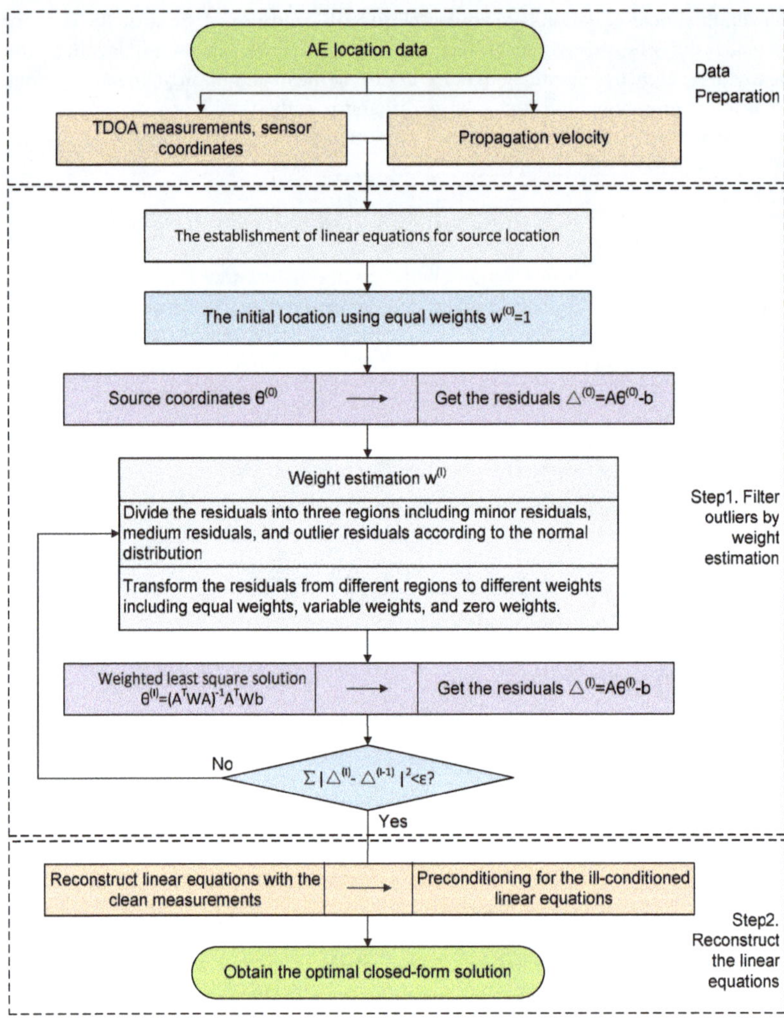

Figure 3. Flow charts of preconditioned closed-form solution based on weight estimation (PCFWE) method location process.

4. Experimental Verification

To verify the feasibility of this method, the AE source location experiment is carried out in this paper. The acoustic wave is collected by a *DS5-16C Holographic Acoustic Emission Signal Analyzer* (made by Beijing SoftLand Scientific and Technology Co., Ltd., Beijing, China). AE sources are generated by pencil-lead breaks. The HB pencil lead is 0.5 mm in diameter, and the pencil lead is broken at 30° on the surface of the specimen, according to the requirements of *Metal Pressure Vessel Acoustic Emission Testing and Result Evaluation (GB/T18182—2000)*. The granite specimen that is 200 mm × 179 mm × 84 mm in size and an average velocity of 4600 m/s is used as a monitoring system. Sixteen RS-2A piezoelectric ceramic resonant sensors with the coordinates of (10, 10, 84), (190, 10, 84), (190, 170, 84), (12, 170, 84), (0, 80, 74), (110, 0, 74), (200, 80, 74), (90, 180, 74), (0, 170, 10), (0, 90, 10), (10, 0, 10), (100, 0, 10), (190, 0, 10), (200, 90, 10), (190, 180, 10), and (100, 180, 10) (in mm) respectively are mounted on the surface of the monitoring system. Sensors are numbered as S_i ($i = 0, 1, \cdots, 15$)

according to the arrival orders of the AE signals, and the nearest sensor S_0 from the AE source is regarded as the reference sensor. Later, 15 TDOA measurements $\triangle t_i$ ($i = 1, 2, \cdots, 15$) between the sensor pairs of S_0 and S_i are obtained. In addition, there are four AE sources generated by pencil-lead breaks, and their coordinates are P (80, 120, 84), Q (160, 60, 84), R (0, 120, 42), and T (120, 0, 42) respectively. Before signal acquisition, the voltage threshold of each channel is set as 10 mv, and the preamplifier gain is 40 dB.

Figure 4 shows the received AE waveforms by two sensors, in which t_0 and t_2 are the arrival times at sensor S_0 and sensor S_2, respectively. Then, TDOA measurement is formed by using the arrival times at the sensor pair S_0 and S_2, i.e., $\Delta t_2 = t_2 - t_0$. The arrival times of the acoustic waves are picked manually according to the waveform diagram to ensure a high picking quality.

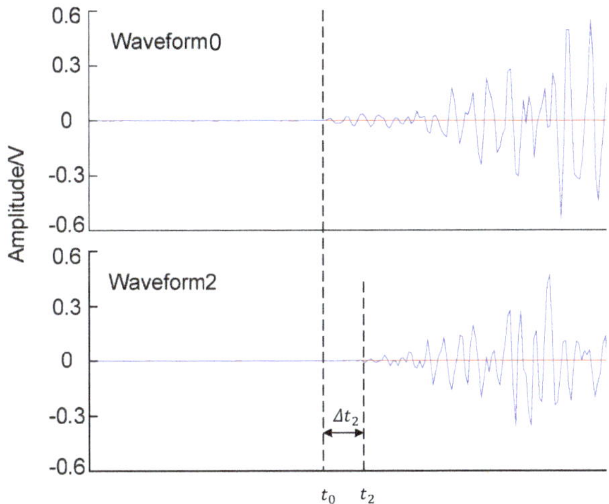

Figure 4. Acoustic emission (AE) signals and arrival times at different sensors.

4.1. The Filtering of Outliers

TDOA measurements, the coordinates of triggered sensors, and average velocity are used to locate the four AE events by the method proposed in this paper. The outlier TDOA measurements are generated factitiously by adding ±30% large errors to the largest two measurements, which are used to simulate the picking deviation caused by automated programs in complex engineering practice. Location results with the additive outliers in measurements are determined by using the LLS method without the outliers filtering process, and the detailed data are listed in upper half of Table 1. It is clear to see that the location errors between the authentic coordinates and the location results are great, where the maximum absolute distance error reaches to 57.61 mm. Obviously, the location results are unreasonable and unreliable, because corrupted measurements dramatically affect the location accuracy. Therefore, it is significant to identify and filter the outliers for the AE source location.

Figure 5 illustrates the process of eliminating the outliers at the location of source T. The initial location using the LLS method with equal weights has a large location error due to the existence of the outliers shown in Step a. The corresponding residuals are obtained and shown in Step b; it can be seen that the absolute residuals of 1, 4, 5, 6, 7, 8, 9, 12, and 13 are less than the threshold λ_m, so the corresponding weights of the measurements $\triangle t_1$, $\triangle t_4$, $\triangle t_5$, $\triangle t_6$, $\triangle t_7$, $\triangle t_8$, $\triangle t_9$, $\triangle t_{12}$, and $\triangle t_{13}$ are set as equal according to Equation (7), while the weights of the other measurements are set as $1/\triangle_i$. Moreover, it should be noted that no outliers can be found after the first location, because the residuals obtained by the initial location result can not reflect the deviations of the true model effectively, especially when outliers exist. Therefore, further iterations between the source location and weight estimation are necessary in order to filter the outliers, which are shown from step *c* to Step *f*. The 14th and 15th residuals exceed the maximum allowable deviation λ_L at step *c* and *d*, respectively. Therefore, they are regarded as outliers, and $\triangle t_{14}$ and $\triangle t_{15}$ are considered outlier measurements that are filtered by setting the weights to zeros. The change of residuals between steps *e* and *f* is small enough, so the iteration stops. After that, linear equations are reconstructed by using the remaining measurements. Finally, the location results can be obtained by solving the new linear equations using the LLS method. It can be seen in Step *g* that the calculated result and theoretical locations respectively denoted by the diamond and spot are close to each other. Therefore, the location accuracy is improved effectively after outlier filtering.

The lower half of Table 1 lists the location results and absolute distance errors of the four AE sources after filtering the outliers. Through the comparison between location results before and after filtering the outliers in Figure 6, it is obvious that the location accuracy is improved significantly. For example, the absolute distance error of the source *R* is reduced to 9.28 mm from 57.61 mm, and the absolute distance error of source *T* is reduced to 10.55 mm from 46.82 mm. Therefore, it can be concluded that the location accuracy is improved effectively through the filtering of outliers. However, it should be noticed that linear equations are always in or near ill condition, which still cause a large deviation with a minor disturbance in the coefficient. Location accuracy is improved using a preconditioning method by reducing the ill condition of linear equations. Location results in the following section, Section 4.2, prove the ability of solving source coordinates with higher accuracy by preconditioning for linear equations.

Table 1. Closed-form solutions before and after filtering using the linear least squares (LLS) method.

Sources		True Source Coordinates			Location Results			Absolute Distance Error (mm)
		x (mm)	*y* (mm)	*z* (mm)	*x* (mm)	*y* (mm)	*z* (mm)	
Without filtering	P	80.00	120.00	84.00	93.94	106.20	76.60	20.96
	Q	160.00	60.00	84.00	121.71	71.02	73.45	41.21
	R	0.00	120.00	42.00	53.29	117.88	63.78	57.61
	T	120.00	0.00	42.00	95.25	38.87	33.73	46.82
filtering	P	80.00	120.00	84.00	78.14	124.71	84.00	5.07
	Q	160.00	60.00	84.00	168.66	54.04	84.00	10.51
	R	0.00	120.00	42.00	7.82	119.33	46.96	9.28
	T	120.00	0.00	42.00	111.22	0.00	36.15	10.55

4.2. Comparing Location Results with and without Outliers

For each event, the mentioned TSWLS and LLS methods in Sections 1 and 2 are applied for comparison with the PCFWE with additive outliers and without additive outliers.

Figure 7 shows the location results in a three-dimensional space. It can be seen that the location results of the PCFWE method are approximate to the true sources, regardless of whether the additive outliers are contained in the TDOA measurements or not. Location results, which are determined by the TSWLS and LLS methods, also achieve desirable location results with minor deviations from true sources, but it deviates dramatically if outliers exist. Figure 8 shows the absolute distance errors of location results for three methods with and without outliers. When there are no outliers in the measurements, the absolute distance errors of the three methods are all small, while the PCFWE is the smallest, because of the applied preconditioning method. When the outliers are contained in measurements, the location errors of the traditional methods are dramatically large, while the location performance of the proposed method always remains stable with a higher location accuracy.

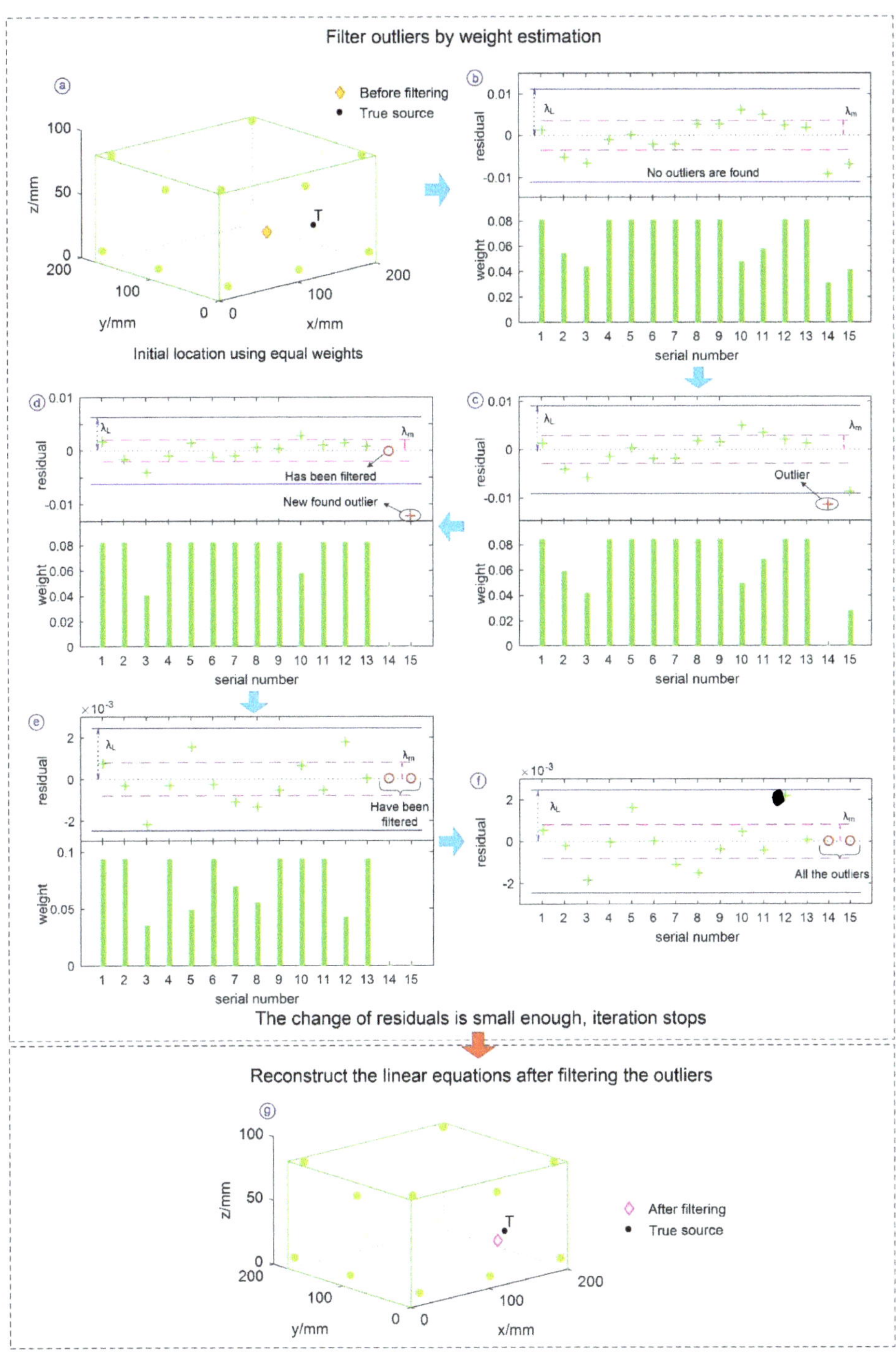

Figure 5. The process of locating the AE source T using the PCFWE method. The weights in this figure have been standardized by $w_i/\mathrm{sum}(w_i)$.

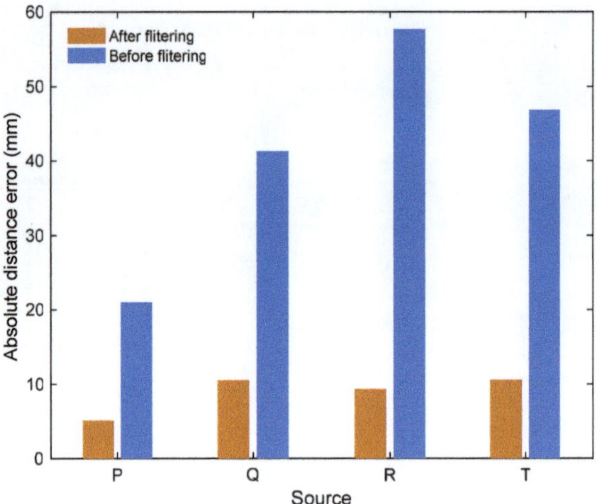

Figure 6. Comparison of location errors before and after filtering using the LLS method.

From Table 2, it is obvious that the biggest absolute distance errors of traditional methods can occasionally exceed 50 mm. There are two main reasons for the poor location results of traditional methods. Firstly, due to the influence of an outlier in the measurement, residuals are significantly intensified by the least squares principle, which results in a large deviation to the final location result for traditional methods. Secondly, traditional methods take no account of the problem of ill-conditioned linear equations, which also affect the location accuracy to some extent. It is worth mentioning that the proposed method in this paper can achieve an accurate location, where the best absolute distance error can reach about 1 mm. The good performance of the new method is attributed to the fact that, it not only reduces the ill condition of linear equations, it also eliminates outliers. Therefore, location accuracy is improved effectively.

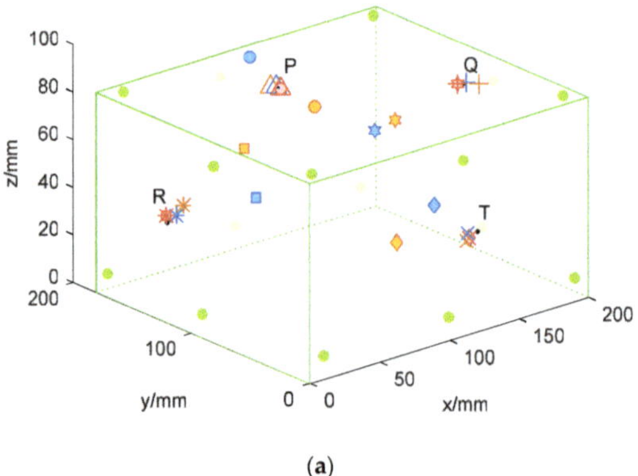

(**a**)

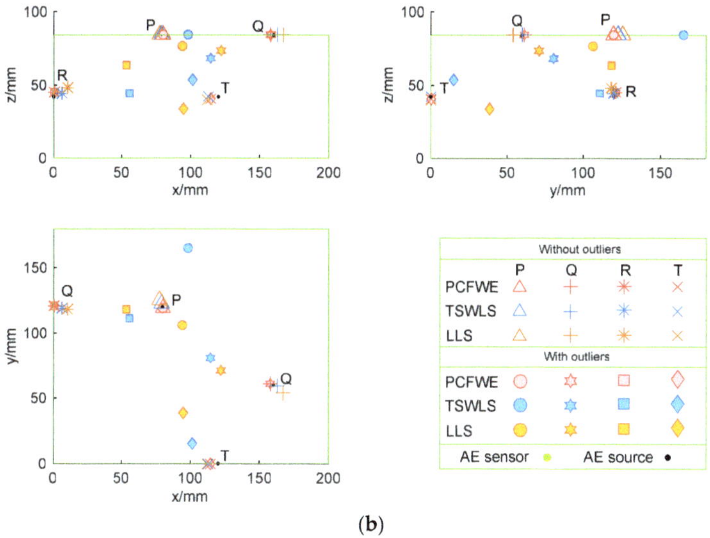

Figure 7. Location results from three methods: (**a**) three-dimensional (3D) schematic diagram of location results and sensor layout; (**b**) projections of AE events on the *x-z*, *y-z*, and *x-y* planes.

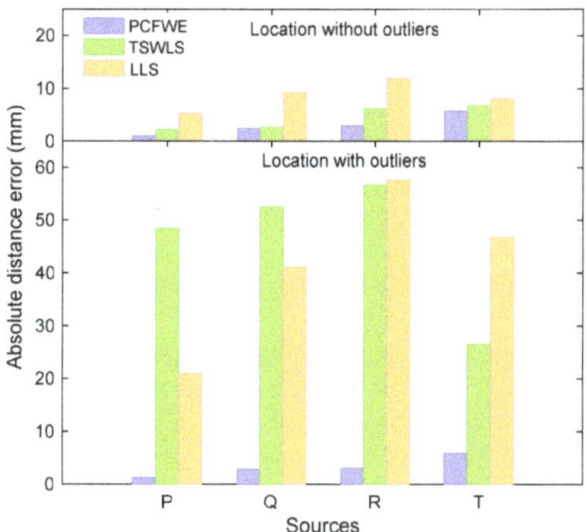

Figure 8. The absolute distance error of the location results for three methods.

Table 2. Coordinates of AE sources solved by three methods.

AE Source	Outliers	PCFWE				TSWLS				LLS			
		x (mm)	y (mm)	z (mm)	Absolute Distance Error (mm)	x (mm)	y (mm)	z (mm)	Absolute Distance Error (mm)	x (mm)	y (mm)	z (mm)	Absolute Distance Error (mm)
P	No	80.46	119.12	84.00	0.99	79.96	122.18	84.00	2.18	77.98	124.97	84.00	5.37
	Yes	80.60	118.86	84.00	1.29	98.24	164.94	84.00	48.50	93.94	106.20	76.60	20.96
Q	No	158.08	61.40	84.00	2.38	162.65	59.57	84.00	2.69	167.06	54.11	84.00	9.20
	Yes	157.80	61.88	84.00	2.89	114.45	80.80	68.23	52.50	121.71	71.02	73.45	41.21
R	No	0.00	121.08	44.75	2.96	5.78	119.36	44.05	6.17	10.11	117.79	47.93	11.93
	Yes	0.00	121.24	44.81	3.07	55.81	110.82	44.08	56.60	53.29	117.88	63.78	57.61
T	No	114.43	0.00	40.53	5.76	113.24	0.00	42.30	6.76	112.17	0.00	39.88	8.12
	Yes	114.28	0.00	40.55	5.90	101.38	14.82	53.56	26.46	95.25	38.87	33.73	46.82

5. Conclusions and Discussion

Considering that the outliers in the complex engineering environment have a serious effect on the location performance, the PCFWE method is proposed to improve the location accuracy in this paper. Firstly, the weight estimation is developed by transforming residuals to weights according to normal distribution. Updated weights are incorporated into the source location to provide a weighted least squares solution. Then, outliers are filtered by iterating between the weight estimation and source location. Linear equations, which are reconstructed with remaining measurements after filtering outliers, are ill-conditioned and need to be preconditioned to lower the condition number. Finally, the optimal source coordinates can be obtained by solving preconditioned linear equations. The proposed location method is verified by the experiment of pencil-lead breaks; the results show that the proposed PCFWE method achieves a stable and accurate location both with outliers and without outliers in measurements. However, how many outliers the proposed method can tolerate, and the sensitivity of the wave velocity, still need further investigation.

5.1. Outlier Tolerance

To investigate the outlier-tolerant ability of PCFWE under different outlier proportions, simulating tests are used because of the admirable repeatability and flexibility. Simulating tests with controllable errors in the input data such as TDOA measurements and the velocity system are described in this section. Figure 9 shows an assumed cubic locating system with a side length of 300 mm, where an AE source O (150, 100, 200) surrounded by 21 sensors is set to generate AE signals. It is assumed that the trigger time is 0 µs, and the average wave velocity of the media is 5000 m/s. In addition, to simulate minor systemic errors in TDOA measurements, the extra random errors from the normal distribution with a mean of zero and a standard deviation of 2% of measurements are added. To simulate the uncertainty of the velocity along different paths, a minor error of 5% of velocity for each path is generated. For measurements with abnormal errors (outliers), dramatic errors of ±30% of TDOA measurements, which are much larger than the systemic error, are added with different proportions. Thus, the simulated location consists of velocity uncertainty, systemic minor errors, and different outlier proportions.

To obtain a reliable statistical result, the AE source location is repeated 100 times by random changes of TDOA measurements and the velocity errors, and then their location coordinates are calculated. The location results of the 100 simulating tests with different outlier proportions are shown in Figure 10. It can be seen that all of the sizes of solid circles are small and close to true sources, when there is no outlier in the input data. However, they have obvious differences as shown in Figure 11, where the average absolute distance error of the PCFWE is smaller than that of traditional methods, due to the use of preconditioning for the linear equations. In other words, the PCFWE performs better than traditional location methods without outliers. The location performance of the PCFWE differs more substantially from traditional methods in the presence of outliers. Solid circles of traditional methods always appear more discrete and have a larger diameter with the increase of outlier proportions, which indicates that the location errors increase with outlier proportions. While the solid circles of the PCFWE always appear more compact and darker under different outlier proportions, which illustrate the PCFWE location method has a higher accuracy and stability than traditional methods.

Figure 11 shows the average absolute distance errors of 100 location results under different outlier proportions determined by three methods. The average absolute distance errors for traditional methods increase dramatically with the increasing of outlier proportions. Whereas, the average absolute distance errors of PCFWE method always keep low and stable under different outlier proportions, due to the filtering of outliers in calculation, which further illustrates the good performance of the PCFWE method.

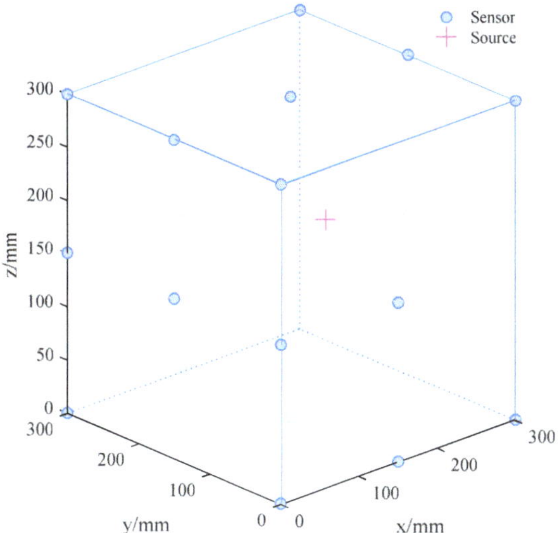

Figure 9. Layout of sensors and location of the AE source.

Moreover, the relationship between valid location ratios and outlier proportions is displayed in Figure 12. The valid location ratio is the location proportion whose location errors are smaller than 2% of the distance between the two farthest sensors in the sensor array. Clearly, the two curves of traditional methods decline dramatically with the increase of outliers, the valid locations of which are less than 30% when there are 25% outliers. Compared with traditional methods, the curve of the PCFWE always keeps stable and achieves valid locations of nearly 100% when the outliers are less than or equal to 25%. Then, the curve falls to 73% at the outlier proportion of 45%, which is still far above the other two curves that have valid locations of less than 10%.

Therefore, the outlier-tolerant ability of the proposed method is far higher than those of traditional methods, which can realize valid locations of more than 90%, even when the outlier proportion reaches 35%. Therefore, the proposed method is suitable for engineering practice where the outliers commonly exist. In addition, when there are too many outliers (more than 35%), the PCFWE method gives a warning to check the environment noise or equipment malfunctions.

5.2. Velocity Sensibility

The premeasured velocity is necessary for most location methods, and measurement errors in velocity are unavoidable due to the complex construction environment and manual operation. Moreover, the wave velocity of the propagation medium always changes with material testing or engineering construction. An effective location method should have a lower velocity sensitivity, i.e., a higher tolerance of velocity error. Therefore, the analysis of the velocity sensitivity of different methods is conducted in this section. This simulating test is still based on the above location system. To obtain a reliable statistical result, 100 AE sources are selected randomly inside the sensor array to generate AE signals. Besides, the 2% minor systemic errors in measurements are applied to simulate minor systemic errors, and 5% to 25% errors are added to velocity, in turn to investigate the velocity sensitivity. Hence, the simulated location consists of systemic minor errors in TDOA measurements and different error scales in velocity.

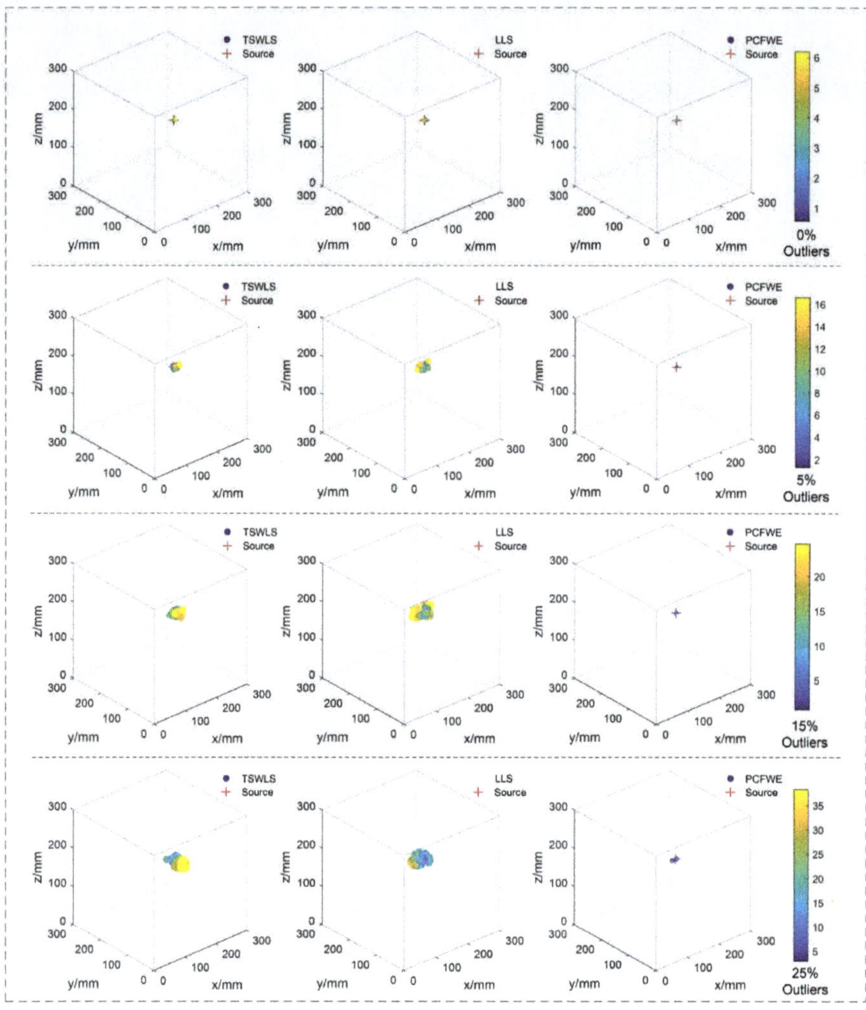

Figure 10. Three-dimensional location maps determined by three methods with outlier proportions of 0%, 5%, 15%, and 25% respectively. Diameters of solid circles indicate the location error, and smaller diameters indicate a higher location accuracy. To further distinguish the location accuracy of different sources, colors in the solid circles also indicate the location accuracy in millimeters.

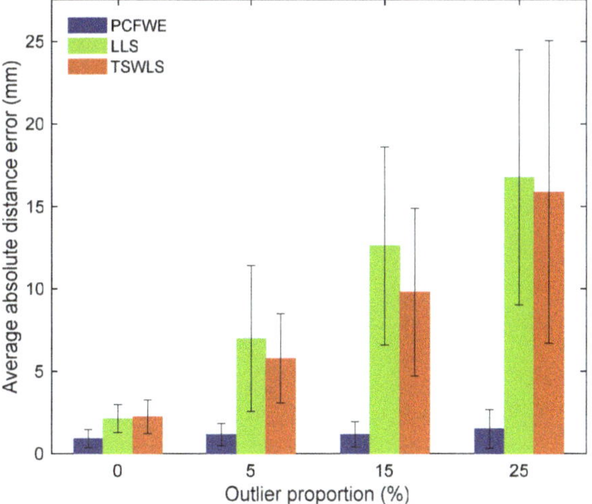

Figure 11. The average absolute errors of the AE sources determined by three methods under different outlier proportions.

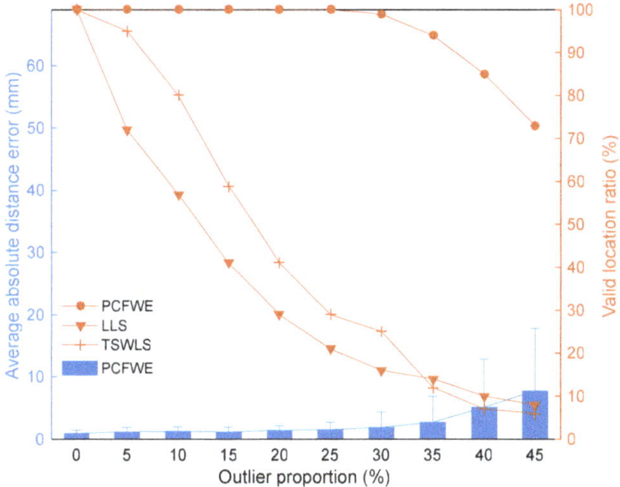

Figure 12. Valid location ratios of three methods and the average location error of the PCFWE under different outlier proportions.

The average absolute distance errors under different error scales in velocity are calculated, which are shown in Figure 13. All three curves show a rising trend with the increase of error scales in velocity, but the curve of the PCFWE is significantly lower than the other two methods. Therefore, the PCFWE has the lowest velocity sensitivity, which indicates that the PCFWE can bear larger errors in velocity than traditional methods.

During material testing, the wave velocity changes with the loading stress. Through the analysis of the velocity sensitivity of different methods, conclusions are gained, as the proposed method has a stronger tolerance to velocity errors, which is suitable for a material test where the wave velocity changes along with the test. However, in the later stage of a test, the wave velocity changes dramatically due to the propagation of cracks and voids. The proposed method also fails to achieve an ideal location result. A possible solution is the dynamic inversion for real-time wave velocity by taking the velocity as an unknown parameter or modifying the velocity data according to the real-time velocity test, which still needs to be investigated in future research [46].

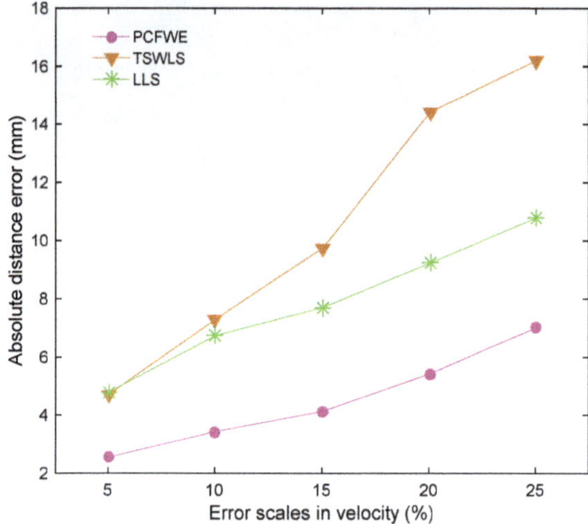

Figure 13. The average absolute distance error determined using three methods under different error scales in velocity.

In practical engineering applications, the proposed method still has the following two limitations. This method is established based on a uniform medium, while most of the materials in engineering practice are anisotropic and heterogeneous, such as rock and concrete. In addition, although the sensitivity of the wave velocity is low, the proposed method is still affected by the velocity error. Besides, it is troublesome to measure the wave velocity in advance or real-time. Therefore, further studies should extend the method to address these issues.

Author Contributions: Z.Z. and Y.R. proposed the location method and performed the experiments; Z.Z., Y.R., J.Z. and R.C. analysed the data and wrote the paper; and L.D., L.C. and X.C. supervised the research and revised the manuscript.

Acknowledgments: The authors would like to acknowledge financial support from the National Basic Research Program of China (973 Program) (No. 2015CB060200), the National Natural Science Foundation of China (Nos. 41772313 and 51478479), and the Fundamental Research Funds for the Central Universities of Central South University (No. 2018zzts716).

Conflicts of Interest: The authors declare no conflict of interest.

Abbreviations

The following abbreviations are used in this manuscript:

AE	Acoustic Emission
PCFWE	Preconditioned closed-form solution based on weight estimation
TDOA	Time difference of arrival
TSWLS	Two-step weighted least squares
LLS	Linear least squares

References

1. Dworakowski, Z.; Kohut, P.; Gallina, A.; Holak, K.; Uhl, T. Vision-based algorithms for damage detection and localization in structural health monitoring. *Struct. Control Health* **2016**, *23*, 35–50. [CrossRef]
2. Topolář, L.; Pazdera, L.; Kucharczyková, B.; Smutný, J.; Mikulášek, K. Using acoustic emission methods to monitor cement composites during setting and hardening. *Appl. Sci.* **2017**, *7*, 451. [CrossRef]
3. Muñoz, C.G.; Márquez, F.G. A new fault location approach for acoustic emission techniques in wind turbines. *Energies* **2016**, *9*, 40. [CrossRef]
4. Martini, A.; Troncossi, M.; Rivola, A. Leak detection in water-filled small-diameter polyethylene pipes by means of acoustic emission measurements. *Appl. Sci.* **2016**, *7*, 2. [CrossRef]
5. Zhou, Z.L.; Cai, X.; Ma, D.; Cao, W.Z.; Chen, L.; Zhou, J. Effects of water content on fracture and mechanical behavior of sandstone with a low clay mineral content. *Eng. Fract. Mech.* **2018**, *193*, 47–65. [CrossRef]
6. Dong, L.J.; Sun, D.Y.; Li, X.B.; Du, K. Theoretical and Experimental Studies of Localization Methodology for AE and Microseismic Sources without Pre-Measured Wave Velocity in Mines. *IEEE Access* **2017**, *5*, 16818–16828. [CrossRef]
7. Lacidogna, G.; Piana, G.; Carpinteri, A. Acoustic Emission and Modal Frequency Variation in Concrete Specimens under Four-Point Bending. *Appl. Sci.* **2017**, *7*, 339. [CrossRef]
8. Li, X.; Deng, Z.D.; Rauchenstein, L.T.; Carlson, T.J. Contributed review: Source-localization algorithms and applications using time of arrival and time difference of arrival measurements. *Rev. Sci. Instrum.* **2016**, *87*, 921–960. [CrossRef] [PubMed]
9. Huang, Y.; Benesty, J.; Elko, G.W.; Mersereati, R.M. Real-time passive source localization: A practical linear-correction least-squares approach. *IEEE Trans. Speech Audio Process.* **2001**, *9*, 943–956. [CrossRef]
10. Roberts, R.G.; Christoffersson, A.; Cassidy, F. Real-time event detection, phase identification and source location estimation using single station three-component seismic data. *Geophys. J. Int.* **2010**, *97*, 471–480. [CrossRef]
11. Al-Jumaili, S.K.; Pearson, M.R.; Holford, K.M.; Eaton, M.J.; Pullin, R. Acoustic emission source location in complex structures using full automatic delta t mapping technique. *Mech. Syst. Signal Process.* **2016**, *72–73*, 513–524. [CrossRef]
12. Dong, L.J.; Shu, W.W.; Han, G.J.; Li, X.L.; Wang, J. A Multi-Step Source Localization Method With Narrowing Velocity Interval of Cyber-Physical Systems in Buildings. *IEEE Access* **2017**, *5*, 20207–20219. [CrossRef]
13. Mostafapour, A.; Davoodi, S. Acoustic emission source locating in two-layer plate using wavelet packet decomposition and wavelet-based optimized residual complexity. *Struct. Control. Health* **2017**. [CrossRef]
14. Kundu, T. Acoustic source localization. *Ultrasonics* **2014**, *54*, 25–38. [CrossRef] [PubMed]
15. Schau, H.; Robinson, A. Passive source location employing spherical surfaces from time-of-arrival differences. *IEEE Trans. Aerosp. Electron. Syst.* **1987**, *35*, 1223–1225.
16. Smith, J.O.; Abel, J.S. Closed-form least-squares source location estimation from range-difference measurements. *IEEE Trans. Acoust. Speech Signal Process.* **1987**, *35*, 1661–1669. [CrossRef]
17. Zhou, Z.L.; Rui, Y.C.; Zhou, J.; Dong, L.J.; Cai, X. Locating an acoustic emission source in multilayered media based on the refraction path method. *IEEE Access* **2018**. [CrossRef]
18. Zhou, Z.L.; Zhou, J.; Dong, L.J.; Cai, X.; Rui, Y.C.; Ke, C.T. Experimental study on the location of an acoustic emission source considering refraction in different media. *Sci. Rep.* **2017**, *7*, 7472. [CrossRef] [PubMed]
19. Drew, J.; White, R.S.; Tilmann, F.; Tarasewicz, J. Coalescence microseismic mapping. *Geophys. J. Int.* **2013**, *195*, 1773–1785. [CrossRef]
20. Enosh, I.; Weiss, A.J. Outlier identification for toa-based source localization in the presence of noise. *Signal Process.* **2014**, *102*, 85–95. [CrossRef]
21. Allen, R.V. Automatic earthquake recognition and timing from single traces. *Bull. Seismol. Soc. Am.* **1978**, *68*, 1521–1532.
22. Kuperkoch, L.; Meier, T.; Lee, J.; Friederich, W.; Grp, E.W. Automated determination of P-phase arrival times at regional and local distances using higher order statistics. *Geophys. J. Int.* **2010**, *181*, 1159–1170. [CrossRef]
23. Niccolini, G.; Xu, J.; Manuello, A.; Lacidogna, G.; Carpinteri, A. Onset time determination of acoustic and electromagnetic emission during rock fracture. *Prog. Electromagn. Res. Lett.* **2012**, *35*, 51–62. [CrossRef]
24. Akaike, H. A new look at the statistical model identification. *Trans. Autom. Control* **1974**, *19*, 716–723. [CrossRef]

25. Ge, M.C. Efficient mine microseismic monitoring. *Int. J. Coal Geol.* **2005**, *64*, 44–56. [CrossRef]

26. Dong, L.J.; Li, X.B.; Ma, J.; Tang, L.Z. Three-dimensional analytical comprehensive solutions for acoustic emission/microseismic sources of unknown velocity system. *Chin. J. Rock Mech. Eng.* **2017**, *36*, 186–197.

27. Taghizadeh, M.J.; Haghighatshoar, S.; Asaei, A.; Garner, P.N.; Bourlard, H. Robust microphone placement for source localization from noisy distance measurements. In Proceedings of the IEEE International Conference on Acoustics, Speech and Signal Processing (ICASSP), Brisbane, QLD, Australia, 19–24 April 2015; pp. 2579–2583.

28. Picard, J.S.; Weiss, A.J. Time difference localization in the presence of outliers. *Signal Process.* **2012**, *92*, 2432–2443. [CrossRef]

29. Li, P.; Ma, X. Robust acoustic source localization with TDOA based RANSAC algorithm. In Proceedings of the 5th International Conference on Intelligent Computing (ICIC), Ulsan, Korea, 16–19 September 2009; pp. 222–227.

30. Weng, Y.; Xiao, W.D.; Xie, L.H. Total least squares method for robust source localization in sensor networks using tdoa measurements. *Int. J. Distrib. Sens. Netw.* **2011**, *7*, 172902. [CrossRef]

31. Lin, L.; So, H.C.; Chan, F.K.W.; Chan, Y.T.; Ho, K.C. A new constrained weighted least squares algorithm for tdoa-based localization. *Signal Process.* **2013**, *93*, 2872–2878. [CrossRef]

32. Mensing, C.; Plass, S. Positioning algorithms for cellular networks using TDOA. In Proceedings of the IEEE International Conference on Acoustics, Speech and Signal Processing (ICASSP), Toulouse, France, 14–19 May 2006; pp. 513–516.

33. Chan, Y.T.; Hang, H.Y.C.; Ching, P.C. Exact and approximate maximum likelihood localization algorithms. *IEEE Trans. Veh. Technol.* **2006**, *55*, 10–16. [CrossRef]

34. Chen, J.C.; Hudson, R.E.; Yao, K. Maximum-likelihood source localization and unknown sensor location estimation for wideband signals in the near-field. *IEEE Trans. Signal Proces.* **2002**, *50*, 1843–1854. [CrossRef]

35. Jiang, Y.; Azimi-Sadjadi, M.R. A Robust Source Localization Algorithm Applied to Acoustic Sensor Network. In Proceedings of the IEEE International Conference on Acoustics, Speech and Signal Processing (ICASSP), Honolulu, HI, USA, 15–20 April 2007; pp. 1233–1236.

36. Li, X.B.; Wang, Z.W.; Dong, L.J. Locating single-point sources from arrival times containing large picking errors (lpes): The virtual field optimization method (vfom). *Sci. Rep.* **2016**, *6*, 19205. [CrossRef] [PubMed]

37. Li, N.; Wang, E.Y.; Sun, Z.Y.; Li, B.L. Simplex microseismic source location method based on l1 norm statistical standard. *J. China Coal Soc.* **2014**, *39*, 2431–2438.

38. Prugger, A.F.; Gendzwill, D.J. Microearthquake location: A nonlinear approach that makes use of a simplex stepping procedure. *Bull. Seismol. Soc. Am.* **1988**, *78*, 799–815.

39. Ciampa, F.; Meo, M. Acoustic emission source localization and velocity determination of the fundamental mode A0 using wavelet analysis and a Newton-based optimization technique. *Smart Mater. Struct.* **2010**, *19*, 045027. [CrossRef]

40. Foy, W.H. Position-location solutions by taylor-series estimation. *IEEE Trans. Aerosp. Electron. Syst.* **2007**, *AES-12*, 187–194. [CrossRef]

41. Chan, Y.T.; Ho, K.C. A simple and efficient estimator for hyperbolic location. *IEEE Trans. Signal Process.* **1994**, *42*, 1905–1915. [CrossRef]

42. Ho, K.C.; Xu, W. An accurate algebraic solution for moving source location using TDOA and FDOA measurements. *IEEE Trans. Signal Process.* **2004**, *52*, 2453–2463. [CrossRef]

43. Ho, K.C. Bias reduction for an explicit solution of source localization using TDOA. *IEEE Trans. Signal Process.* **2012**, *60*, 2101–2114. [CrossRef]

44. Ho, K.C.; Lu, X.; Kovavisaruch, L.O. Source localization using TDOA and FDOA measurements in the presence of receiver location errors: Analysis and solution. *IEEE Trans. Signal Process.* **2007**, *55*, 684–696. [CrossRef]

45. Xu, B.; Qi, W.D.; Wei, L.; Liu, P. Turbo-TSWLS: Enhanced two-step weighted least squares estimator for TDOA-based localisation. *Electron. Lett.* **2012**, *48*, 1597–1598. [CrossRef]

46. Dong, L.J.; Shu, W.W.; Li, X.B.; Han, G.J.; Zou, W. Three dimensional comprehensive analytical solutions for locating sources of sensor networks in unknown velocity mining system. *IEEE Access* **2017**, *5*, 11337–11351. [CrossRef]

47. Dong, L.J.; Zou, W.; Li, X.B.; Shu, W.W.; Wang, Z.W. Collaborative localization method using analytical and iterative solutions for microseismic/acoustic emission sources in the rockmass structure for underground mining. *Eng. Fract. Mech.* **2018**. [CrossRef]

48. Leighton, F.; Duvall, W.I. *Least Squares Method for Improving Rock Noise Source Location Techniques*; USBM RI 7626; Bureau of Mines: Washington, DC, USA, 1972; pp. 26–69.

49. Fang, W.H.; Liu, L.C.; Yang, J.P.; Shui, A.S. A non-iterative ae source localization algorithm with unknown velocity. *J. Log. Eng. Univ.* **2016**, *32*, 1–7.

50. Li, X.B.; Dong, L.J. An efficient closed-form solution for acoustic emission source location in three-dimensional structures. *AIP Adv.* **2014**, *4*, 152–160. [CrossRef]

51. Dong, L.J.; Li, X.B.; Xie, G.N. An analytical solution for acoustic emission source location for known P wave velocity system. *Math. Probl. Eng.* **2014**, *2014*, 290686. [CrossRef]

52. Strutz, T. *Data Fitting and Uncertainty: A Practical Introduction to Weighted Least Squares and Beyond*; Vieweg and Teubner Verlag: Wiesbaden, Germany, 2010.

53. Chauvenet, W. *A Manual of Spherical and Practical Astronomy*, 4th ed.; J.B. Lippincott Co.: Philadelphia, PA, USA, 1871; pp. 558–568.

54. Holland, P.W.; Welsch, R.E. Robust regression using iteratively reweighted least-squares. *Commun. Stat. Theory Mothods* **1977**, *6*, 813–827. [CrossRef]

Article

Multiple Signal Classification-Based Impact Localization in Composite Structures Using Optimized Ensemble Empirical Mode Decomposition

Yongteng Zhong [1,2,3], Jiawei Xiang [2,*], Xiaoyu Chen [3], Yongying Jiang [2] and Jihong Pang [2]

[1] School of Mechanical Engineering, Zhejiang University, Hangzhou 310058, China;
    zhongyongteng@wzu.edu.cn
[2] College of Mechanical and Electrical Engineering, Wenzhou University, Wenzhou 325035, China;
    yyjiang2003@126.com (Y.J.); pangjihong@163.com (J.P.)
[3] Zhejiang Linuo Fluid Control Technology Co., Ltd., Wenzhou 325200, China; lvyanping2010@163.com
*   Correspondence: wxw8627@163.com

Received: 31 July 2018; Accepted: 22 August 2018; Published: 24 August 2018

Abstract: Multiple signal classification (MUSIC) algorithm-based structural health monitoring technology is a promising method because of its directional scanning ability and easy arrangement of the sensor array. However, in previous MUSIC-based impact location methods, the narrowband signals at a particular central frequency had to be extracted from the wideband Lamb waves induced by each impact using a wavelet transform. Additionally, the specific center frequency had to be obtained after carefully analyzing the impact signal, which is time consuming. Aiming at solving this problem, this paper presents an improved approach that combines the optimized ensemble empirical mode decomposition (EEMD) and two-dimensional multiple signal classification (2D-MUSIC) algorithm for real-time impact localization on composite structures. Firstly, the impact signal at an unknown position is obtained using a unified linear sensor array. Secondly, the fast Hilbert Huang transform (HHT) with an optimized EEMD algorithm is introduced to extract intrinsic mode functions (IMFs) from impact signals. Then, all IMFs in the whole frequency domain are directly used as the input vector of the 2D-MUSIC model separately to locate the impact source. Experimental data collected from a cross-ply glass fiber reinforced composite plate are used to validate the proposed approach. The results show that the use of optimized EEMD and 2D-MUSIC is suitable for real-time impact localization of composite structures.

Keywords: optimized EEMD; 2D-MUSIC; composite structure; impact localization

---

1. Introduction

Composite structures have been increasingly used in various aircraft structures due to their superior stiffness and weight characteristics. However, composite structures are susceptible to low velocity impacts, which can cause internal damage that will lead to a significant reduction of the local structural strength [1]. Therefore, impact monitoring has become an important task of structural health monitoring (SHM) to ensure the safety of these applications.

Recently, sensor array signal processing-based methods have emerged as a set of new promising SHM methods. Yin et al. constructed two Z-shaped clusters at different positions for acoustic source localization in anisotropic plates [2]. Xiao et al. proposed an acoustic emission source localization approach based on beamforming with two uniform linear arrays [3]. He developed a method for localizing two acoustic emission sources simultaneously based on beamforming and singular value decomposition. Agarwal and Macháň proposed a statistical super-resolution technique called the multiple signal classification (MUSIC) algorithm published in Nature Communications, and demonstrated that the MUSIC algorithm had a comparable or better performance in comparison with other localization techniques [4]. Because it has such advantages, the MUSIC algorithm was developed for the impact monitoring of plate-like structures. By processing the whole sensor array outputs using the MUSIC algorithm, the transfer function and material properties of the structure are not needed [5]. Since the MUSIC algorithm belongs to an eigen-structure and sub-space approach, it can effectively extract the key features of signals received and successfully detect the source signal under a low signal-to-noise ratio [6]. Some developments have been published by applying this method to process the sensor array signals in the structural health monitoring area. Stepinski and Engholm [7] presented a far-field MUSIC algorithm-based method to estimate the direction of arrival (DOA) of single elastic waves on an aluminum plate using a uniform circular array. Yang et al. [8] employed the far-field MUSIC algorithm to estimate the DOA of one impact on an aluminum plate. Yuan et al. presented a single frequency component-based re-estimated MUSIC algorithm to reduce the localization error caused by the anisotropy of the complex composite structure [9]. To deal with the near-field monitoring problem, the authors in previous work proposed the near-field 2D-MUSIC algorithm-based impact localization method for a composite plate using Gabor wavelet transform [10,11]. However, in previous MUSIC-based impact location methods, the narrowband signals at a particular central frequency had to be extracted from the wideband Lamb waves induced by each impact, and the specific center frequency had to be obtained after carefully analyzing the impact signal, which is time consuming.

EMD is a new time–frequency analysis technique which can decompose the complicated signal into a set of complete and almost orthogonal components named intrinsic mode functions (IMFs) [12]. The IMFs represent the natural oscillatory mode embedded in the signal and work as the basis functions, which are determined by the signal itself [13]. Thus, it is a self-adaptive signal processing method that can be applied to a nonlinear and non-stationary process to decompose it into stationary signals. Generally, EMD cannot accurately extract fault features because of the mode mixing phenomena. Wu and Huang proposed a new ensemble EMD (EEMD) method to reduce mode mixing [14]. However, the EMD and its improved version EEMD are computation intensive methods, which are not suitable for on-line detection. A noise-assisted approach in conjunction with a multivariate empirical mode decomposition (MEMD) algorithm has been proposed for the computation of EMD, in order to produce localized frequency estimates at the accuracy level of instantaneous frequency [15–17]. Leo developed Bivariate EMD to detect defective areas in composite materials [18]. Wang et al. [19] proposed a fast HHT with an optimized EEMD algorithm to speed up the computational efficiency by 1000 times. They also proved that the computational complexity of the EMD is equivalent to fast Fourier transform (FFT). Hence, the optimized EEMD method might be a good choice to be applied to the real-time impact localization of composite structures.

This paper presents an improved approach that combines the optimized EEMD and 2D-MUSIC algorithm for the on-line impact localization on composite structures without selecting the center frequency. The layout of the paper is structured as follows: In Section 2, an optimized EEMD-based 2D-MUSIC method for impact monitoring is presented. An impact monitoring experiment on a cross-ply glass fiber reinforced composite plate is performed in Section 3 to verify the proposed method. Finally, the conclusion and future works are given in Section 4.

2. Optimized EEMD Based 2D-MUSIC Method

2.1. 2D-MUSIC Algorithm for Impact Localization

In this section, the 2D-MUSIC algorithm in [10,11] is briefly introduced. As seen in Figure 1, a uniform linear array (ULA) consists of M piezoelectric (PZT) sensors on the structure, which are arranged uniformly along the x axis and asymmetric with the y axis. The distance between two sensors is d.

$\mathbf{A}(r,\theta)$ is defined as the array steering vector

$$\mathbf{A}(r,\theta) = [a_1(r,\theta), a_2(r,\theta), \cdots, a_M(r,\theta)] \tag{1}$$

where

$$a_i(r,\theta) = \frac{r}{r_i}\exp(-j\omega_0\tau_i)$$

$$\tau_i = \frac{(-d\sin\theta)}{c_{AV}}(i-1) + \left(-\frac{d^2}{c_{AV}r}\cos^2\theta\right)(i-1)^2$$

r is defined as the distance between the impact source and the ULSA, which is the distance from the source to the reference sensor labeled as 1 in the sensor array. θ denotes the wave propagating direction caused by the impact with respect to the coordinate y axis. c_{AV} is the average velocity of the Lamb wave.

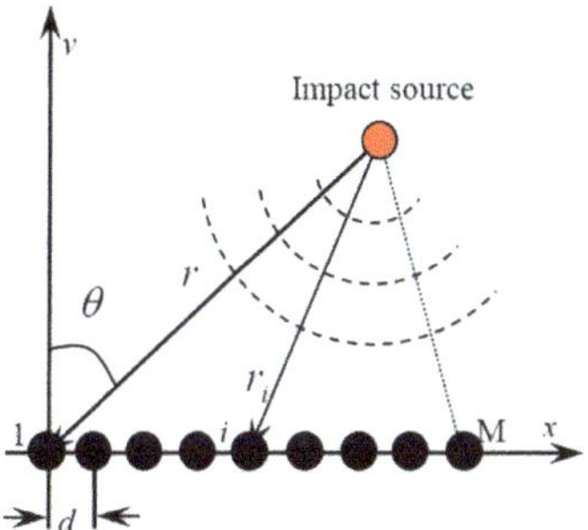

Figure 1. Near-field impact signal model.

The basic idea of the MUSIC algorithm is to obtain the signal subspace and noise subspace through eigenvalue decomposition of the signal covariance matrix, and then to estimate the signal parameter using the orthogonality of two spaces. To describe the orthogonality between the signal subspace and noise subspace, the spatial spectrum is used, which can be calculated by

$$\mathbf{P}_{\text{MUSIC}}(r,\theta) = \frac{1}{\mathbf{A}^H(r,\theta)\mathbf{U}_N\mathbf{U}_N^H\mathbf{A}(r,\theta)} \tag{2}$$

where $\mathbf{U}_N$ denotes the noise subspace spanned by the eigenvector matrix corresponding to those small eigenvalues. Based on Equation (2), by varying r and θ to realize a scanning process, $\mathbf{A}(r,\theta)$ is steered to scan the whole structure area. The peak point on the spatial spectrum corresponds to the impact source point. Both the distance and direction of the impact source can be obtained.

2.2. Fast EEMD for Impact Signal Extraction

In 1998, Huang et al. introduced the EMD, which is able to adaptively and effectively decompose a complicated signal into a collection of stationary IMFs [12]. Therefore, it has often been used in nonlinear and non-stationary signal processing. In the EMD method, the data $x(t)$ is decomposed in terms of IMFs c_j as

$$x(t) = \sum_{j=1}^{n} c_j + r_n \tag{3}$$

where r_n is the residual of data $x(t)$, after n number of IMFs are extracted. IMFs are simple oscillatory functions that have the following two properties:

1. Throughout the whole length of a single IMF, the number of extrema N_e (maxima and minima) and the number of zero-crossings N_z must either be equal or differ at most by one, i.e.,

$$(N_z - 1) \leq N_e \leq (N_z + 1) \tag{4}$$

2. At any time point t_i, the mean value of the envelope $f_{max}(t_i)$ and $f_{min}(t_i)$ respectively defined by the local maxima and the local minima are zero, i.e.,

$$[f_{max}(t_i) + f_{min}(t_i)]/2 = 0 t_i \in [t_a, t_b] \tag{5}$$

where $[t_a, t_b]$ is the time interval.

In practice, the EMD decomposition procedures (sifting procedure) are as follows:

1. Identify all the local maxima and minima and connect all of them using a cubic spline as the upper and lower envelope $f_{max}(t)$ and $f_{min}(t)$, respectively. Then, calculate the mean value $m(t)$ of $f_{max}(t)$ and $f_{min}(t)$ as

$$m(t) = [f_{max}(t) + f_{min}(t)]/2 \tag{6}$$

2. Obtain the first component h_1 by taking the difference between the data $x(t)$ and the local mean $m(t)$ as

$$h_1(t) = x(t) - m(t) \tag{7}$$

3. Treat $h_1(t)$ as the data and repeat steps 1 and 2 as many times as is required until the two properties of IMF as shown in Equations (4) and (5) are satisfied. Then, the final $h_1(t)$ is designated as an IMF $c_1(t)$.
4. Treat $r_i(t) = x(t) - c_i(t)$, $(i = 1, 2, \cdots, n-1)$ as the data and repeat steps 1–3. Finally, we obtain additional IMFs $c_2(t), c_3(t), \cdots c_n(t)$ and the final residual $r_n(t)$, which are represented by Equation (3).

Generally, the stop criterion of the sifting procedure is restrained by

$$S_d = \sum_{t=0}^{T} \frac{|h_{k-1}(t) - h_k(t)|^2}{h_k^2(t)} \tag{8}$$

where T is the signal length; $h_{k-1}(t)$ and $h_k(t)$ are the neighbor components in sifting procedures for one IMF; and S_d is the standard deviation, which is suggested to be 0.2–0.3.

However, mode mixing appears to be the most significant drawback of EMD [19]. Therefore, in 2009, a new artificial noise-excited EMD method was proposed by Wu and Huang, which was called EEMD [19]. The procedures are similar to EMD, except only one series of white noise with a finite amplitude are added into the original signals and the procedures are summarized as follows:

1. Add a white noise $n_i(t)$ series (noise level is N_l) to the targeted data and decompose the data with added white noise into IMFs as

$$x(t) + n_i(t) = \sum_{j=1}^{n} c_j^i(t) + r_n^i(t) \tag{9}$$

where $i = 1, 2, \cdots, q$ and q is the average time (ensemble number).

2. Repeat step1 q times with different white noise series $n_i(t)$.
3. Obtain the (ensemble) means of corresponding IMFs of the decompositions as the final result, that is

$$x(t) + n_i(t) = \frac{1}{q}\sum_{i=1}^{q}\sum_{j=1}^{n} c_j^i(t) + \frac{1}{q}\sum_{i=1}^{q} r_n^i(t) = \sum_{j=1}^{n} d_j(t) + r_n(t) \tag{10}$$

where $d_j(t)$ is the jth IMFs of EEMD decomposition as

$$d_j(t) = \frac{1}{q} \sum_{i=1}^{q} c_j^i(t) \tag{11}$$

and $r_n(t)$ is the final residual of EEMD decomposition as

$$r_n(t) = \frac{1}{q} \sum_{i=1}^{q} r_n^i(t) \tag{12}$$

2.3. Impact Localiaztion Process

The implementing process of the impact location method based on the proposed method is shown in Figure 2. The fast HHT with an optimized EEMD algorithm is introduced to extract IMFs from impact signals. The spatial spectrum $\mathbf{P}_{\text{MUSIC}}$ of each IMF is evaluated versus all (r, θ) in the $r - \theta$ plane to find the steering vector $\mathbf{A}(r, \theta)$ which matches the actual impact signal vectors. When the right steering vector is found, the denominator of Equation (2) approaches zero due to the orthogonal properties, resulting in a peak in the spatial spectrum which corresponds to the estimated impact position.

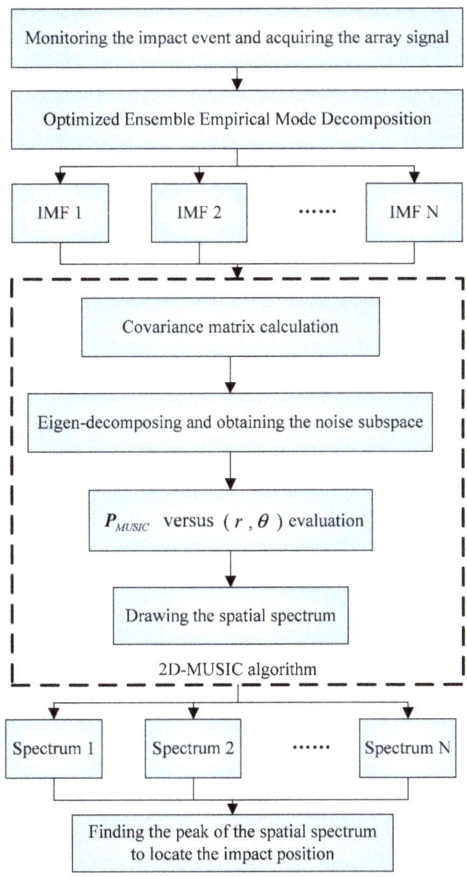

Figure 2. The impact localization process.

3. Experimental Investigations

3.1. Experiment Setup

The experiment setup is shown in Figure 3, including a 600 mm × 600 mm × 2 mm cross-ply glass fiber reinforced composite plate with unknown material properties. The thickness of each ply is 0.125 mm and the ply sequence is $[0_2/90_4/0_2]_S$. The integrated structural health monitoring scanning system (ISHMS) is adopted as the monitoring system. This system is developed to control the excitation and sensing of the PZT sensor array. As seen in Figure 3, the array used in the experiment is a ULSA bonded on the structure surface with seven PZT sensors. The diameter of the PZT sensor is 8 mm. These sensors are arranged with a space of 10 mm and are labeled as PZT1–7, respectively, from the left to the right, where the length of the sensor array L is 60 mm.

Ten low-velocity impacts with a 2 J energy are performed at various positions on the structure shown in Figure 3. These impacts are induced by an impact hammer. The sampling rate is set to be 2 MHz and the trigger voltage is set to 3 V in the experiments. The sampling length is 10,000, including 2000 pre-trigger samples.

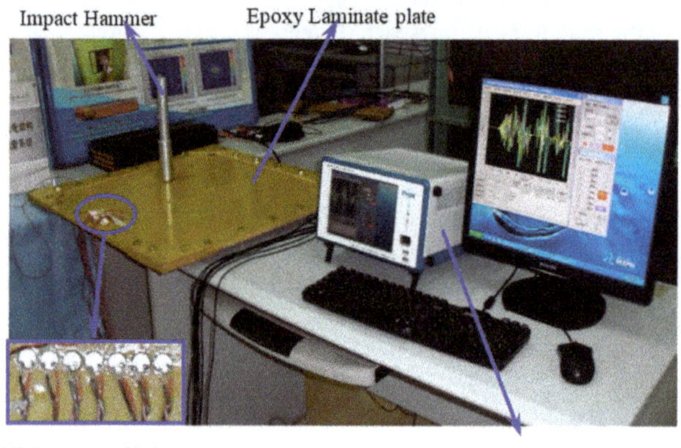

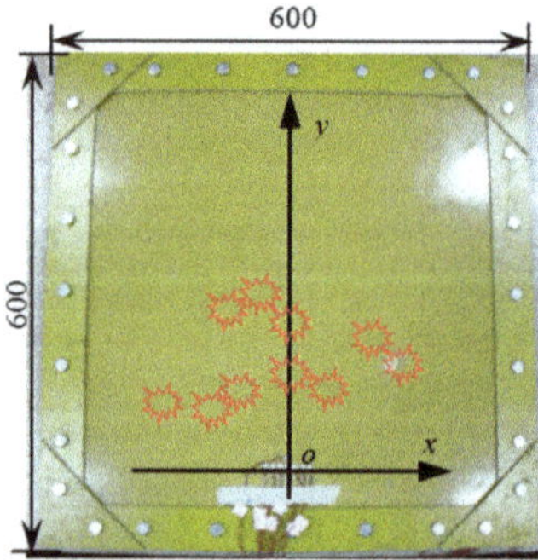

Figure 3. Experiment setup, ULSA arrangement, and impact applied positions (mm).

3.2. Impact Localization Results

A near-field impact at the point (200 mm, 90°) is chosen as a typical case to be analyzed first. The impact signals extracted from each sensor of each IMF are shown in Figure 4. From Figure 4, the wave fronts cannot been seen in the high frequency region, including IMF1, IMF2, and IMF3.

Additionally, much more noise signal, boundary reflection, and pattern aliasing appear in the IMF5, IMF6, IMF7, and IMF8. Only the wave fronts of impact signals can be easily found in the IMF4.

All the IMFs extracted from the impact signals are used to form the input vector of the near-field 2D-MUSIC model. The measured average velocity on this plate is 1154 m/s [11]. According to the structure dimension, the scanned area is set to be the distance from 0 to 450 mm and the direction from $0°$ to $180°$, and the scanning step length of the distance and direction is 1 mm and $1°$, respectively. The near-field impact obtained at the point (200 mm, $90°$) spatial spectrum $\mathbf{P}_{\mathrm{MUSIC}}(r, \theta)$ is shown in Figure 5. The figure represents the spatial spectrum magnitudes of each scanned point (r, θ), and the highest pixel point of the figure represents the impact point localized by the presented 2D-MUSIC algorithm. Seen from Figure 5, the spatial spectrum of IMF4 located the impact source, and the direction and distance error are $0°$ and 0.4 cm, respectively.

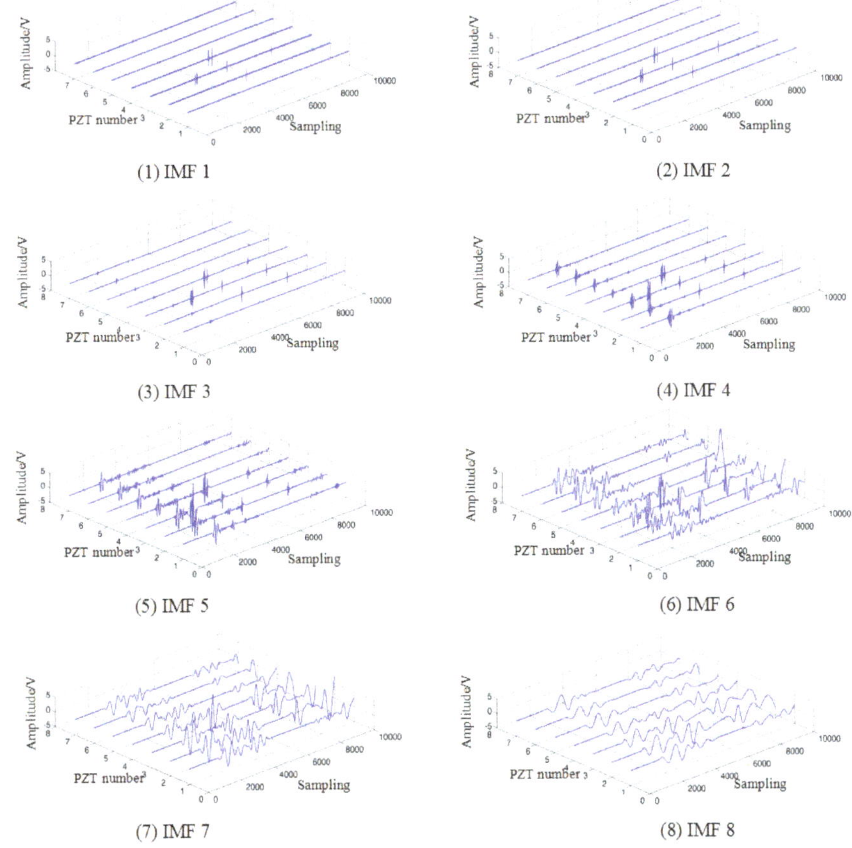

Figure 4. The IMFs using the optimized EEMD to decompose the impact array signals.

Ten estimated results of impact points and the errors compared with actual impact points are listed in Table 1, including seven located results at the spatial spectrum of IMF4, and three located results at the spatial spectrum of IMF5. It shows that ten near-field impacts are in good agreement with the actual impacts. The maximum error is at the impact position (255 mm, $150°$), whose direction and distance error are $3°$ and 2 cm, respectively.

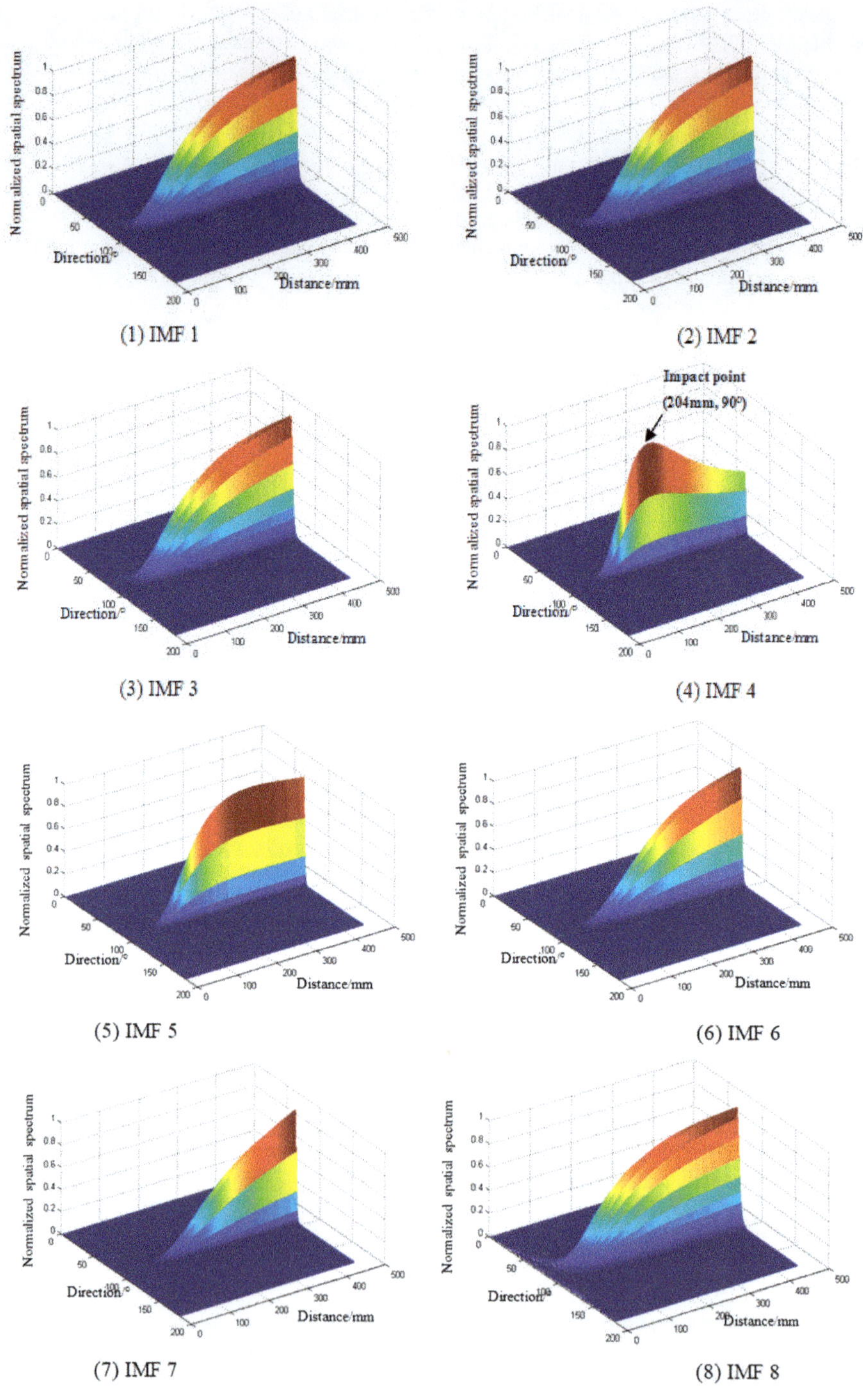

Figure 5. The spatial spectrums of each IMF estimated by the 2D-MUSIC model.

Table 1. The localization results.

Impacts	Parameter	Actual Positions		Estimated Positions		Errors	
		r/mm	θ/°	IMF4	IMF5	*r*/mm	θ/°
1		150	45	(135,43)	-	15	2
2		200	56	(208,54)	-	8	2
3		124	75	(109,76)	-	15	1
4		150	90	(155,90)	-	5	0
5		200	90	(194,90)	-	6	0
6		255	100	-	(242,100)	13	0
7		255	110	-	(245,112)	10	2
8		150	124	(140,123)	-	10	1
9		150	135	(146,133)	-	4	2
10		255	150	-	(235,147)	20	3

4. Conclusions

This paper presents an improved approach that combines the EEMD and 2D-MUSIC algorithm for the real-time impact localization on composite structures. The fast Hilbert Huang transform with an optimized EEMD algorithm is introduced to extract IMFs from impact signals in the whole frequency domain. Then, all IMFs in the whole frequency domain are directly used as the input vector of the 2D-MUSIC model separately to locate the impact source. The main advantage of the proposed method is its computational efficiency, which requires less time in comparison to the previous MUSIC-based impact location method, in order to achieve the same location accuracy. From experimental results on a cross-ply glass fiber reinforced composite plate, the wave fronts of impact signals can be easily found in the IMF4 or IMF5, and the corresponding spatial spectrum peak also easily located the impact source. Ten near-field impacts are in good agreement with the actual impacts. The maximum direction and distance error are 3° and 2 cm, respectively. The experiment on a cross-ply glass fiber reinforced composite plate proved that the use of the 2D-MUSIC algorithm improved by EEMD is a suitable approach for the real-time impact localization of composite structures.

Author Contributions: Y.Z. and J.X. developed the method. Y.Z. and X.C. conceived of, designed the experiments. Y.J. and J.P. performed the experiments. Y.Z. wrote the paper.

Funding: This work is supported by the National Natural Science Foundation of China (No. 51505339, No. 51575400), the Zhejiang Provincial Natural Science Foundation of China (No. LQ16E050005), the Postdoctoral Foundation of Zhejiang Province, China.

Acknowledgments: The authors would like to acknowledge the Natural Science Foundation of China, the Zhejiang Provincial Natural Science Foundation of China, and the Postdoctoral Foundation of Zhejiang Province, China.

Conflicts of Interest: The authors declare no conflicts of interest.

References

1. Kersemans, M.; Martens, A.; Degrieck, J.; Abeele, K.V.D.; Delrue, S.; Pyl, L.; Zastavnik, F.; Sol, H.; Paepegem, W.V. The ultrasonic polar scan for composite characterization and damage assessment: Past, present and future. *Appl. Sci.* **2016**, *6*, 58. [CrossRef]

2. Yin, S.; Cui, Z.; Kundu, T. Acoustic source localization in anisotropic plates with "z" shaped sensor clusters. *Ultrasonics* **2018**, *84*, 34–37. [CrossRef] [PubMed]

3. Xiao, D.C.; He, T.; Pan, Q.; Liu, X.D.; Wang, J.; Shan, Y.C. A novel acoustic emission beamforming method with two uniform linear arrays on plate-like structures. *Ultrasonics* **2014**, *54*, 737–745. [CrossRef] [PubMed]

4. Agarwal, K.; Macháň, R. Multiple signal classification algorithm for super-resolution fluorescence microscopy. *Nat. Commun.* **2016**, *7*, 13752. [CrossRef] [PubMed]

5. Schmidt, R.O. Multiple emitter location and signal parameter estimation. *IEEE Trans. Antennas Propag.* **1986**, *34*, 276–280. [CrossRef]

6. Fan, Y.B.; Gu, F.S.; Ball, A. Acoustic emission monitoring of mechanical seals using MUSIC algorithm based on higher order statistics. *Key Eng. Mater.* **2009**, *413*, 811–816. [CrossRef]

7. Engholm, M.; Stepinski, T. Direction of arrival estimation of Lamb waves using circular arrays. *Struct. Health Monit.* **2011**, *10*, 467–480. [CrossRef]

8. Yang, H.; Lee, Y.J.; Lee, S.K. Impact source localization in plate utilizing multiple signal classification. *Proc. Inst. Mech. Eng. Part C J. Mech. Eng. Sci.* **2013**, *227*, 703–713. [CrossRef]

9. Yuan, S.F.; Bao, Q.; Qiu, L.; Zhong, Y.T. A single frequency component-based re-estimated music algorithm for impact localization on complex composite structures. *Smart Mater. Struct.* **2015**, *24*, 105021. [CrossRef]

10. Zhong, Y.T.; Yuan, S.F.; Qiu, L. Multiple damage detection on aircraft composite structures using near-field MUSIC algorithm. *Sens. Actuators A* **2014**, *214*, 234–244. [CrossRef]

11. Yuan, S.F.; Zhong, Y.T.; Qiu, L.; Wang, Z.L. Two-dimensional near-field multiple signal classification algorithm–based impact localization. *J. Intell. Mater. Syst. Struct.* **2015**, *26*, 400–413. [CrossRef]

12. Huang, N.E.; Shen, Z.; Long, S.R.; Wu, M.C.; Shih, H.H.; Zheng, Q.; Yen, N.C.; Tung, C.C.; Liu, H.H. The empirical mode decomposition and the Hilbert spectrum for nonlinear and non-stationary time series analysis. *Proc. R. Soc. A Math. Phys. Eng. Sci.* **1998**, *454*, 903–995. [CrossRef]

13. Wang, Y.X.; He, Z.J.; Zi, Y.Y. A comparative study on the local mean decomposition and empirical mode decomposition and their applications to rotating machinery health diagnosis. *J. Vib. Acoust. Trans.* **2010**, *132*, 021010. [CrossRef]

14. Wu, Z.; Huang, N.E. Ensemble empirical mode decomposition: A noise assisted data analysis method. *Adv. Adapt. Data Anal.* **2009**, *1*, 1–41. [CrossRef]

15. Looney, D.; Hemakom, A.; Mandic, D.P. Intrinsic multi-scale analysis: A multi-variate empirical mode decomposition framework. *Proc. Math. Phys. Eng. Sci.* **2015**, *471*, 20140709. [CrossRef] [PubMed]

16. Rehman, N.U.; Park, C.; Huang, N.E.; Mandic, D.P. EMD via MEMD: Multivariate noise-aided computation of standard EMD. *Adv. Adapt. Data Anal.* **2013**, *5*, 1350007. [CrossRef]

17. Mandic, D.P.; Rehman, N.U.; Wu, Z.; Mandic, D.P. Empirical Mode Decomposition-Based Time-Frequency Analysis of Multivariate Signals: The Power of Adaptive Data Analysis. *IEEE Signal Process. Mag.* **2013**, *30*, 74–86. [CrossRef]

18. Leo, M.; Looney, D.; D'Orazio, T.; Mandic, D.P. Identification of Defective Areas in Composite Materials by Bivariate EMD Analysis of Ultrasound. *IEEE Trans. Instrum. Meas.* **2011**, *61*, 221–232. [CrossRef]

19. Wang, Y.H.; Yeh, C.H.; Young, H.W.V.; Hu, K.; Lo, M.T. On the computational complexity of the empirical mode decomposition algorithm. *Phys. A Stat. Mech. Appl.* **2014**, *400*, 159–167. [CrossRef]

*applied
sciences*

Article

Structural Reliability Prediction Using Acoustic Emission-Based Modeling of Fatigue Crack Growth

Azadeh Keshtgar *, Christine M. Sauerbrunn and Mohammad Modarres

Center for Risk and Reliability, University of Maryland, College Park, MD 20742, USA;
csauerbr@terpmail.umd.edu (C.M.S.); modarres@umd.edu (M.M.)
* Correspondence: keshtgar@umd.edu; Tel.: +1-301-789-3106

Received: 29 May 2018; Accepted: 23 July 2018; Published: 25 July 2018

Abstract: In this paper, AE signals collected during fatigue crack-growth of aluminum and titanium alloys (Al7075-T6 and Ti-6Al-4V) were analyzed and compared. Both the aluminum and titanium alloys used in this study are prevalent materials in aerospace structures, which prompted this current investigation. The effect of different loading conditions and loading frequencies on a proposed AE-based crack-growth model were studied. The results suggest that the linear model used to relate AE and crack growth is independent of the loading condition and loading frequency. Also, the model initially developed for the aluminum alloy proves to hold true for the titanium alloy while, as expected, the model parameters are material dependent. The model parameters and their distributions were estimated using a Bayesian regression technique. The proposed model was developed and validated based on post processing and Bayesian analysis of experimental data.

Keywords: reliability; acoustic emission; structural integrity; crack growth; fatigue life prediction; uncertainty analysis; nondestructive testing

---

1. Introduction

Acoustic emission (AE) technology has the potential for on-line structural health monitoring; a desired procedure for evaluating material degradation in aircrafts. Acoustic emissions are stress waves that propagate through a material as a result of applied stresses. When a material is subjected to cyclic fatigue loading, AE signals may be generated frequently with cracking within the material. These waves can be detected by piezoelectric sensors when placed on the surface of the material. The characteristics of the AE signal are determined by the mechanism that generated the signal, and the means by which it travels through the material and the sensor that transforms the emission into the signal [1].

The most commonly used AE feature for fatigue is the AE counts, which is defined as the number of times that an AE signal amplitude exceeds a predefined subjective threshold value. Other common characteristics used to describe features of an AE hit include the signal's peak amplitude, risetime, and duration. The amplitude is defined as the maximum voltage value of the signal often reported in decibels, the risetime is the time between the first count and the count with peak amplitude, and the duration is the time between the first and last AE count. These features are illustrated in Figure 1, which shows a typical waveform from an AE hit with the associated features.

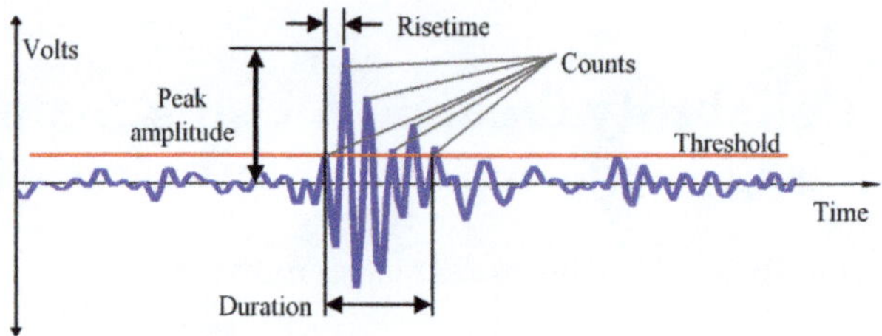

Figure 1. Characteristics of AE Signal.

Besides relating features from the AE signals to fatigue crack growth, some researchers supplement the AE signals with other crack growth indicators. For example, surface temperature mapping known as thermography has been used to enhance detection of crack initiation and growth. The driving factor for this development is that in addition to the release and propagation of stress waves that are detectable by AE sensors during fatigue crack growth, small amounts of thermal energy are also dissipated. Recent promising researches [2–5], built on the previous pioneering works on infrared t1hermography and AE signals, have also quantified fatigue crack growth. This paper, however, uses features of AE signals alone to describe fatigue crack growth behavior.

Mechanical structures typically operate under a wide range of loading scenarios. Previously proposed fatigue life models based on the AE signal properties reported by Talebzadeh and Roberts, [6]; Bianocolini [7]; and later by Rabiei & Modarres [8] have not been validated with respect to different loading conditions. These studies were also focused only on studying aluminum alloys. Different material and loading features, such as frequency and loading ratio, may also affect model predictions and can be verified through specially designed experiments.

Several attempts have been made to relate different AE parameters such as AE count, energy, and amplitude to fatigue crack growth, stress intensity factor range (ΔK), maximum stress intensity factor (K_{max}) and crack growth rates [9–13]. These studies showed that the relationship between the stress intensity factor range and the number of AE counts can be captured by an equation with a form similar to the Paris-Erdogan equation [14–16]. A leading general model that relates the AE count rate to the crack growth rate was proposed by Bianocolini [7], and has the following form:

$$\frac{dc}{dN} = \alpha_2 \left(\frac{da}{dN} \right)^{\alpha_1} \tag{1}$$

where, c is the AE count, a represents crack size, N is the load cycle, (da/dN) is the crack growth rate, (dc/dN) is the AE count rate, and α_1 and α_2 are the model parameters. Rabiei and Modarres [11] used a variation of Equation (1) in the form of the linear regression in Equation (2):

$$\log\left(\frac{da}{dN} \right) = \alpha_1 \log\left(\frac{dc}{dN} \right) + \alpha_2 + \varepsilon \tag{2}$$

where, the error term, ε, in Equation (2) accounts for the difference between the model prediction and the observed AE count rate. The model described by Equation (2) assumes that a small crack may be difficult to measure, and as the crack becomes larger, the measurement of crack length becomes more accurate [17]. In order to capture any changes in the error distribution, it was assumed that the error follows a normal probability density function with zero mean and a standard deviation, s (Equation (3)).

Based on a single experiment, Rabiei [17] concluded that s was independent of crack growth rate. To capture the dependence of the crack length on s, Rabiei [17] assumed that s follows a two-parameter exponential distribution that changes as a function of AE count rates. However, these assumptions were not supported by the results of this research. The experimental results of this study suggested that s follows a normal distribution and is not a function of AE count rates; $s \approx N(\mu_s, \sigma_s)$.

$$\varepsilon \approx N(0, s), \ s \approx N(\mu_s, \sigma_s) \tag{3}$$

The significance of the proposed model represented by Equations (2) and (3) is that once the model parameters are estimated using experimental data, this equation can be used to estimate crack growth rates of structures by monitoring AE signals and extracting the AE count rate from the observed signals.

Rabiei's model [17] was derived based on the results from one experiment with sinusoidal loading conditions of a maximum load of P_{max} = 300 lbf (1334 kN), loading ratio of R = 0.1, and loading frequency of f = 10 Hz. The results have not been validated with respect to different loading conditions and were limited to one experiment on Al7075-T6. Therefore, the first step in this research was to study the consistency and variability of the model by performing several standard fatigue experiments using compact tension test specimens of the same aluminum alloy (Al7075-T6) subjected to cyclic loading with varied loading ratios and frequencies. This part of the research addressed model development and validation, with respect to changes in loading frequency and loading ratio, through statistical analysis of the crack growth data. In the second step, the material effect was studied based on three identical crack growth tests with Ti-6Al-4V titanium alloy (known as Ti-6-4). Experimental procedures and methods of analyzing the data were similar to those used for the Al7075-T6 samples.

This paper first discusses the details of the performed fatigue experiments, followed by the description of the data analysis methods, and finally presents results including model development and validation. In the final section, conclusions are summarized and suggestions are made for future work.

2. Experimental Approach

The goal of the experiments performed was to monitor fatigue crack propagation in standardized specimens. The test samples for both Al7075-T6 and Ti-6-4 were supplied in the form of compact tension (CT) specimens, based on ASTM standard E647 [18]. The test specimens were manufactured from 0.125 in (3.175 mm) thick plates. The geometry and dimensions of the specimens are shown in Figure 2.

A servo-hydraulic Material Testing System (MTS) fatigue machine retrofitted with an Instron 8800 controller was used to define and control the loading conditions. An advanced DiSP-4 AE system supplied by the Physical Acoustics Corporation (now MISTRAS Group), was used to record the AE signals. The setup included one 100–900 kHz wideband differential AE sensor (WD) to collect the signal, a 40 dB preamplifier to amplify them, a data acquisition module to perform primary filtration and record the signals, and a software module to visualize the data and to perform feature extraction.

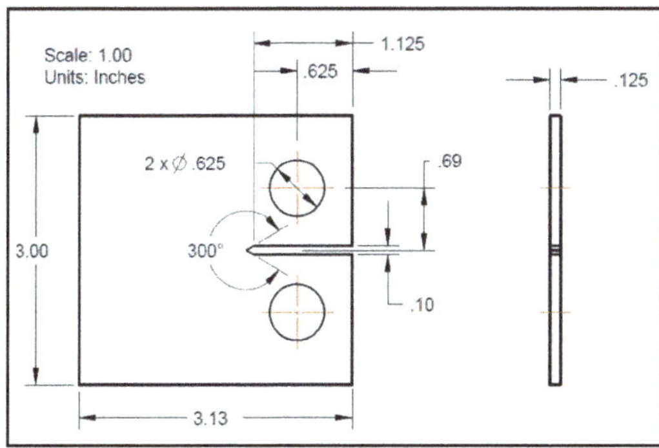

Figure 2. Technical drawing of the CT specimen (dimensions in inches).

A ruler with smallest divisions of 0.01 in. (0.254 mm) was placed on the specimen so the actual crack length could be measured during post-processing of the images. The AE sensor was attached to the upper left-hand side of the specimen with silicone grease and held in place with a C-clamp (Figure 3). The instrumented CT specimens were mounted on the 5 kip MTS machine which applied sinusoidal loading cycles until a fatigue pre-crack of adequate length and straightness (according to ASTM E647) could be detected. Then, the standard crack growth tests were conducted using the same MTS machine. All the experiments were continued until fracture.

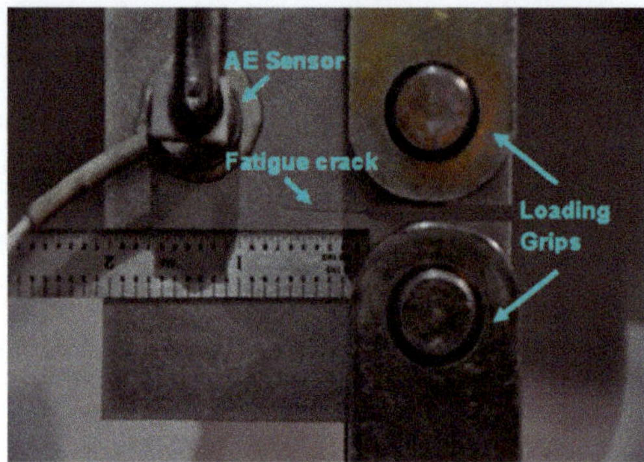

Figure 3. Standard CT specimen with a mounted AE sensor.

A 40 dB preamplifier amplified the AE signals received from the sensor. A band pass filter was used in the amplifier, and the amplified signals were analyzed using the DiSP-4 system. The AE features that were measured and recorded included AE counts, energy, and the time of the event. A digital close-up camera took time-lapsed high-resolution pictures of the crack growth, and the crack length could be measured based on the ruler attached onto the specimen within the accuracy of 0.01 inch (0.254 mm). As a result, considerable amounts of data were captured and stored for further processing.

Experiments performed on several CT specimens were subjected to cyclic loading with different loading ratios. For each case, the loading ratio was changed, with all other test parameters remaining constant. Table 1 lists the experiments performed for the corresponding load ratios:

Table 1. Details of experiments and load parameters.

Test Reference	Material	Loading Frequency (Hz)	Loading Ratio	Maximum Force (lbf)	Maximum Force (kN)
CT1	Al7075-T6	10	0.1	500	2.22
CT2	Al7075-T6	10	0.1	500	2.22
CT3	Al7075-T6	10	0.3	500	2.22
CT4	Al7075-T6	10	0.3	500	2.22
CT5	Al7075-T6	10	0.5	500	2.22
CT6	Al7075-T6	10	0.5	500	2.22
CT7	Al7075-T6	2	0.1	500	2.22
CT8	Al7075-T6	2	0.1	500	2.22
CT9	Al7075-T6	7	0.1	500	2.22
CT10	Al7075-T6	7	0.1	500	2.22
CT11	Al7075-T6	10	0.1	500	2.22
CT12	Al7075-T6	10	0.1	500	2.22
CT13	Ti-6Al-4V	5	0.1	900	4
CT14	Ti-6Al-4V	5	0.1	900	4
CT15	Ti-6Al-4V	5	0.1	900	4

For the three Ti-6-4 tests, the loading was increased because Ti-6-4 has a higher yield strength of about 120 ksi (830 MPa) compared to the yield strength of Al7075-T6 of about 67 ksi (460 MPa) [19].

Additionally, the frequency was decreased to 5 Hz in order to better distinguish and process AE events during post-processing [20].

Noise Reduction

AE signals may be generated from a number of possible sources including background noise, micro-crack generation, or plastic deformation. In order to reduce uncertainties and determine the AE signals corresponding to crack growth, applying noise reduction techniques on the captured data was required. Various de-noising techniques have been proposed to filter AE signals due to extraneous events during crack growth [10,21,22].

To filter out noise from AE signals associated with fatigue crack propagation, the recorded AE data was filtered using the acquisition system's band pass filter (200 kHz–3 MHz). In addition to the band pass filter that restricts the captured signals based on frequency, an amplitude threshold was determined and filtered out low-amplitude background noise. It should be noted that the lack of significant AE activity in the initial stages of fatigue loading makes it more difficult to distinguish background noise from crack-related acoustic events. For each set of materials, a dummy specimen was tested beforehand under the same conditions as the main experiments to determine the background noise. The dummy specimen was simply the first sample from the test specimen batch and had the same geometry and characteristics. The dummy specimen was installed in the testing machine, and the background noise was captured as a cyclic load was applied for a few cycles. Based on this test, the AE detection threshold was set to 45 dB for Al7075-T6 samples to eliminate the background noises. This value was found to be 35 dB for the Ti-6-4 samples.

It has also been observed that AE events occurring during the loading portion of a cycle are related to crack growth [6,17,21]. Therefore, the AE data taken only during the loading portion of each cycle were used for data analysis, while AE events during the unloading portion of the cycle were ignored. In addition, many researchers have assumed that only events occurring close to the maximum or peak load are associated directly with crack growth [12,21,22]. So, the filtered AE events were separated for different percentages of the applied load range, and it was determined that the AE counts occurring within the top 20% of peak load showed the closest correlation with crack propagation rates [23].

3. Analysis of Experimental Data and Results

3.1. Crack Growth Measurement

The lengths of the pictured cracks were measured using an image processing software called ImageJ [24]. Crack measurement with the image processing software was calibrated using the scale ruler attached to the specimen. The accuracy of the ruler was 0.01 in. (0.254 mm); therefore, the scale error was estimated as ± 0.005 in. (0.127 mm). One example of the process for measuring the crack length and matching the crack length with the cumulative AE counts is shown in Figure 4. By knowing the crack length and AE cumulative counts at numerous instances during the experiment, values for crack growth rate and AE count rate were estimated.

When the crack lengths were determined, the fatigue crack growth rates were approximated at different cycles using Equation (4):

$$\frac{da}{dN} = lim_{\Delta N \to 0} \frac{\Delta a}{\Delta N} \tag{4}$$

According to the observed data, the correlation between the AE count rates and stable crack growth rates follow the log-linear model of Equation (2). To properly compare the correlations estimated from all observed data obtained from different fatigue loading conditions on a similar scale, the correlation between the logarithm of crack growth rate and the normalized logarithm of dc/dN values was found. That is, the logarithm of every count rate was normalized to the range of the logarithms of minimum and maximum count rates observed in each test. With this normalization, the estimated parameters of Equation (2) were nearly identical for each test regardless of the loading ratio or loading frequency and thus allowed all tests data to be combined to estimate one correlation model, as discussed in Section 3.3. Therefore, every dc/dN data point can be reversely expressed in terms of the corresponding normalized value and the two fixed minimum and maximum $log(dc/dN)$ values. The correlation between normalized $log(dc/dN)$ and $log(da/dN)$ for different experiments at different loading ratios is shown in Figure 5a. Similarly, results for tests at different loading frequencies is shown in Figure 5b.

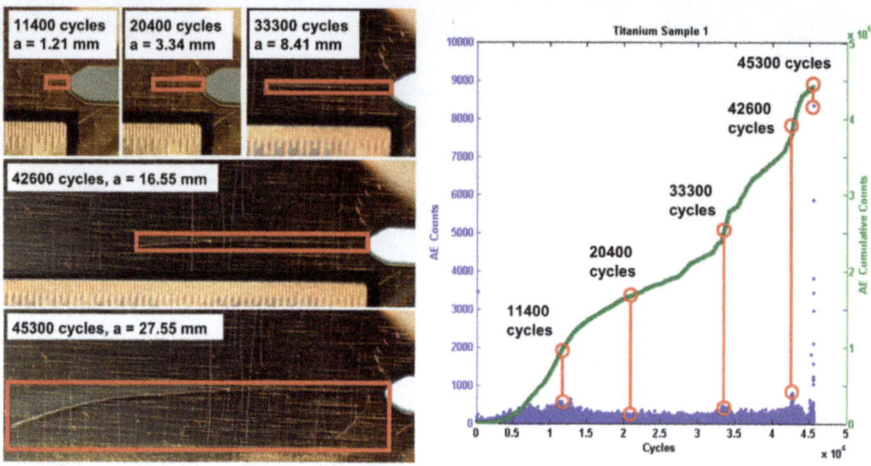

Figure 4. Example of crack length measurements paired with cumulative AE counts.

After the post processing of the AE data, the experimental results were used to determine the point estimate unknown parameters α_1 and α_2 in Equation (2) using regression analysis. The observed correlation of AE count rates versus crack growth rates and the regression analysis of the experimental data corresponding with different frequencies and different loading ratios showed that estimated parameters of the regression line are in the same range and do not have a considerable difference. It should be noted that the mean value of the regression parameters are calculated using all the experimental data and the estimation of the error term was implemented later in the probabilistic model development section. The point estimates of the linear regression model parameters for the different experiments are listed in Table 2. It should be noted that the fatigue lives (cycles-to-fracture) for Al7075-T6 samples were in the range of 3500 and 12,000 cycles. This range was determined to be 45,000 to 109,000 cycles for the Ti-6-4 samples.

Table 2. Regression parameters for individual test data, Al7075-T6.

Test	Frequency	R	α_1	α_2
CT1	10	0.1	0.03	-11.46
CT2	10	0.1	0.03	-11.17
CT3	10	0.3	0.03	-11.09
CT4	10	0.3	0.02	-12.01
CT5	10	0.5	0.03	-11.97
CT6	10	0.5	0.02	-11.77
CT7	2	0.1	0.02	-10.88
CT8	2	0.1	0.05	-13.86
CT9	7	0.1	0.02	-10.86
CT10	7	0.1	0.02	-10.74
CT11	10	0.1	0.04	-12.91
CT12	10	0.1	0.04	-13.41

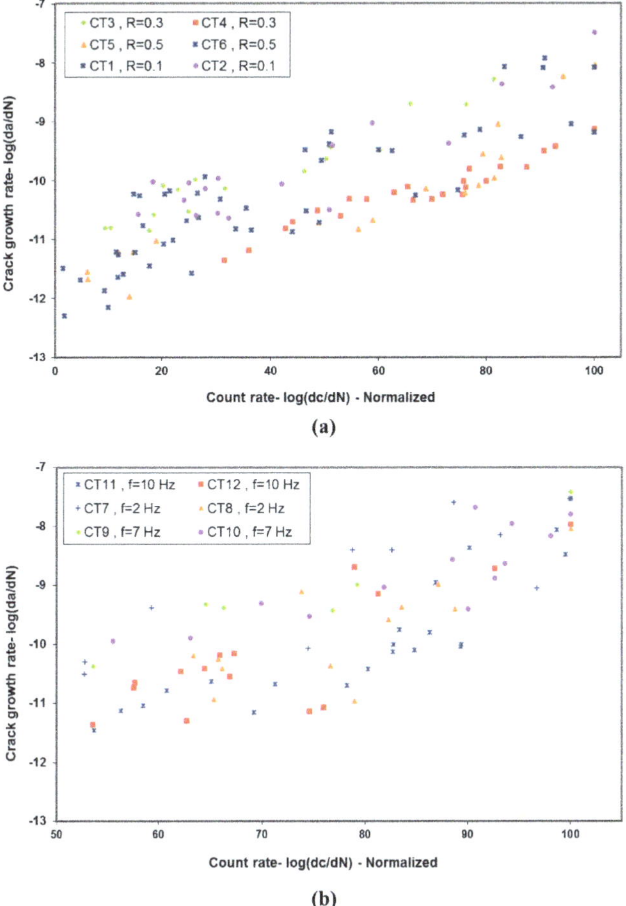

Figure 5. Crack growth rate versus AE count rate for Al7075-T6 tests at (**a**) different loading ratios; (**b**) different loading frequencies—log scale. The base scale of *da/dN* is in/cycle.

3.2. Test of Homogeneity

To determine if there is any significant statistical difference in the regression lines derived from the different test data, an Analysis of Covariance (ANCOVA) was used [25]. The topic of interest is whether the regression lines have similar slopes. This method allows for testing the null hypothesis that the regression model parameters were derived from samples estimating populations which all had equal slopes.

For this hypothesis analysis, the confidence limit level was selected to be 90% (i.e., $P_{critical} = 0.05$). The F statistic was calculated as 0.82 and the corresponding probability (*p*-value) was estimated to be 0.618. Since F statistic is less than unity and *p*-value is greater than $P_{critical}$, the null hypothesis could not be rejected. It was concluded that there is no statistically significant difference between the slopes of regression lines and the test of ANCOVA confirmed that the model is not influenced by the loading ratio and loading frequencies. For more information about ANCOVA, refer to References [26,27].

3.3. Probabilistic Model Development

Since the results show that changes in certain loading conditions do not result in any significant influences on the linear model of AE count rate versus crack growth rate, all the experimental data from different tests could be used to develop the crack length vs. AE count rate model. To do this, the experimental data set for Al7075-T6 was divided into three different sets. One set was used for parameter estimation when defining the model, one set was used for evaluating the error term in the model, and the final data set was used for model validation. The first set was of four experiments that were selected from different loading conditions to arrive at a more generic model in the phase of model development. Those experiments were CT1, CT3, CT5 and CT7 from Table 1. The second set of data used to estimate the uncertainties and capture the error term in the model were CT2, CT4, CT6, CT10 and CT12 from Table 1. The last step was to validate the developed model. The experiments used for validation were CT9 and CT11 from Table 1. The procedure and results of model development and validation are reported in the remainder of this section.

3.3.1. Bayesian Data Analysis

A Bayesian regression approach was implemented to estimate the model parameters and error. This technique is used to estimate and update the posterior distribution of the unknown model parameters [28]. In the Bayesian inference, a prior probability distribution function (pdf) of the model parameters is combined with observed data (evidence) in the form of a likelihood function of an unknown parameter (θ). The result is an updated state of knowledge in the form of the posterior joint distribution of the model parameters, $f(\theta \,|\, x)$. The posterior pdf of the model parameter can be assessed as described according to the Bayes' Theorem [29] as:

$$f(\theta|x) = \frac{f(x\,|\,\theta)}{f(x)} \propto f(x\,|\,\theta)f(\theta) \tag{5}$$

where, $f(\theta \,|\, x)$ is the posterior pdf of the vector of parameters (θ) *with* $\theta^T = [\alpha_1, \alpha_2, \sigma]$ given the observed data (x); $f(x)$ is the marginal pdf of the random variable x; $f(\theta)$ is the prior pdf of the model parameters; and $f(x\,|\,\theta)$ is the likelihood of the model and contains the available information provided by the observed data:

$$f(\theta|x) = \prod_{i=1}^{n} f(x_i\,|\,\theta) \tag{6}$$

The results of the analyzed experimental data were used to develop a probabilistic linear model for the estimation of the crack growth rate as the dependent variable, and using the AE count rate as the independent variable. So, based on the observed correlation, the unknown parameters of the linear model described in Equation (2) are updated. An error term was added later to assess the model error.

After the first steps of analyzing the experimental data (crack growth measurement, AE data filtration, and AE count rate calculations) were completed, the results were used to estimate the marginal and joint distribution of the unknown parameters in Equations (2) and (3) using the Bayesian regression using Equations (5) and (6). The software package WinBUGS [28] was used to perform the Bayesian inference. In this Bayesian inference, the likelihood function in Equation (6) for the observed independent crack data points (x_i, y_i) can be expressed as a normal distribution:

$$p(D|\alpha_1, \alpha_2, s) = \prod_{i=1}^{n} \frac{1}{s\sqrt{2\pi}} exp\left(-\frac{1}{2}\left(\frac{y_i - (\alpha_1 x_i + \alpha_2)}{s} \right)^2 \right) \tag{7}$$

where $p(.)$ is the likelihood of all data in form of $x_i = log(dc/dN)_i$, $y_i = log(da/dN)_i$, and D is the data set of all pairs (x_i, y_i) for $i = 1$ to n of the tests.

3.3.2. Error and Uncertainty

Uncertainties associated with the model have various sources which may be grouped as follows:

(a) Aleatory uncertainty: Also known as inherent uncertainty, aleatory uncertainty is a natural randomness of a quantity such as uncertainty in the material features. Generally, there are different factors during manufacturing of a material that cause random variation of the material properties from experiment to experiment [30]. To minimize aleatory uncertainty, test data of specimens from the same material lot (test samples coming from the same sheet of Al7075-T6 or Ti-6-4) were used. It should be noted that this physical variation is inherent to experiments and cannot be eliminated.

(b) Epistemic uncertainties: This uncertainty is the result of limited information or incomplete information due to finite experimental data or a limited number of observed data points. The addition of experiments corresponding with each loading condition in the parameter estimation process can help reduce epistemic uncertainty bounds for the estimated model parameters. Two types of epistemic uncertainty are discussed below:

1. Measurement error: This is the error caused by imperfect measuring equipment and/or human observation errors. There are some inherent variations resulting from the method of observation. The image processing method used for measuring the specimen's crack length based on the high-resolution photography with the close-up lens carried some measurement errors due to difficulties in finding the crack tip in images.

Some methods were applied to reduce the crack length measurement error. If the measurement value is known to be constant from one measurement to another and if the measurement can be made several times, information about the uncertainty of the measuring method can be obtained. In this case, this type of uncertainty can be reduced through averaging. It was attempted in this research to reduce the measurement uncertainty using the measurement data produced by two different individual testers on each data point. Since these measurements are mutually independent the mean value of the measured crack length was used for each data point.

2. Modeling error: Besides the error of crack length measurement, filtration method and insufficient data, there is an important model uncertainty that must be captured. This model uncertainty is related to the formulation of the proposed probabilistic model. There might be some other sources of uncertainty that can be considered such as the uncertainty resulting from the de-noising technique. For example, the noise reduction method may filter out some crack growth related signals and contribute to the model uncertainty. Improvement can be made by exploring alternate methods of classification of AE-related data and filtration techniques.

In Section 3.3.2.1, the AE model error is estimated, which expresses the sum of aleatory and epistemic uncertainties described above. Later in Section 3.4, the model is validated and the uncertainties are estimated.

3.3.2.1. Model Error Estimation

After establishing a linear relationship between the explanatory variable, dc/dN, and the response variable, da/dN, the relationship can be used to make predictions of the next value of crack growth rate given the next value of AE count rate. Predicting future observations for certain dc/dN values is one of the main goals of this linear regression modeling. Better prediction can be made considering the uncertainties and errors in the model. True values for model parameters are unknown and using estimated values in the prediction adds to the uncertainty that should be captured by the error term. The error term showed by ε in Equation (2) includes all the uncertainties discussed in the previous section and follows a normal distribution with the mean of zero and standard deviation that needs to be captured (s). The results of this study showed that parameter s of the error term can be described by another normal distribution: $s \sim N(\mu_s, \sigma_s)$.

The point estimates of the model parameters, α_1 and α_2, and the model error, ε, described by Equations (2) and (3) were computed for Al7075-T6 and Ti-6-4 experiments separately. As discussed, the error term is considered to follow a normal distribution with the mean of zero and a standard deviation of s which accounts for the model uncertainties after the data were normalized. The model parameters and error term distributions were determined through Bayesian analysis for each set of experiments. The software WinBUGS was used to capture the error term distribution in the model in which uninformative, or uniform, prior distributions were chosen for all variables. The posterior distributions of the error term and the corresponding standard deviations were determined and are listed in Table 3.

Table 3. Regression parameters and error term distribution for Al7075-T6 and Ti-6Al-4V tests.

Material	α_1	α_2	$\varepsilon \sim N(\mu_\varepsilon, s)$	$s \sim N(\mu_s, \sigma_s)$	
			μ_ε	μ_s	σ_s
Al7075-T6	0.03	-11.319	0	0.86	0.057
Ti-6Al-4V	0.006	-4.971	0	0.24	0.010

The error estimation for Al7075-T6 tests was performed by comparing experimental measurements with estimated model predictions obtained through Bayesian analysis (using parameter estimation for Al7075-T6 in Table 3). Subsequently, it was found that for Al7075-T6, the parameter s of the error term follows a normal distribution as well and was estimated as $s \sim N(0.86, 0.057)$, where $\varepsilon \sim N(0, s)$. The results are shown in Figure 6.

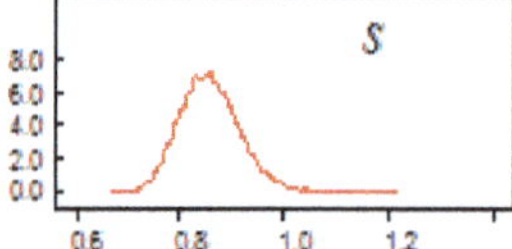
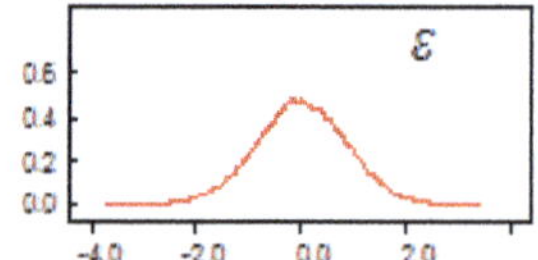

Figure 6. Estimated error terms.

The result of the posterior predictive distribution for da/dN as a function of dc/dN for Al7075-T6 is plotted in Figure 7. The posterior distribution is shown by its median and the 95% confidence level (2.5% and 97.5% prediction bounds). The data used to fit the model is also plotted in Figure 7.

This finding suggests that since almost all procedures were maintained between the two sets of experiments, the model parameters must be updated to account for material variation from material to material, as expected. In addition, both model parameters are material-dependent since both vary significantly between the material testing.

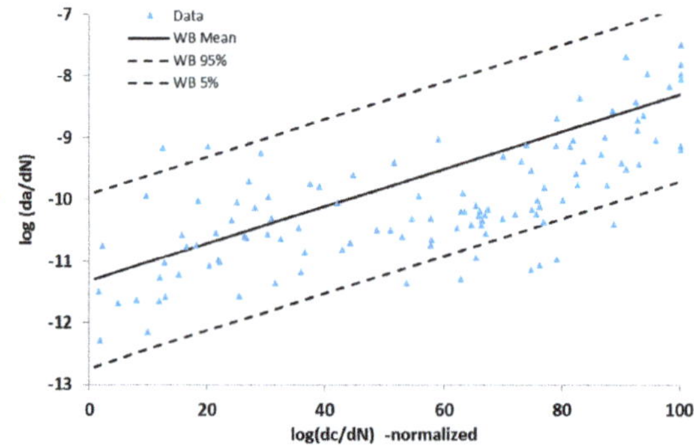

Figure 7. Posterior predictive AE model with the uncertainty bounds, material: Al7075-T6; WB indicates Bayesian regression analysis results using WinBugs.

Each of the three Ti-6-4 experiments demonstrated a good correlation between AE count rate and crack growth rate. These results suggest that the proposed linear model based on Al7075-T6 holds for the Ti-6-4 as well. The regression between $log(da/dN)$ and $log(dc/dN)$ for the experiments and confidence interval are depicted in Figure 8. As expected, the model parameters are considerably different between Al7075-T6 and Ti-6-4 testing, but the form of the model remains unchanged. To provide a visual comparison, the linear regressions between crack growth rate and AE count rate and the raw data for Al7075-T6 and Ti-6-4 are displayed in Figure 9.

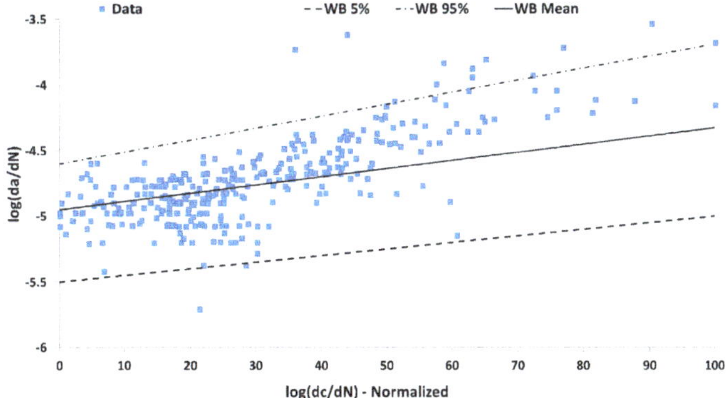

Figure 8. Posterior predictive AE model with the uncertainty bounds, material: Ti-6Al-4V, WB indicates Bayesian regression analysis results using WinBugs.

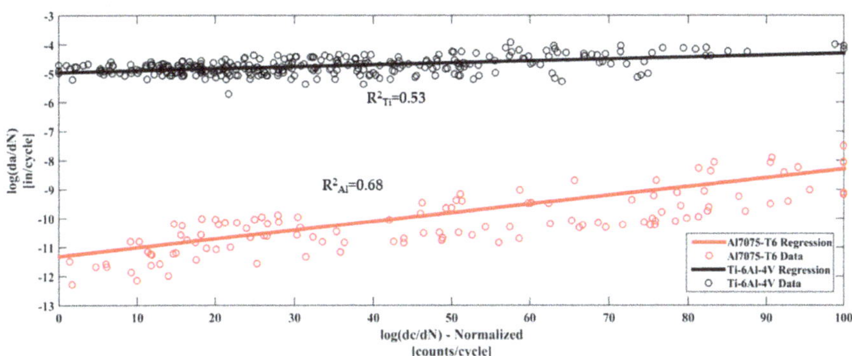

Figure 9. Crack growth rate versus AE count rate with linear regressions for Al7075-T6 and Ti-6Al-4V.

Some inferences about AE phenomena in different materials can be made by comparing the regression parameters listed in Table 3. The slope parameter, α_1, for the Ti-6-4 is smaller than the reported α_1 value for Al7075-T6. This relationship suggests that as the AE count rate increases, less increase is observed in the crack growth rate in the Ti-6-4 as compared to Al7075-T6. The intercept parameter, α_2, for the Ti-6-4 is larger than Al7075-T6. Physically, this means that a crack will generally grow at a faster rate in the Ti-6-4 than Al7075-T6 for the same relative AE count rate. In other words, as a crack propagates, higher crack growth rates are detected for the Ti-6-4 than Al7075-T6. The tradeoff when detecting crack growth through means of observing AE count rate is a crack will grow more quickly but at a relatively consistent rate for Ti-6-4, compared to a slower but relatively variable crack growth rate in Al7075-T6. Finally, it should be noted that the differences in the crack growth models are simplified into two dimensions, the two model parameters, α_1 and α_2. If the relationships between

$log(da/dN)$ and $log(dc/dN)$ for Ti-6-4 were not linear, at least one other material-dependent dimension would need to be considered.

3.4. Model Validation

In order to validate the developed AE model, model predictions of crack growth rate were compared against the validation experimental data set. For a given value of AE count rate, a prediction of crack growth rate was estimated based on the developed AE-based model. Available information captured from Al7075-T6 experimental data was used as the input to the Bayesian estimation procedure, as the model was developed based on Al7075-T6 data. The Bayesian estimation approach updated the model prediction with the experimental results. With this Bayesian estimation inference, uncertainties in the experimental values were propagated in the model and resulted in model prediction uncertainty assessment. The prediction results were then compared against the true crack growth rates obtained by validation experiments.

In the proposed model validation methodology [31], both model prediction and experimental results are considered to be estimations and representations of the true values, given some error as it is shown in Equations (8) and (9):

$$\frac{X_i}{X_{e,i}} = F_{e,i}; \; F_e \sim LN(b_e, s_e) \tag{8}$$

and

$$\frac{X_i}{X_{m,i}} = F_{m,i}; \; F_m \sim LN(b_m, s_m) \tag{9}$$

where X_i is the true value, $X_{e,i}$ and $X_{m,i}$ indicate the experimental results and model prediction, respectively. F_e is the multiplicative error of the experiment with respect to true value, and F_m is the multiplicative error of the model prediction with respect to true value. A multiplicative error of the experiment with respect to the model prediction is defined by Equation (10):

$$\frac{X_{e,i}}{X_{m,i}} = \frac{F_{m,i}}{F_{e,i}} = F_{t,i} \tag{10}$$

Since both $F_{m,i}$ and $F_{e,i}$ distributions are lognormal, the distribution of $F_{t,i}$ would also be lognormal with mean and standard deviation of $(b_m - b_e)$ and $\sqrt{s_m^2 + s_e^2}$, respectively. In this approach, the likelihood used is shown in Equation (11):

$$L(X_{e,i}/X_{m,i}, b_e, s_e | b_m, s_m) = \prod_{i=1}^{n} \frac{1}{\sqrt{2\pi}\left(\frac{X_{e,i}}{X_{m,i}}\right)\sqrt{s_m^2 + s_e^2}} exp\left(-\frac{1}{2} \times \frac{\left[ln\left(\frac{X_{e,i}}{X_{m,i}}\right) - (b_m - b_e)\right]^2}{s_m^2 + s_e^2}\right) \tag{11}$$

Using the validation sets of data, the validation approach was implemented and the results are discussed in this section. For more information about the validation method, refer to the paper by Ontiveros [31]. For simplicity, the distributions of model predictions were reduced to a mean value and compared one-to-one with the experimental results. The mean and standard deviation of F_e, which are b_e and s_e, were determined from the unbiased experimental error of $\pm 1\%$. The values determined were -0.00002 for b_e and 0.002 for s_e. The summary statistics for the marginal posterior *pdf* of parameters b_m and s_m as well as the distribution of F_m are presented in Table 4.

Table 4. Model validation statistic summary.

Parameter	Mean	Standard Deviation	2.5%	Median	97.5%
b_m	0.009	0.011	−0.013	0.009	0.031
s_m	0.059	0.008	0.045	0.058	0.078
F_m	1.011	0.061	0.894	1.009	1.141

The model uncertainty bounds for the crack length estimation can be determined from the 2.5 and 97.5 percentile of the multiplicative error of F_m. The resulting upper bound on reality was calculated as 14%, while the lower bound is -10%. These results are presented graphically in Figure 10. It can be noticed that the mean values of the model prediction are standing on the upper bounds of the experimental error, and the value of F_m for the model prediction is greater than 1. The value of F_m around 1.01 suggests a very small bias in the AE model to under predict the true crack growth rate. Therefore, the estimation of the true crack growth rate given by the model prediction is expected to be slightly higher. The results show that to correct the model, the predictions must be increased by a factor of 1.01.

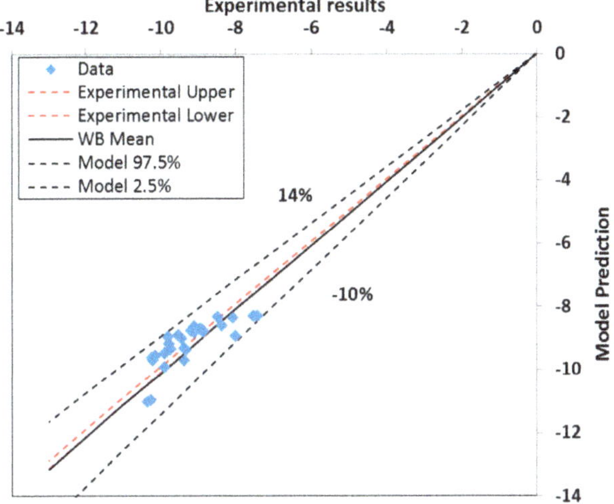

Figure 10. Comparison of AE model prediction and experimental results (log da/dN). WB indicates Bayesian regression analysis results using WinBugs.

Using the experimental data points for $log(da/dN)$ and assuming an experimental error of $\pm1\%$, the 45-degree solid line shows the difference between the model prediction and the "true" $log(da/dN)$ (with the 45-degree line showing the perfect match). However, because of the experimental data scatter and a slight bias expressed by F_m (showing systematic model error) between the true $log(da/dN)$ and regression models of Figure 9, the true $log(da/dN)$, with a 95% confidence level, will be somewhere as high as 14% above the model prediction and as low as 10% below this prediction.

As expected, the validation results showed good agreement between model predictions and experimental observations. The validated AE model can be used in real time monitoring of large cracks that subsequently improves the structural health management.

4. Conclusions

This research focused on the AE model for large crack growth assessment. In order to establish the AE signal feature versus the fatigue crack growth model and study the consistency and accuracy of the model, several standard fatigue experiments were performed. A set of tests were performed using standard Al7075-T6 test specimens subjected to cyclic loading with different amplitude and frequencies. A previously proposed relationship between the crack growth rate and AE signal features generated during crack growth was modified and validated. The AE-based crack growth model was found to be independent of the loading condition and loading frequency. These findings validate the previous work by Rabiei [17] as the proposed model is independent of the loading ratio and frequency.

Based on three identical tests with Ti-6-4, it was concluded that while the model parameters are material-dependent, the linear model form depicting the relationship between the crack growth rate and AE count rate remains valid when the material is changed.

The obtained experimental data was uncertain in nature due to considerable uncertainties in the optical crack detection method and measurement errors associated with the utilized crack sizing technique. The procedure of probabilistic model development and validation were discussed, and uncertainties of the model were investigated. To deal with uncertainties, a Bayesian approach was used to consider systematic and random errors in the model by capturing the effect of uncertainties. This approach provided a framework for updating the distribution of the model parameters.

Development of the proposed AE monitoring technique reported in this paper facilitates for the prognostics and life predictions of the structure. The developed methodology can be utilized for continuous in-service monitoring of structures and has proven to be promising for use in life predictions and assessment for structures subject to fatigue loading. Ultimately, these predictions can be used to define the appropriate inspection policies and maintenance schedules.

To update the model for any material variation, quantitative material properties that correlate to the AE count rate and crack growth rate could be identified. This process would be extensive due to the numerous material properties that affect a material's failure mechanisms including fracture toughness, modulus of elasticity, and yield strength. Once properties are identified, numerous types of materials could be tested to validate and update the model to account for material variations. In addition, the model is based on tests using the same specimen geometry of a specific plate thickness. Since crack growth behavior can also be dependent on the thickness of the material, the AE response and subsequently the model would be dependent on material thickness and specimen geometry. Future research may also focus on experimental studies to account for the effects of material thickness.

Author Contributions: A.K. developed and performed the Al7075-T6 acoustic emission and fatigue crack growth experiments while C.S. performed the Ti-6-4 experiments. A.K. and C.S. each performed the initial data analysis steps including crack growth measurements, de-noising, and AE count rate calculations for each of their respective experiments. A.K. performed the Bayesian analysis and model parameter development for all tests. A.K. wrote the majority of the paper, while C.S. contributed to several sections and assisted in editing. Finally, M.M. was the academic research advisor in this work and provided guidance throughout the research and edited the paper.

Conflicts of Interest: The authors declare no conflict of interest.

References

1. Beattie, A.G. Acoustic emission, principles and instrumentation. *J. Acoust. Emiss.* **1983**, *2*, 95–128.
2. Vanniamparambil, P.A.; Guclu, U.; Kontsos, A. Identification of crack initiation in aluminum alloys using acoustic emission. *Exp. Mech.* **2015**, *55*, 837–850. [CrossRef]
3. Barile, C.; Casavola, C.; Pappalettera, G.; Pappalettere, C. Analysis of crack propagation in stainless steel by comparing acoustic emissions and infrared thermography data. *Eng. Fail. Anal.* **2016**, *69*, 35–42. [CrossRef]
4. Kordatos, E.Z.; Aggelis, D.G.; Matikas, T.E. Monitoring mechanical damage in structural materials using complimentary NDE techniques based on thermography and acoustic emission. *Compos. Part B Eng.* **2012**, *43*, 2676–2686. [CrossRef]
5. Barile, C.; Casavola, C.; Pappalettera, G.; Pappalettere, C. Acoustic sources from damage propagation in Ti grade 5. *Measurement* **2016**, *91*, 73–76. [CrossRef]
6. Roberts, T.M.; Talebzadeh, M. Acoustic emission monitoring of fatigue crack propagation. *J. Constr. Steel Res.* **2003**, *59*, 695–712. [CrossRef]
7. Biancolini, M.E.; Brutti, C.; Paparo, G.; Zanini, A. Fatigue cracks nucleation on steel, Acoustic emission and fractal analysis. *Int. J. Fatigue* **2006**, *28*, 1820–1825. [CrossRef]
8. Rabiei, M.; Modarres, M.; Hoffman, P. Quantitative Methods for Structural Health Management using in-situ AE Monitoring. In Proceedings of the Annual Conference of the Prognostics and Health Management (PHM) Society, Montreal, QC, Canada, 25–29 September 2011.
9. Bassim, M.N.; Hamel, F.; Bailon, J.P. Acoustic emission mechanism during high-cycle fatigue. *Eng. Fract. Mech.* **1981**, *14*, 853–860.

10. Roberts, T.M.; Talebzadeh, M. Fatigue life prediction based on crack propagation and acoustic emission count rates. *J. Constr. Steel Res.* **2003**, *59*, 679–694. [CrossRef]

11. Rabiei, M.; Modarres, M. Quantitative methods for structural health management using in situ acoustic emission monitoring. *Int. J. Fatigue* **2013**, *49*, 81–89. [CrossRef]

12. Han, Z.; Luo, H.; Cao, J.; Wang, H. Acoustic emission during fatigue crack propagation in a micro-alloyed steel welds. *Mater. Sci. Eng. A* **2011**, *528*, 7751–7756. [CrossRef]

13. Strantza, M.; Hemelrijck, D.; Guillaume, P.; Aggelis, D.G. Acoustic emission monitoring of crack propagation in additively manufactured and conventional titanium components. *Mech. Res. Commun.* **2017**, *84*, 8–13. [CrossRef]

14. Gong, Z.; Nyborg, E.O.; Oommen, G. Acoustic emission monitoring of steel railroad bridges. *Mater. Eval.* **1992**, *50*, 883–887. [CrossRef]

15. Bassim, M.N.; St. Lawrence, S.; Liu, C.D. Detection of the onset of fatigue crack growth in rail steels using acoustic emission. *Eng. Fract. Mech.* **1994**, *47*, 207–214. [CrossRef]

16. Paris, P.; Erdogan, F. A critical analysis of crack propagation laws. *J. Basic Eng.* **1963**, *85*, 528–533. [CrossRef]

17. Rabiei, M. A Bayesian Framework for Structural Health Management Using Acoustic Emission Monitoring and Periodic Inspections. Ph.D. Dissertation, Department of Mechanical Engineering, University of Maryland, College Park, MD, USA, 2011. Available online: https://search.proquest.com/docview/880410901?pq-origsite=gscholar (accessed on 15 July 2013).

18. ASTM. *Constant Load Amplitude Fatigue Crack Growth Rate above 10^{-8} m/cycle*; American Society for Testing and Materials: Washington, DC, USA, 2008; pp. 321–339.

19. Gauthier, M.M. *Engineered Materials Handbook—Desk Edition*; ASM International: Materials Park, OH, USA, 1995.

20. Sauerbrunn, C.M.; Modarress, M. Effects of Material Variation on Acoustic Emissions-Based, Large-Crack Growth Model. In Proceedings of the 25th American Society of Nondestructive Testing (ASNT) Research Symposium, New Orleans, LA, USA, 11–14 April 2016.

21. Morton, T.M.; Harrington, R.M.; Bjeletich, J.G. Acoustic Emission of fatigue crack growth. *Eng. Fract. Mech.* **1973**, *5*, 691–697. [CrossRef]

22. Wang, Z.F.; Li, J.; Ke, W.; Zhu, Z. Characteristics of acoustic emission for A537 structural steel during fatigue crack propagation. *Scr. Metall. Mater.* **1992**, *27*, 641–646. [CrossRef]

23. Keshtgar, A.; Modarres, M. Acoustic emission-based fatigue crack growth prediction. In Proceedings of the 2013 Reliability and Maintainability Symposium (RAMS), Orlando, FL, USA, 28–31 January 2013; pp. 1–5.

24. Ferreria, T.; Rasband, W. *ImageJ User Guide*; U.S. National Institutes of Health: Bethesda, MD, USA, 2012.

25. Keshtgar, A. Acoustic Emission-Based Structural Health Management and Prognostics Subject to Small Fatigue Cracks. Ph.D. Dissertation, University of Maryland, College Park, MD, USA, 2013.

26. Miller, G.A.; Chapman, J.P. Misunderstanding analysis of covariance. *J. Abnorm. Psychol.* **2001**, *110*, 40–48. [CrossRef] [PubMed]

27. Smith, F. Interpretation of adjusted treatment means and regression in analysis of covariance. *Biometrics* **1975**, *13*, 282–308. [CrossRef]

28. Ntzoufras, I. *Bayesian Modeling Using WinBUGS*; John Wiley and Sons: Hoboken, NJ, USA, 2009.

29. Bolstad, W.M. *Introduction to Bayesian Statistics*, 2nd ed.; John Wiley and Sons: Hoboken, NJ, USA, 2007.

30. Ditlevsen, O.; Madsen, H.O. *Structural Reliability Methods*; John Wiley and Sons: Hoboken, NJ, USA, 2007.

31. Ontiveros, V. An Integrated Methodology for Assessing Fire Simulation Code Uncertainty. Master's Thesis, University of Maryland, College Park, MD, USA, 2010.

Article

Transverse Vibration of Clamped-Pinned-Free Beam with Mass at Free End

Jonathan Hong [1,2,*,†], Jacob Dodson [3], Simon Laflamme [2] and Austin Downey [4]

1   Applied Research Associates, Emerald Coast Division, Niceville, FL 32578, USA
2   Department of Civil, Construction, and Environmental Engineering, Iowa State University, Ames, IA 50011, USA
3   Air Force Research Laboratory, Munitions Directorate, Eglin AFB, FL 32542, USA
4   Department of Mechanical Engineering, University of South Carolina, Columbia, SC 29201, USA
*   Correspondence: jhong@ara.com
†   Current address: 956 W. John Sims Pkwy, Niceville, FL 32578, USA.

Received: 25 May 2019; Accepted: 12 July 2019; Published: 26 July 2019

Abstract: Engineering systems undergoing extreme and harsh environments can often times experience rapid damaging effects. In order to minimize loss of economic investment and human lives, structural health monitoring (SHM) of these high-rate systems is being researched. An experimental testbed has been developed to validate SHM methods in a controllable and repeatable laboratory environment. This study applies the Euler-Bernoulli beam theory to this testbed to develop analytical solutions of the system. The transverse vibration of a clamped-pinned-free beam with a point mass at the free end is discussed in detail. Results are derived for varying pin locations and mass values. Eigenvalue plots of the first five modes are presented along with their respective mode shapes. The theoretical calculations are experimentally validated and discussed.

Keywords: beam; vibration; structural health monitoring; high-rate dynamics

---

1. Introduction

High-rate dynamics are defined as events having amplitudes greater than 100 g over durations less than 100 ms [1]. Some examples of high-rate systems may include civil structures exposed to blast, passenger vehicles experiencing collisions, and aerial or spacecraft vehicles subjected to ballistic impacts. Such systems have the potential to experience rapid changes in mechanical configuration through damage. Economic investments and lives could be saved if fast detection of parameter changes can be accurately quantified [2]. A variable input observer has been studied by the authors as a potential solution to increasing convergence times through richer inputs [3]. However, there is a need to validate high-rate structural health monitoring (SHM) methods [4].

An experimental testbed has been designed and built to test and validate SHM methods systems experiencing high-rate dynamics. The development of an experimental testbed is critical, because the experimentation on real-life high-rate systems would be complex, difficulty to verify, and potentially very costly. This testbed design incorporates a cantilever beam with a roller that restrains the displacement in the vertical direction and is allowed to move freely along the length of the beam. Additionally, the mass at the free end of the beam can be dropped through the de-energizing of the electromagnet that detaches the mass from the beam. The roller is a moving cart that provides a changing boundary condition while the mass drop provides a sudden change in mechanical configuration. This system is easily controllable and repeatable in a laboratory setting.

To develop analytical solutions for this beam structure, the Euler-Bernoulli beam theory is applied. The system is modeled as clamped-pinned-free with a point mass at free end.

To the best knowledge of the authors, this configuration has not been previously studied analytically nor experimentally. There is no mention of this beam configuration in the book authored by Blevins [5]. The clamped-pinned-free without mass [6] and the clamped-free with mass at free end without pin [7] have been studied. In more recent years, researchers have investigated the free vibration of multi-span beams with flexible constraints [8], axial vibrations of multi-span beams with concentrated masses [9], multi-span beams with moving masses [10], and multi-span beams carrying spring-mass systems [11,12].

Using beam theory, Section 2 derives the transcendental equation. A generalized form is presented that is applicable to any pin location and any mass value. Derived results are verified through comparison between well known cases in literature. Section 3 calculates the eigenvalues for normalized pinned location for various mass ratios. Section 4 calculates the mode shapes for several different pinned locations, while Section 5 compares the results from the theoretical calculations to experimental data.

2. Frequency Calculations

The transverse vibrations of a slender clamped-pinned-free beam with a mass at free end of interest is shown in Figure 1. The governing equation for the beam using Euler-Bernoulli's beam theory [13] can be written:

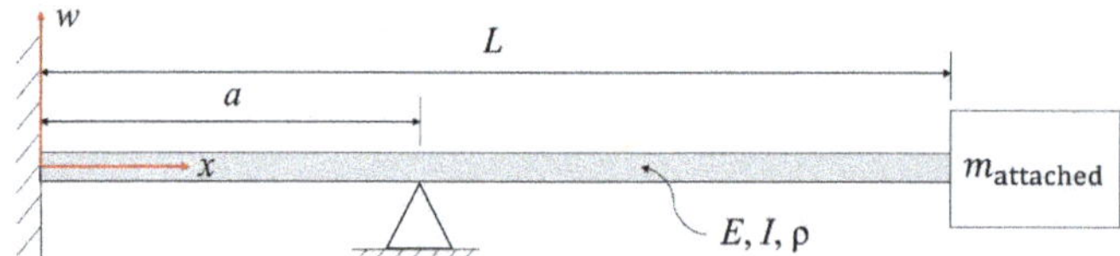

Figure 1. Schematic of a clamped-pinned-free beam with mass at free end.

$$\rho A \frac{\partial^2 w}{\partial t^2} + EI \frac{\partial^4 w}{\partial x^4} = 0 \tag{1}$$

where E is the Young's modulus, I is the cross-sectional moment of inertia, w is the vertical deflection, x is the axial coordinate, ρ is the density of the beam, A is the cross-sectional area, and t is time. Equation (1) can be solved assuming a separation-of-variables solution in the standard form:

$$w(x,t) = X(x)T(t) \tag{2}$$

where X is the spatial solution and T is the temporal solution. The spatial solution for a two-span beam then is expressed:

$$X(x) = \begin{cases} X_1(x), & 0 \le x \le a \\ X_2(x), & a \le x \le L \end{cases} \tag{3}$$

The sub-functions in Equation (3) can be written as the following general solutions:

$$X_1(x) = a_1 \sin(\beta x) + a_2 \cos(\beta x) + a_3 \sinh(\beta x) + a_4 \cosh(\beta x) \tag{4}$$

$$X_2(x) = b_1 \sin(\beta x) + b_2 \cos(\beta x) + b_3 \sinh(\beta x) + b_4 \cosh(\beta x) \tag{5}$$

where β is the beam vibration eigenvalue. Parameter β and seven of the eight coefficients can be solved by applying the boundary conditions of the system. For the clamped-pinned-free, the displacement and slope at the clamped end are zero [7]:

$$X_1(0) = 0 \tag{6}$$

$$\frac{dX_1(0)}{dx} = 0 \tag{7}$$

while at the free end, the bending moment and the shear vanish such that:

$$\frac{d^2 X_2(L)}{dx^2} = 0 \tag{8}$$

$$EI\frac{d^3 X_2(L)}{dx^3} = m_{\text{attached}}\frac{d^2 T_2(L,t)}{dt^2} \tag{9}$$

where m_{attached} is the mass attached to the beam at the free end. In addition to these four boundary conditions, four more boundary conditions (displacement and rotation) are found at the pin location a:

$$X_1(a) = 0 \tag{10}$$

$$X_2(a) = 0 \tag{11}$$

$$\frac{dX_1(a)}{dx} = \frac{dX_2(a)}{dx} \tag{12}$$

$$\frac{d^2 X_1(a)}{dx^2} = \frac{d^2 X_2(a)}{dx^2} \tag{13}$$

Substituting the first transverse displacement (Equation (4)) into the clamped boundary condition (Equations (6) and (7)) gives:

$$a_2 + a_4 = 0 \tag{14}$$

$$a_1 + a_3 = 0 \tag{15}$$

Substituting the second transverse displacement (Equation (5)) into the free end boundary condition (Equation (8)) yields:

$$- b_1 \sin(\beta L) - b_2 \cos(\beta L) + b_3 \sinh(\beta L) + b_4 \cosh(\beta L) = 0 \tag{16}$$

Additionally, inserting the second transverse displacement (Equation (5)) into Equation (1) and applying the boundary condition at the free end (Equation (9)) results in:

$$b_1\left(- \cos(\beta L) + \beta L\frac{m_{\text{attached}}}{m_{\text{beam}}} \sin(\beta L)\right) + b_2\left(\sin(\beta L) + \beta L\frac{m_{\text{attached}}}{m_{\text{beam}}} \cos(\beta L)\right)$$
$$+ b_3\left(\cosh(\beta L) + \beta L\frac{m_{\text{attached}}}{m_{\text{beam}}} \sinh(\beta L)\right) + b_4\left(\sinh(\beta L) + \beta L\frac{m_{\text{attached}}}{m_{\text{beam}}} \cosh(\beta L)\right) = 0 \tag{17}$$

where m_{beam} is the mass of the beam.

Substituting the first transverse displacement (Equation (4)) into the pinned boundary condition (Equations (10) and (11)) results in:

$$a_1 \sin(\beta L\frac{a}{L}) + a_2 \cos(\beta L\frac{a}{L}) + a_3 \sinh(\beta L\frac{a}{L}) + a_4 \cosh(\beta L\frac{a}{L}) = 0 \tag{18}$$

and

$$b_1 \sin(\beta L\frac{a}{L}) + b_2 \cos(\beta L\frac{a}{L}) + b_3 \sinh(\beta L\frac{a}{L}) + b_4 \cosh(\beta L\frac{a}{L}) = 0 \tag{19}$$

After, substituting the second transverse displacement (Equation (5)) into the boundary conditions defined by Equations (12) and (13) provides the following expressions:

$$a_1 \cos(\beta L\frac{a}{L}) - a_2 \sin(\beta L\frac{a}{L}) + a_3 \cosh(\beta L\frac{a}{L}) + a_4 \sinh(\beta L\frac{a}{L})$$
$$- b_1 \cos(\beta L\frac{a}{L}) + b_2 \sin(\beta L\frac{a}{L}) - b_3 \cosh(\beta L\frac{a}{L}) - b_4 \sinh(\beta L\frac{a}{L}) = 0 \tag{20}$$

$$-a_1 \sin(\beta L\frac{a}{L}) - a_2 \cos(\beta L\frac{a}{L}) + a_3 \sinh(\beta L\frac{a}{L}) + a_4 \cosh(\beta L\frac{a}{L})$$

$$+b_1 \sin(\beta L\frac{a}{L}) + b_2 \cos(\beta L\frac{a}{L}) - b_3 \sinh(\beta L\frac{a}{L}) - b_4 \cosh(\beta L\frac{a}{L}) = 0 \tag{21}$$

Aggregating Equations (14)–(21) into an 8×8 matrix (see Appendix A) and solving for the determinant leads to the transcendental equation expressed:

$$4\cos(\beta L(\frac{a}{L} - 1)) \sinh(\beta L(\frac{a}{L} - 1)) - 4\cosh(\beta L(\frac{a}{L} - 1)) \sin(\beta L(\frac{a}{L} - 1))$$

$$+2\cos(\beta L(2\frac{a}{L} - 1)) \sinh(\beta L) - 2\cosh(\beta L(2\frac{a}{L} - 1)) \sin(\beta L)$$

$$+4\cos(\frac{a}{L}\beta L) \sinh(\frac{a}{L}\beta L) - 4\cosh(\frac{a}{L}\beta L) \sin(\frac{a}{L}\beta L)$$

$$+2\cos(\beta L) \sinh(\beta L) - 2\cosh(\beta L) \sin(\beta L) + 8\beta L\frac{m_{\text{attached}}}{m_{\text{beam}}} \sin(\beta L(\frac{a}{L} - 1)) \sinh(\beta L(\frac{a}{L} - 1))$$

$$+2\beta L\frac{m_{\text{attached}}}{m_{\text{beam}}} \cos(\beta L(2\frac{a}{L} - 1)) \cosh(\beta L) - 2\beta L\frac{m_{\text{attached}}}{m_{\text{beam}}} \cosh(\beta L(2\frac{a}{L} - 1)) \cos(\beta L)$$

$$+2\beta L\frac{m_{\text{attached}}}{m_{\text{beam}}} \sin(\beta L(2\frac{a}{L} - 1)) \sinh(\beta L) + 2\beta L\frac{m_{\text{attached}}}{m_{\text{beam}}} \sinh(\beta L(2\frac{a}{L} - 1)) \sin(\beta L)$$

$$-4\beta L\frac{m_{\text{attached}}}{m_{\text{beam}}} \sin(\beta L) \sinh(\beta L) = 0 \tag{22}$$

where the natural frequencies (in Hz) are given by:

$$f_n = \frac{(\beta_n L)^2}{2\pi L^2} \sqrt{\frac{EI}{\rho A}} \tag{23}$$

To verify Equation (22), the first five natural frequencies were calculated for three well known cases:

- Case 1: Clamped-free [14]: $\frac{a}{L} = \frac{m_{\text{attached}}}{m_{\text{beam}}} = 0$
- Case 2: Clamped-free with mass at free end [7]: $\frac{a}{L} = 0$
- Case 3: Clamped-pinned-free [6]: $\frac{m_{\text{attached}}}{m_{\text{beam}}} = 0$

The results are tabulated in Tables 1–3. The frequencies of the first five modes (β_1–β_5) are compared between what is found in literature against the results from Equation (22) (proposed model). The small differences are due to rounding errors of the beam vibration eigenvalues β, which cause large changes in the calculated frequency ($f_n \propto \beta_n^2$). The precision for β in this paper is ± 0.0002.

Table 1. Comparison of analytical results: clamped-free (Case 1).

Mode	Literature [14] (Hz)	Proposed Model (Hz)	Difference (%)
1	19.64	19.63	0.051
2	123.07	123.02	0.041
3	344.64	344.45	0.055
4	675.31	674.97	0.050
5	1116.33	1115.79	0.048

Table 2. Comparison of analytical results: Clamped-free with mass at free end ($\frac{m_{\text{attached}}}{m_{\text{beam}}} = 0.2$) (Case 2).

Mode	Literature [7] (Hz)	Proposed Model (Hz)	Difference (%)
1	14.56	14.58	0.14
2	101.47	101.64	0.17
3	298.59	299.00	0.14
4	603.06	604.02	0.16
5	1017.07	1018.48	0.14

Table 3. Comparison of analytical results: Clamped-pinned-free (pinned at $a = 200$ mm) (Case 3).

Mode	Literature [6] (Hz)	Proposed Model (Hz)	Difference (%)
1	41.70	41.59	0.26
2	279.25	278.46	0.28
3	635.80	635.19	0.10
4	899.94	897.69	0.25
5	1650.85	1646.48	0.26

3. Calculations of Eigenvalues

The beam vibration eigenvalues are calculated in terms of βL for different mass ratios, $\frac{m_{\text{attached}}}{m_{\text{beam}}}$. The eigenvalues are plotted as a function of the normalized pinned location, $\frac{a}{L}$ in Figures 2–6. Note, the βL values corresponding to $\frac{a}{L} = 0$ is equivalent to the clamped-free system with a mass at the free.

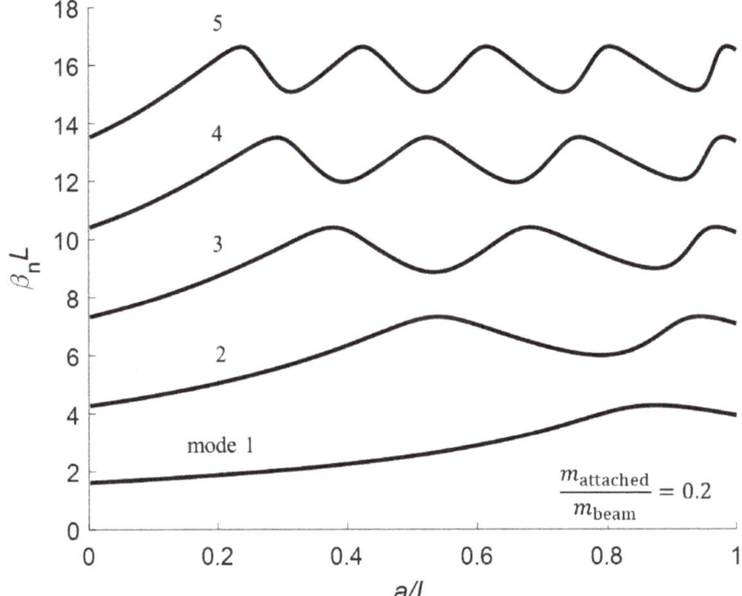

Figure 2. Eigenvalues of first 5 modes, $\frac{m_{\text{attached}}}{m_{\text{beam}}} = 0.2$.

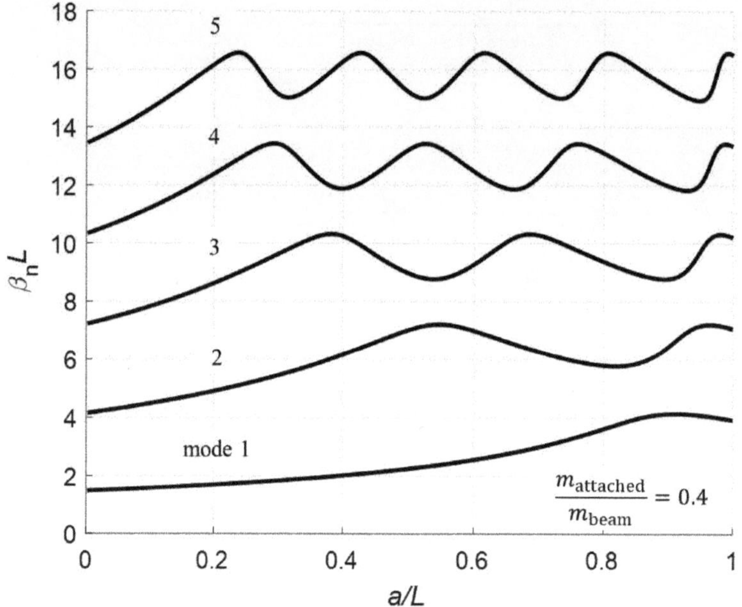

Figure 3. Eigenvalues of first 5 modes, $\frac{m_{\text{attached}}}{m_{\text{beam}}} = 0.4$.

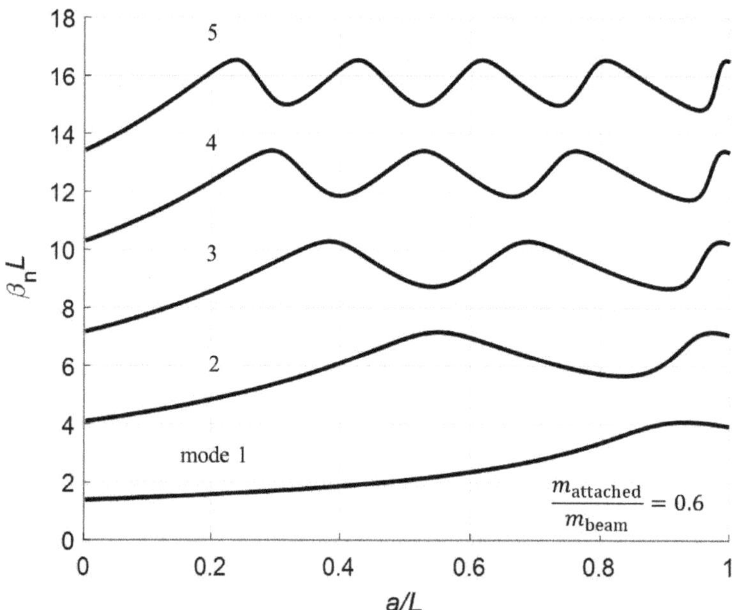

Figure 4. Eigenvalues of first 5 modes, $\frac{m_{\text{attached}}}{m_{\text{beam}}} = 0.6$.

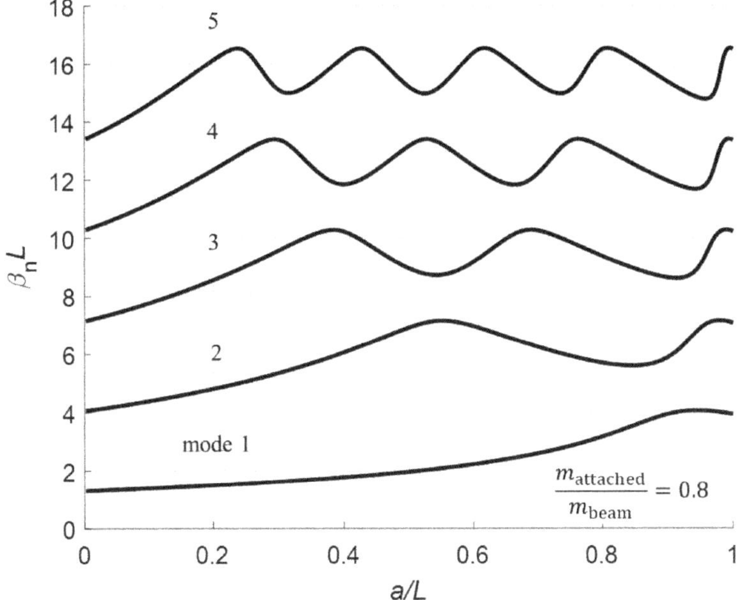

Figure 5. Eigenvalues of first 5 modes, $\frac{m_{\text{attached}}}{m_{\text{beam}}} = 0.8$.

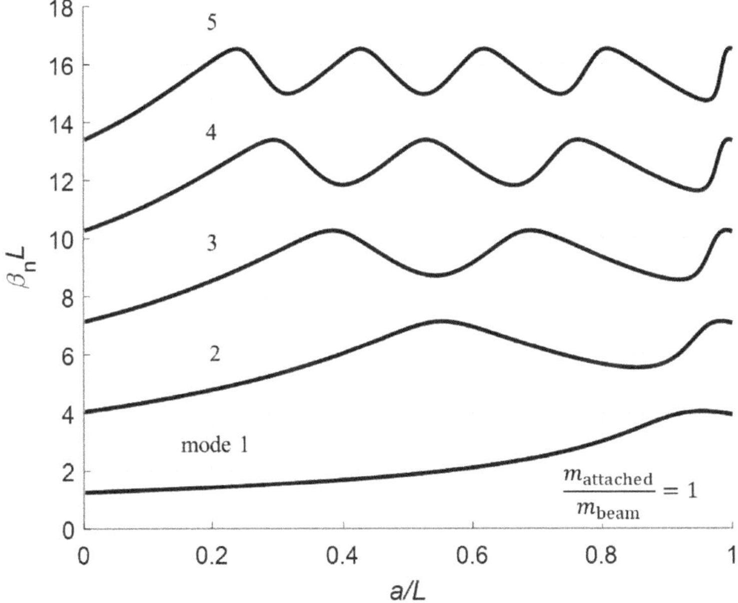

Figure 6. Eigenvalues of first 5 modes, $\frac{m_{\text{attached}}}{m_{\text{beam}}} = 1$.

4. Mode Shapes

The mode shapes are calculated for the two different sections of the beam corresponding to the clamped-pinned and pinned-free sections. To calculate the mode shapes, the boundary condition

(Equations (6)–(13)) are used to find a relationship between the coefficients. The method used here consists of solving all coefficients in terms of a_4. Note, there are not enough equations to determine a unique solution for each coefficient. The solutions for the a coefficients are:

$$a_1 = -a_3 \tag{24}$$

$$a_2 = -a_4 \tag{25}$$

$$a_3 = \frac{\cos(\beta L \frac{a}{L}) - \cosh(\beta L \frac{a}{L})}{\left(\sinh(\beta L \frac{a}{L}) - \sin(\beta L \frac{a}{L})\right)} a_4 \tag{26}$$

and for the b coefficients:

$$b_1 = -b_2 \cot(\beta L) + b_3 \frac{\sinh(\beta L)}{\sin(\beta L)} + b_4 \frac{\cosh(\beta L)}{\sin(\beta L)} \tag{27}$$

$$b_2 = b_3 \frac{z_1}{z_3} + b_4 \frac{z_2}{z_3} \tag{28}$$

$$b_3 = \frac{\frac{z_2 z_4}{z_3} + z_6}{\frac{z_1 z_4}{z_3} + z_5} \tag{29}$$

where

$$z_1 = \frac{\sinh(\beta L)}{\sin(\beta L)} \left(\cos(\beta L) + \beta L \frac{m_{\text{attached}}}{m_{\text{beam}}} \sin(\beta L)\right) - \left(\cosh(\beta L) + \beta L \frac{m_{\text{attached}}}{m_{\text{beam}}} \sinh(\beta L)\right) \tag{30}$$

$$z_2 = \frac{\cosh(\beta L)}{\sin(\beta L)} \left(\cos(\beta L) + \beta L \frac{m_{\text{attached}}}{m_{\text{beam}}} \sin(\beta L)\right) - \left(\sinh(\beta L) + \beta L \frac{m_{\text{attached}}}{m_{\text{beam}}} \cosh(\beta L)\right) \tag{31}$$

$$z_3 = \cot(\beta L)\left(\cos(\beta L) + \beta L \frac{m_{\text{attached}}}{m_{\text{beam}}} \sin(\beta L)\right) + \left(\sin(\beta L) + \beta L \frac{m_{\text{attached}}}{m_{\text{beam}}} \cos(\beta L)\right) \tag{32}$$

$$z_4 = \cos(\beta L \frac{a}{L}) - \cot(\beta L) \sin(\beta L \frac{a}{L}) \tag{33}$$

$$z_5 = \frac{\sinh(\beta L)}{\sin(\beta L)} \sin(\beta L \frac{a}{L}) + \sinh(\beta L \frac{a}{L}) \tag{34}$$

$$z_6 = \frac{\cosh(\beta L)}{\sin(\beta L)} \sin(\beta L \frac{a}{L}) + \cosh(\beta L \frac{a}{L}) \tag{35}$$

Substituting the equations for the coefficients (Equations (24)–(35)) into the boundary condition from Equation (12), a relationship between a_4 and b_4 is obtained. For brevity, this expression is not presented here. The mode shapes are determined for the multi-span beam by substituting all coefficient expressions in terms of a_4 into Equation (3).

Normalizing at $a_4 = 1$, the first five mode shapes for the clamped-pinned-free beam with a mass at the free end are plotted in Figures 7–10 for $a = 100$, 200, 300, and 400 mm with $\frac{m_{\text{attached}}}{m_{\text{beam}}} = 0.2$. The red triangle on the plots denotes the pin location. Note that for $a = 100$ (Figure 7), the mode shapes are as expected for a fixed-pinned-free cantilever beam with a mass on the free end. However, when $a = 200$ (Figure 8), mode shape 4 is highly non-symmetric because the constraint point (pin) is just past the node and in combination with the effect of the mass this mode shape flattens out for the remainder of the beam. For $a = 300$ (Figure 9) and $a = 400$ (Figure 10) the more expected sinusoidal shape dominates the mode shapes.

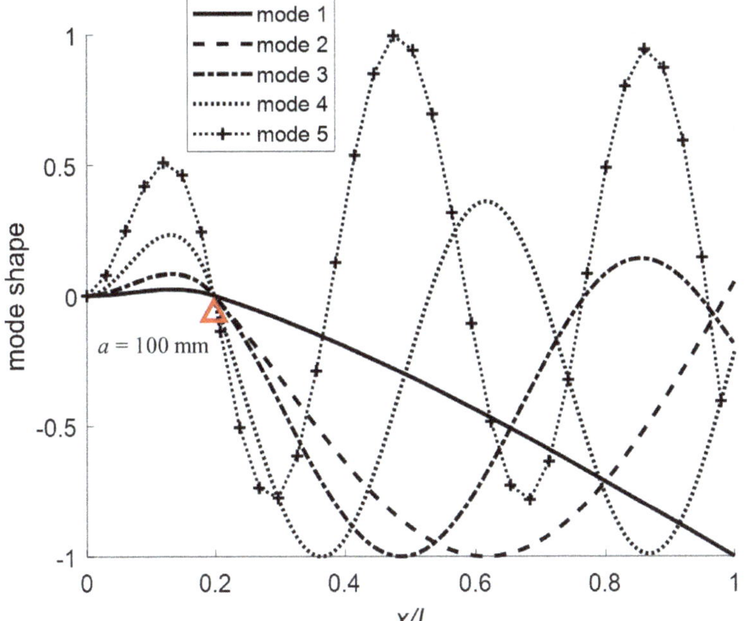

Figure 7. Mode shapes for pinned at $a = 100$ mm and $\frac{m_{\text{attached}}}{m_{\text{beam}}} = 0.2$.

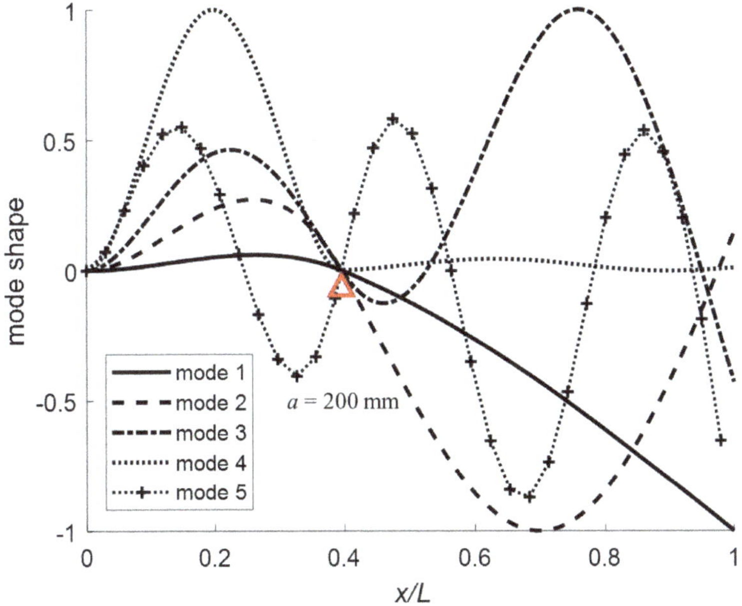

Figure 8. Mode shapes for pinned at $a = 200$ mm and $\frac{m_{\text{attached}}}{m_{\text{beam}}} = 0.2$.

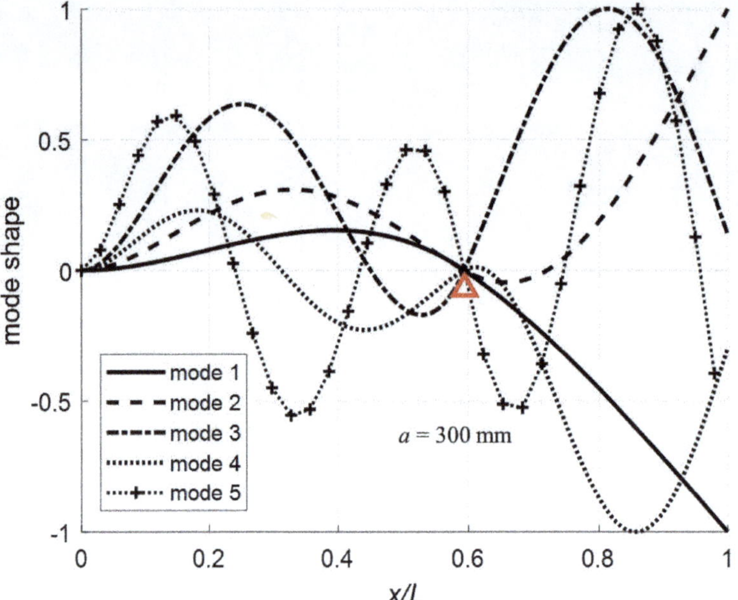

Figure 9. Mode shapes for pinned at $a = 300$ mm and $\frac{m_{\text{attached}}}{m_{\text{beam}}} = 0.2$.

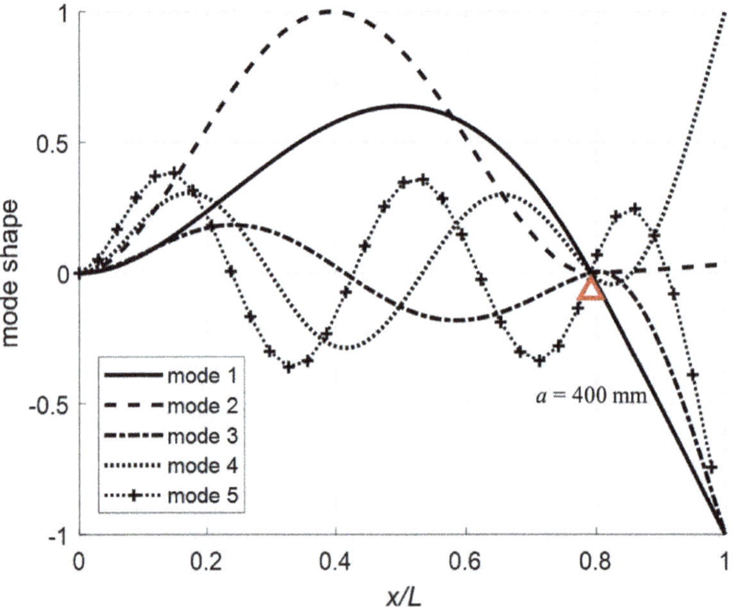

Figure 10. Mode shapes for pinned at $a = 400$ mm and $\frac{m_{\text{attached}}}{m_{\text{beam}}} = 0.2$.

5. Experimental Validation

In this section, the theoretical results are compared with experimental data. The experimental setup is illustrated in Figure 11. A cart with rollers is used as a moving pin along the beam.

Accelerometers are attached at locations 300 mm and 400 mm. The mass of the accelerometers is assumed to have a negligible effect. Each accelerometer weighs 1.7 gm, not including cables, which is 0.2% of the weight of the beam. At the free end, an electromagnet is used to simulate the point mass. The specifications of the experiment are listed in Table 4.

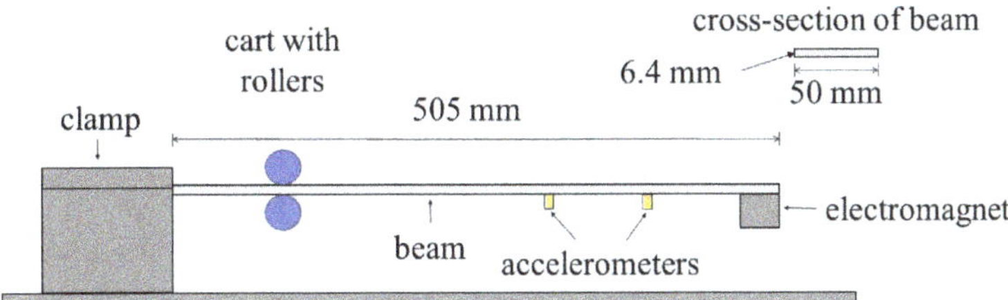

Figure 11. Illustration of the experimental setup.

Table 4. Specifications of the experimental setup.

Parameter	Value
L	505 mm
base	50 mm
height	6.4 mm
ρ	7970 kg/m^3
E	190 GPa
$m_{attached}$	0.259 Kg

The accelerometers are single axis PCB 353B17. They are connected to a NI-9234 IEPE analog input module seated in an NI cDAQ-9172 chassis. A PCB 086C01 modal hammer with a white ABS plastic tip is used to excite the beam at 300 mm. Five tests under each condition are conducted and averaged in the frequency domain to generate frequency response functions (FRFs) using the H_v algorithm [15]. The FRFs for the different tests are plotted in Figures 12–15. The vertical red dashed lines represent the theoretical modes computed from the proposed model. To better understand the differences, the modes are extracted and tabulated in Tables 5–8.

For the four test conditions evaluated, the difference between the theoretical calculations and experimental results for modes 4 and 5 are non-trivial. Three possible explanations for these differences are (1) the electromagnet vibrates separate from the beam, (2) the beam vibrates within the rollers, and (3) the rotational inertia from a large mass at the end of a long beam impacts the higher frequencies. In Figures 13 and 15, the coherence for mode 5 drops significantly such that it cannot be said with certainty that the frequencies are correct. Percent difference is used to quantify how different the theoretical frequencies are from the experimental. All frequencies fall below 20% difference with the exception of mode 4 in Table 5.

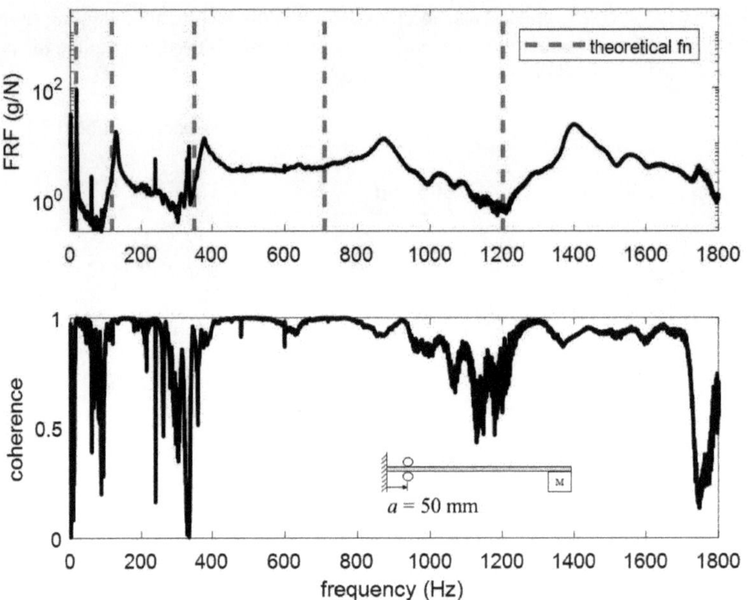

Figure 12. FRF for pinned at $a = 50$ mm and $\frac{m_{\text{attached}}}{m_{\text{beam}}} = 0.2$.

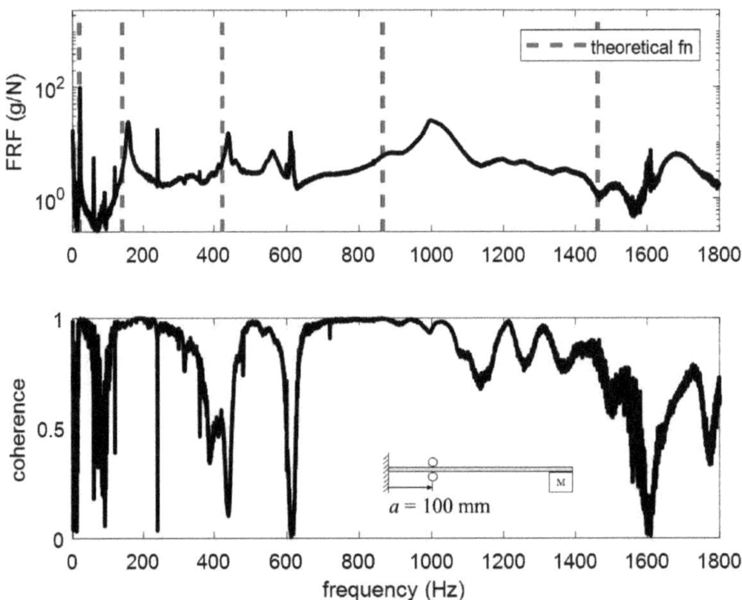

Figure 13. FRF for pinned at $a = 100$ mm and $\frac{m_{\text{attached}}}{m_{\text{beam}}} = 0.2$.

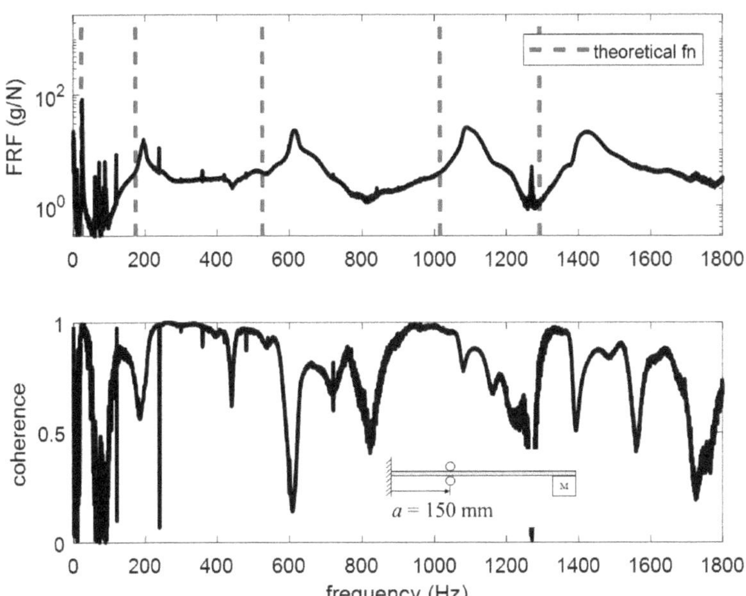

Figure 14. FRF for pinned at $a = 150$ mm and $\frac{m_{\text{attached}}}{m_{\text{beam}}} = 0.2$.

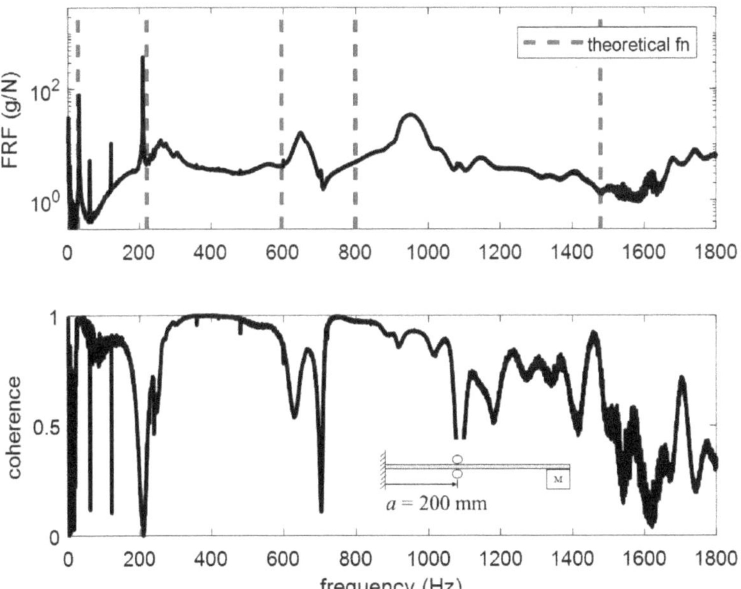

Figure 15. FRF for pinned at $a = 200$ mm and $\frac{m_{\text{attached}}}{m_{\text{beam}}} = 0.2$.

Table 5. Pinned at $a = 50$ mm and $\frac{m_{\text{attached}}}{m_{\text{beam}}} = 0.2$.

Mode	Proposed Model (Hz)	Experiment (Hz)	Difference (%)
1	16.73	17.75	6.1
2	118.28	128.88	9.0
3	350.24	378.68	8.1
4	710.71	872.01	22.7
5	1202.46	1400.19	16.4

Table 6. Pinned at $a = 100$ mm and $\frac{m_{\text{attached}}}{m_{\text{beam}}} = 0.2$.

Mode	Proposed Model (Hz)	Experiment (Hz)	Difference (%)
1	19.48	21.37	9.7
2	141.29	157.37	11.4
3	423.53	440.11	3.9
4	864.86	996.16	15.2
5	1462.39	1680.39	14.9

Table 7. Pinned at $a = 150$ mm and $\frac{m_{\text{attached}}}{m_{\text{beam}}} = 0.2$.

Mode	Proposed Model (Hz)	Experiment (Hz)	Difference (%)
1	23.09	25.87	12.0
2	173.89	195.62	12.5
3	526.12	614.09	16.7
4	1015.61	1088.84	7.2
5	1292.51	1421.86	10.0

Table 8. Pinned at $a = 200$ mm and $\frac{m_{\text{attached}}}{m_{\text{beam}}} = 0.2$.

Mode	Proposed Model (Hz)	Experiment (Hz)	Difference (%)
1	28.02	31.39	12.0
2	221.20	259.65	17.4
3	595.97	646.99	8.6
4	798.79	950.15	18.9
5	1479.27	1741.16	17.7

6. Conclusions

A high-rate experimental testbed is studied. The testbed is characterized as being a clamped-pinned-free beam with a mass at the free end. Euler-Bernoulli beam theory is applied to derive the transcendental equation for a general case applicable to the system pinned at an arbitrary location and with an arbitrary mass. The eigenvalues and mode shapes were presented under various test conditions. Experimental tests were conducted and results compared with the theoretical calculations of the first five natural frequencies. The comparison of results exhibited a good match in frequency values for the first three modes. The errors increase with the higher modes. The difference in higher modes can be attributed to the electromagnet vibrating separate to the beam, the beam vibrating within the rollers, and the rotational inertia of the mass not taken into consideration. Nevertheless, the percent difference of all modes between the theoretical and experimental values fell below 20% except for one case. These results confirm that within reason, the theory matches the experimental results.

The analytical model developed here can be useful in the design and numerical assessment of structural health monitoring solutions designed for systems operating in high-rate dynamic environments. Furthermore, the experimental quantification of the error in the higher modes validates and defines the bounds for which the high-rate experimental testbed is best utilized. Future work will include the evaluation of algorithms and methodologies for the SHM of structures experiences high-rate dynamic events.

Author Contributions: J.H. performed the analytical work, numerical simulations, and experimental activities, and led the write-up; J.D. and S.L. co-led the investigation; A.D. assisted in the collection of experimental data and validation of the analytical work. All authors contributed to preparing and proofreading the manuscript.

Funding: The material in this paper is based upon work supported by the Air Force Office of Scientific Research (AFOSR) award numbers FA9550-17-1-0131 and FA9550-17RWCOR503, and AFRL/RWK contract number FA8651-17-D-0002. Opinions, interpretations, conclusions and recommendations are those of the authors and are not necessarily endorsed by the United States Air Force. Additionally, this work was partially supported by the National Science Foundation Grant No. 1850012. Any opinions, findings, and conclusions or recommendations expressed in this material are those of the authors and do not necessarily reflect the views of the National Science Foundation.

Conflicts of Interest: The authors declares no conflict of interest.

Appendix A. Boundary Conditions Applied to Transverse Displacement Equations

$$
\begin{bmatrix}
0 & 1 & 0 & 1 & 0 & 0 & 0 & 0 \\
1 & 0 & 1 & 0 & 0 & 0 & 0 & 0 \\
0 & 0 & 0 & 0 & -\sin(\beta L) & -\cos(\beta L) & \sinh(\beta L) & \cosh(\beta L) \\
0 & 0 & 0 & 0 & -\cos(\beta L)+\beta L\frac{m_{attached}}{m_{beam}}\sin(\beta L) & \sin(\beta L)+\beta L\frac{m_{attached}}{m_{beam}}\cos(\beta L) & \cosh(\beta L)+\beta L\frac{m_{attached}}{m_{beam}}\sinh(\beta L) & \sinh(\beta L)+\beta L\frac{m_{attached}}{m_{beam}}\cosh(\beta L) \\
\sin(\beta L\frac{a}{L}) & \cos(\beta L\frac{a}{L}) & \sinh(\beta L\frac{a}{L}) & \cosh(\beta L\frac{a}{L}) & 0 & 0 & 0 & 0 \\
0 & 0 & 0 & 0 & \sin(\beta L\frac{a}{L}) & \cos(\beta L\frac{a}{L}) & \sinh(\beta L\frac{a}{L}) & \cosh(\beta L\frac{a}{L}) \\
\cos(\beta L\frac{a}{L}) & -\sin(\beta L\frac{a}{L}) & \cosh(\beta L\frac{a}{L}) & \sinh(\beta L\frac{a}{L}) & -\cos(\beta L\frac{a}{L}) & \sin(\beta L\frac{a}{L}) & -\cosh(\beta L\frac{a}{L}) & -\sinh(\beta L\frac{a}{L}) \\
-\sin(\beta L\frac{a}{L}) & -\cos(\beta L\frac{a}{L}) & \sinh(\beta L\frac{a}{L}) & \cosh(\beta L\frac{a}{L}) & \sin(\beta L\frac{a}{L}) & \cos(\beta L\frac{a}{L}) & -\sinh(\beta L\frac{a}{L}) & -\cosh(\beta L\frac{a}{L})
\end{bmatrix}
\begin{bmatrix} a_1 \\ a_2 \\ a_3 \\ a_4 \\ b_1 \\ b_2 \\ b_3 \\ b_4 \end{bmatrix}
=
\begin{bmatrix} 0 \\ 0 \\ 0 \\ 0 \\ 0 \\ 0 \\ 0 \\ 0 \end{bmatrix}
$$

References

1. Hong, J.; Laflamme, S.; Dodson, J.; Joyce, B. Introduction to state Estimation of High-Rate System Dynamics. *Sensors* **2018**, *18*, 217. [CrossRef] [PubMed]

2. Lowe, R.; Dodson, J.; Foley, J. Microsecond Prognostics and Health Management. *IEEE Reliab. Soc. Newsl.* **2014**, *60*, 1–5.

3. Hong, J.; Laflamme, S.; Dodson, J. Study of Input Space for State Estimation of High-Rate Dynamics. *Struct. Control Health Monit.* **2018**, *25*, e2159. [CrossRef]

4. Joyce, B.; Dodson, J.; Laflamme, S.; Hong, J. An Experimental Test Bed for Developing High-Rate Structural Health Monitoring Methods. *Shock Vib.* **2018**, *2018*, 3827463. [CrossRef]

5. Blevins, R. *Formulas for Natural Frequency and Mode Shape*; Krieger: Hellerup, Denmark, 2001.

6. Gorman, D.J. Free lateral vibration analysis of double-span uniform beams. *Int. J. Mech. Sci.* **1973**, *16*, 345–351. [CrossRef]

7. Laura, P.A.A.; Pombo, J.L.; Susemihl, E.A. A note on the vibration of a clamped-free beam with a mass at the free end. *J. Sound Vib.* **1974**, *37*, 161–168. [CrossRef]

8. Lin, H.P.; Chang, S. Free vibration analysis of multi-span beams with intermediate flexible constraints. *J. Sound Vib.* **2005**, *281*, 155–169. [CrossRef]

9. Kukla, S. Free Vibrations of Axially Loaded Beams With Concentrated Masses and Intermediate Elastic Supports. *J. Sound Vib.* **1994**, *172*, 449–458. [CrossRef]

10. Ichikawa, M.; Miyakawa, Y.; Matsuda, A. Vibration Analysis Of The Continuous Beam Subjected to A Moving Mass. *J. Sound Vib.* **2000**, *230*, 493–506. [CrossRef]

11. Yesilce, Y.; Demirdag, O.; Catal, S. Free vibrations of a multi-span Timoshenko beam carrying multiple spring-mass systems. *Sadhana* **2008**, *33*, 385–401. [CrossRef]

12. Wang, Y.; Wei, Q.; Shi, J.; Long, X. Resonance characteristics of two-span continuous beam under moving high speed trains. *Lat. Am. J. Solids Struct.* **2010**, *7*, 185–199. [CrossRef]

13. Timoshenko, S.; Young, D. *Vibration Problems in Engineering*; Wiley: New York, NY, USA, 1974.

14. Chang, T.C.; Craig, R.R. Normal Modes of uniform beams. *J. Eng. Mech.* **1969**, *95*, 1027–1031.

15. Avitabile, P. *Modal Testing*; Wiley: New York, NY, USA, 2018.

MDPI

St. Alban-Anlage 66

4052 Basel

Switzerland

Tel. +41 61 683 77 34

Fax +41 61 302 89 18

www.mdpi.com

Applied Sciences Editorial Office

E-mail: applsci@mdpi.com

www.mdpi.com/journal/applsci

www.ingramcontent.com/pod-product-compliance
Lightning Source LLC
LaVergne TN
LVHW071509180726
843512LV00014B/1045